Waves, Oscillations and Acoustics

Second Edition

for **BSc (Pass), BSc (Hons) and BTech Students**

W0081067

Waves, Oscillations and Acoustics

Second Edition

for **BSc (Pass), BSc (Hons) and BTech Students**

SL Kakani

MSc (Physics), PhD

Former Executive Director
Institute of Technology & Management
(Affiliated to Rajasthan Technical University, Kota)
Bhilwara, Rajasthan, India
E-mail: slkakani28@gmail.com

C Hemrajani

MSc (Physics), PhD

Former Head
Department of Physics
MLV Government Postgraduate College
Bhilwara, Rajasthan, India

CBS

CBS Publishers & Distributors Pvt Ltd

New Delhi • Bengaluru • Chennai • Kochi • Kolkata • Mumbai
Hyderabad • Jharkhand • Nagpur • Patna • Pune • Uttarakhand

Waves, Oscillations and Acoustics
Second Edition

ISBN: 978-93-86827-81-4

Second Edition: 2018
First Edition: 2002

Published by Satish Kumar Jain and produced by Varun Jain for

CBS Publishers & Distributors Pvt Ltd
4819/XI Prahlad Street, 24 Ansari Road, Daryaganj, New Delhi 110 002, India.
Ph: 23289259, 23266861, 23266867   Website: www.cbspd.com
Fax: 011-23243014   e-mail: delhi@cbspd.com; cbspubs@airtelmail.in.
Corporate Office: 204 FIE, Industrial Area, Patparganj, Delhi 110 092
Ph: 4934 4934   Fax: 4934 4935   e-mail: publishing@cbspd.com; publicity@cbspd.com

Branches

- **Bengaluru:** Seema House 2975, 17th Cross, K.R. Road,
  Banasankari 2nd Stage, Bengaluru 560 070, Karnataka
  Ph: +91-80-26771678/79   Fax: +91-80-26771680   e-mail: bangalore@cbspd.com
- **Chennai:** 7, Subbaraya Street, Shenoy Nagar, Chennai 600 030, Tamil Nadu
  Ph: +91-44-26680620/26681266   Fax: +91-44-42032115   e-mail: chennai@cbspd.com
- **Kochi:** Ashana House, No. 39/1904, AM Thomas Road, Valanjambalam, Eranakulam 682 018,
  Kochi Kerala
  Ph: +91-484-4059061-65   Fax: +91-484-4059065   e-mail: kochi@cbspd.com
- **Kolkata:** 6/B, Ground Floor, Rameswar Shaw Road, Kolkata-700 014, West Bengal
  Ph: +91-33-22891126, 22891127, 22891128   e-mail: kolkata@cbspd.com
- **Mumbai:** 83-C, Dr E Moses Road, Worli, Mumbai-400018, Maharashtra
  Ph: +91-22-24902340/41   Fax: +91-22-24902342   e-mail: mumbai@cbspd.com

Representatives

Hyderabad	0-9885175004	**Jharkhand**	0-9811541605
Nagpur	0-9021734563	**Patna**	0-9334159340
Pune	0-9623451994	**Uttarakhand**	0-9716462459

Printed by Rashtriya printers, Delhi

Preface to the Second Edition

The book has been thoroughly revised, corrected and significantly enlarged in the light of the revised syllabi of various universities and UGC. A good number of new problems, multiple choice questions and short answer questions have been added. A new chapter on "Acoustics" has also been added.

We hope that this revised and updated edition will definitely cater to the needs of BSc (Pass), BSc (Hons) and BTech students preparing for various entrance examinations.

We are extremely thankful and grateful to Mr SK Jain, Chairman and Managing Director, CBS Publishers & Distributors, New Delhi, for his continued support. We are also thankful to Mr YN Arjuna, Sr Vice President—Publishing, Editorial and Publicity, and the staff involved in this project for taking the trouble in processing the manuscript.

Suggestions and constructive criticism for further improvement are coordially invited.

SL Kakani
C Hemrajani

Preface to the First Edition

This textbook is written for BSc (Pass), BSc (Hons) and BE students of all Indian and other universities. The salient features of this book are:

- Basic principles and fundamental concepts are explained in simple and lucid language.
- A large number of solved problems of different types that appeared in the examinations of various universities have been included to make the students understand the applications of various principles discussed in the chapter.
- A large number of self-explanatory accurate diagrams and tables have been used to supplement the text.
- A good number of typical exercises and problems are provided at the end of each chapter.
- A large number of short-answer questions and objective type questions have been given at the end of each chapter. This will make the book also useful for various competitive entrance examinations.

It is hoped that with these unique features, the book will fulfil the genuine requirements of students and teachers.

Much originality cannot be claimed in a book of this kind. The authors take this opportunity to place on record their indebtedness to the large number of books and journals that they have freely consulted in the preparation of this book. The authors heartily thank the publisher, CBS Publishers & Distributors, for taking keen interest in getting the book printed well in time.

In spite of all precautions and care taken to avoid errors and misprints, there might have crept some due to oversight and the authors will feel highly obliged to those fellow teachers and students who will bring them to their notice. Any suggestions for improvement will be thankfully acknowledged.

SL Kakani
C Hemrajani

Contents

1

Free Oscillations of Systems with One Degree of Freedom: Simple Harmonic Motion

1.1 INTRODUCTION

In a wide range of physical problems, one comes across motions that are periodic, i.e. the motion is repeated after a regular interval of time called *time period*. In general, there are two types of periodic motion:

1. *Rotational motion*, e.g. the motion of the hands of a clock, the movement of the moon around the earth, and that of the earth and other plantets round the sun, are the familiar examples of rotational motion.

2. *Oscillatory motion*, e.g. the motion of the bob of a pendulum, and that of a mass attached to a spring are both oscillatory. The atoms in a solid are vibrating. Similarly, the atoms in molecules are vibrating relative to each other. An understanding of vibrational motion is essential for the discussion of wave phenomenon.

In this chapter, we shall study the simplest and smoothest type of oscillatory motion, namely, *simple harmonic motion* (SHM) of systems having one degree of freedom. A system is said to have one degree of freedom if it is completely specified by a single physical quantity. Some simple examples are, a pendulum oscillating in a plane, mass attached to a spring, and an electrical circuit involving a capacitance and inductance.

Among the various varieties of oscillatory motions, SHM is of central importance for three basic reasons: (i) The oscillations of all physical systems are simple harmonic or a close approximation to it, (ii) We shall learn later that complex oscillation can be expressed as a superposition of harmonic motions, and (iii) The study of SHM is essential for understanding of the wave motion.

1.2 CONDITIONS FOR OSCILLATORY MOTION

Two basic properties of the system, namely, elasticity and inertia, are responsible for the oscillations of a physical system. Consider a body in equilibrium so that the net force on it is zero. Let us displace it from its position of equilibrium (by doing work on it, i.e. by applying a force) by a distance x. When it is released, a force comes into play which tries to restore its equilibrium position, i.e. tries to restore its original value which is zero, by imparting to it an appropriate negative velocity $\dfrac{dx}{dt}$. This force is called *restoring force*. *The magnitude of this restoring force is determined by the elastic properties of the system.* Inertia, on the other hand, tries to oppose any change in velocity. When the body reaches its equilibrium position ($x = \pm\, 0$), the negative velocity is maximum which produces a negative displacement. The body then overshoots its position of equilibrium. The restoring force now becomes positive and it must now overcome the inertia of the negative velocity. Consequently, the velocity keeps on decreasing until it is zero but by that time the displacement has become large and negative and the process is reversed. This process of the restoring force trying to bring x to zero by imparting a velocity and inertia preserving the velocity and making x to overshoot, repeats itself and the body oscillates.

In terms of energy, we can say that a particle undergoing oscillatory motion passes back and forth through a point (its equilibrium position) at which its potential energy u is a minimum. A swinging pendulum is a good example, its potential energy being a minimum at the bottom of the swing, i.e. at the equilibrium position. The force acting on the particle at any position is given by

$$F = -\frac{du}{dx} \tag{1.1}$$

and is zero at the equilibrium position and the force on the particle when displaced from this position towards either side is such as to bring the particle back to this position. This force is called restoring force because it always acts to accelerate the particle towards its position of equilibrium.

Hence, in an oscillatory motion the position of equilibrium is always one of the stable equilibrium and that the potential energy of the particle has a minimum at that point. The particle cannot move outside the limits x_1 and x_2. Thus, *the necessary condition for a particle to be able to have oscillatory motion about a point is that the potential energy of the particle has a minimum at that point.*

1.3 OSCILLATIONS IN AN ARBITRARY POTENTIAL WELL

In all conservative force fields, the potential energy of a particle is continuous function of position, i.e.

$$u = u\ (x, y, z)$$

and the force acting on the particle is given by

$$F = -\nabla u = -\left(\hat{i}\frac{\partial u}{\partial x} + \hat{j}\frac{\partial u}{\partial y} + \hat{k}\frac{\partial u}{\partial z}\right) \tag{1.2}$$

In one dimensional system, i.e. when particle is moving along x-direction

$$F = -\frac{\partial u}{\partial z} \tag{1.2a}$$

Let the variation of the potential energy of a particle with its displacement be represented by full line as shown in Fig. 1.1. It has minimum at point A where its value is $u(x_0)$. The potential energy at any point in the neighbourhood of this equilibrium position will be $u(x_0 + x)$. Using Taylor expansion of this potential energy function, we have

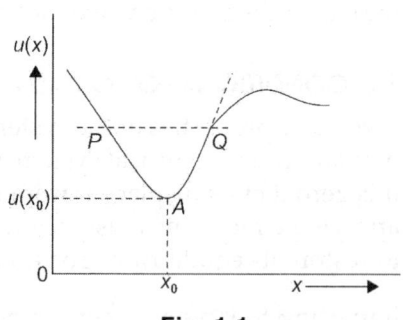

Fig. 1.1

$$u(x_0 + x) = u(x_0) + x\left\{\frac{du}{dx}\right\}_{x=x_0} + \frac{x^2}{2!}\left(\frac{d^2u}{dx^2}\right)_{x=x_0} + \frac{x^3}{3!}\left(\frac{d^3u}{dx^3}\right)_{x=x_0} + \dots$$

Since potential energy has minimum at A, i.e. $\left(\dfrac{du}{dx}\right)_{x=x_0} = 0$

$$\therefore \qquad u(x_0 + x) = u(x_0) + \frac{x^2}{2!}\left(\frac{d^2u}{dx^2}\right)_{x=x_0} + \frac{x^3}{3!}\left(\frac{d^3u}{dx^3}\right)_{x=x_0} + \dots$$

For small displacements from x_0 on either side, higher terms x^3, x^4, etc. can be neglected; then we have

$$\therefore \qquad u(x_0 + x) = u(x_0) + \frac{1}{2}x^2\left(\frac{d^2u}{dx^2}\right)_{x=x_0} \qquad\qquad (1.3)$$

If the potential energy is denoted by $u(x)$ when referred to zero as the PE at $x = x_0$, i.e. $u(x_0) = 0$ and displacement x (small) of the particle from x_0, then we have

$$u(x) = \frac{1}{2}kx^2 \qquad\qquad (1.4)$$

where, $k = \left(\dfrac{d^2u}{dx^2}\right)_{x=0} = $ a constant.

Equation (1.4) represents the equation of a parabola represented by dotted curve in Fig. 1.1. Thus, for small displacements about a potential energy minima the shape of potential energy curve can be considered parabolic. For a moving particle the region of potential energy–displacement curve bound between P and Q (Fig. 1.1) near equilibrium position (minimum potential energy position) A (on both sides for small x) is called *potential well*. The difference between maximum PE and minimum PE of the potential well is called *binding energy*. When the energy of the particle is less than the binding energy, the particle remains inside the potential well. This state of particle is called *bound state*.

The force on the particle for displacement x is

$$f(x) = -\frac{du}{dx} = -kx \qquad\qquad (1.5)$$

i.e. the force on the particle is proportional to its displacement from the mean position. The negative sign indicates that the force opposes the displacement as such is a restoring force. The particle then executes the motion called *simple harmonic motion* (SHM). The constant k which is a measure of the restoring force per unit displacement from the mean position is called *force constant*. The equation of the motion of a particle then is

$$m\frac{d^2x}{dt^2} = -kx$$

or $\qquad\qquad \dfrac{d^2x}{dt^2} = -\dfrac{k}{m}x = -\omega^2 x \qquad$ where $\omega^2 = k/m$

or $\qquad\qquad \dfrac{d^2x}{dt^2} + \omega^2 x = 0 \qquad\qquad (1.6)$

Equation (1.6) is called the *differential equation of SHM* and the particle executing SHM is called *simple harmonic oscillator*.

1.4 KINEMATICS OF SIMPLE HARMONIC MOTION

The solution of Eq. (1.6) which gives the displacement x of the particle at any time t, can be written as

$$x = A \sin(\omega t + \phi) \qquad (1.7)$$

where A and ϕ are a set of arbitrary constants.

If the time t in Eq. (1.7) is increased by $\dfrac{2\pi}{\omega}$ the function becomes

$$x' = A \sin\left\{\omega\left(t + \frac{2\pi}{\omega}\right) + \phi\right\} = A \sin(2\pi + \omega t + \phi)$$
$$= A \sin(\omega t + \phi) = x$$

i.e. the function merely repeats itself after time $\dfrac{2\pi}{\omega}$ and is called the *period of motion, T*.

Thus, we have

$$T = \frac{2\pi}{\omega} = 2\pi\sqrt{\frac{m}{k}} \qquad (1.8)$$

The quantity ω is called *angular frequency* and is related to the physical properties of the system by the relation

$$\omega^2 = \frac{k}{m} = \text{restoring force per unit displacement per unit mass.}$$

The constant A has a simple physical meaning. The sine function takes on values from -1 to $+1$. The displacement x from the central equilibrium position $x = 0$, therefore, has a maximum value of A and is called *amplitude of the motion*.

The quantity $(\omega t + \phi)$ is called the *phase of the motion*. The constant ϕ is called the *phase constant* being the value of phase at time $t = 0$.

Once the motion has started, the particle will continue to oscillate with a constant amplitude and phase constant (ϕ) at a fixed frequency unless other forces disturb the system.

Velocity and Acceleration in Simple Harmonic Motion

It is instructive to learn how velocity and acceleration in a SHM vary with time and displacement. We know that displacement $x(t)$ is given by

$$x = A \sin(\omega t + \phi)$$

then the velocity at any time t is given by

$$v = \frac{dx}{dt} = \dot{x} = A\omega \cos(\omega t + \phi)$$

$$= A\omega \sin\left(\omega t + \phi + \frac{\pi}{2}\right) \qquad (1.9)$$

and velocity when displacement is x, is given by

$$v = \pm A\omega\left(1 - \frac{x^2}{A^2}\right)^{1/2} \qquad \because \sin(\omega t + \phi) = \frac{x}{A}$$

$$= \pm \omega(A^2 - x^2)^{1/2} \qquad (1.9a)$$

Equation (1.9) gives us that velocity and displacement are not in same phase and velocity is $\pi/2$ ahead of displacement in phase. We also notice that the displacement is maximum (+A or −A) the velocity $v = 0$ because now the oscillator has to return, and velocity changes its direction. The velocity of the oscillator is maximum equal to $A\omega$ when $x = 0$, i.e. while crossing the mean position.

Similarly, the acceleration at any time is given by

$$a = \frac{dv}{dt} = \frac{d^2x}{dt^2} = \ddot{x} = -\omega^2 A \sin(\omega t + \phi)$$

$$= \omega^2 A \sin(\omega t + \phi + \pi) \tag{1.10}$$

and acceleration when displacement is x, is given by

$$a = -\omega^2 x \tag{1.10a}$$

Equation (1.10) gives us that acceleration, velocity and displacement are not in same phase. Acceleration is $\pi/2$ ahead in phase with velocity and π ahead in phase with displacement. The acceleration is maximum ($\mp\omega^2 A$) when the displacement x is maximum $\pm A$ and is always directed opposite to the displacement and when $x = 0$ the acceleration is also zero. Figure 1.2(a) gives a qualitative plot of the nature of variation of the displacement x, velocity v, and acceleration a of a particle in SHM with time t.

Here for simplicity we have taken $\phi = 0$.

Similarly, Fig. 1.2(b) gives the variation of velocity (v) and acceleration (a) with displacement of a particle in SHM. In this case, if $\omega = 1$, the graph between velocity and displacement will be a circle.

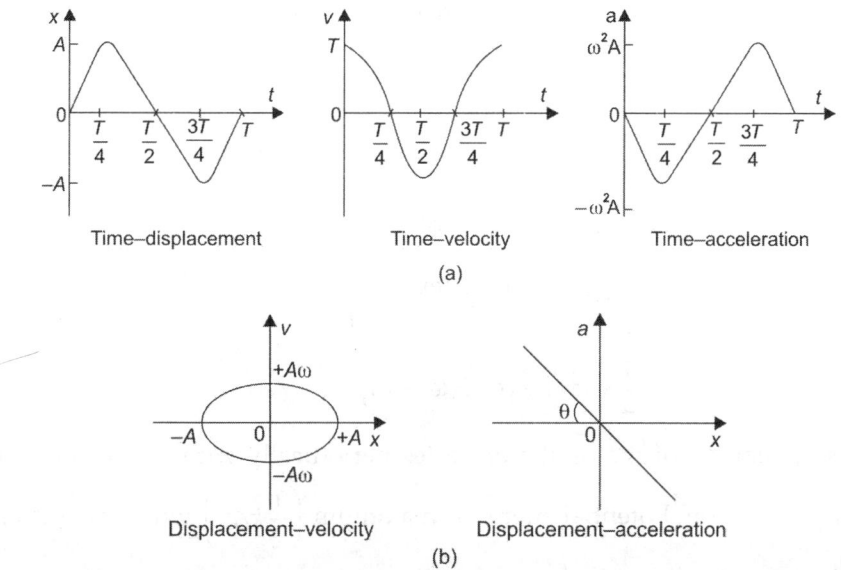

Time–displacement Time–velocity Time–acceleration

(a)

Displacement–velocity Displacement–acceleration

(b)

Fig. 1.2

Energy of a Simple Harmonic Oscillator

Consider a system at rest at its position of equilibrium. When it is displaced from this position, it acquires potential energy. When the system is released, it begins to move with a velocity, thus acquiring kinetic energy. At any instant of time, the kinetic energy of a system of mass m executing SHM is given by

$$\text{Kinetic energy (KE)} = \frac{1}{2}mv^2 = \frac{1}{2}m\omega^2 A^2 \cos^2(\omega t + \phi)$$

$$= \frac{1}{2}kA^2 \left\{ \frac{1 + \cos 2(\omega t + \phi)}{2} \right\}$$

$$= \frac{1}{4}kA^2 \left\{ 1 + \cos 2(\omega t + \phi) \right\} \tag{1.11}$$

here $k = m\omega^2$; and in terms of displacement

$$\text{KE} = \frac{1}{2}k(A^2 - x^2). \tag{1.11a}$$

The kinetic energy of the oscillator varies periodically with frequency (2ω) twice the frequency of motion. It is maximum $\left(\frac{1}{2}m\omega^2 A^2 = \frac{1}{2}kA^2 \right)$ when the velocity is maximum $(= \pm A\omega)$ and displacement is zero. When the displacement is maximum $(= \pm A)$, velocity $v = 0$ and KE = 0. At these extreme positions, the energy is all potential. At intermediate positions (x lying between 0 and $\pm A$), the energy is partly kinetic and partly potential.

Let us now compute potential energy at any instant of time t. Let x be the displacement at time t. The potential energy is given by the amount of work required to move the system from $x = 0$ to x by applying a force.

The force must be just enough to oppose the restoring force $F = -kx$, i.e. the force to be applied must be kx. Work required to give an infinitesimal displacement dx is $kxdx$. Hence, the total work done to displace the system from 0 to x is

$$W = \int_0^x kxdx = \frac{1}{2}kx^2.$$

Thus, potential energy $= \frac{1}{2}kx^2 \tag{1.12}$

$$= \frac{1}{2}kA^2 \sin^2(\omega t + \phi)$$

$$= \frac{1}{2}kA^2 \left[\frac{1 - \cos 2(\omega t + \phi)}{2} \right]$$

$$= \frac{1}{4}kA^2 \left[1 - \cos 2(\omega t + \phi) \right] \tag{1.12a}$$

The potential energy of the oscillator varies periodically with frequency (2ω) twice the frequency of motion. Potential energy is maximum $\left(\frac{1}{2}kA^2 \right)$ when $x = \pm A$ and zero when $x = 0$.

The total energy E of the simple harmonic oscillation, therefore, is

$$E = \text{KE} + \text{PE}$$

$$= \frac{1}{4}kA^2 \left[1 + \cos 2(\omega t + \phi) \right] + \frac{1}{4}kA^2 \left[1 - \cos 2(\omega t + \phi) \right]$$

$$= \frac{1}{2}kA^2 \tag{1.13}$$

which is constant. It is also apparent from the fact that maximum displacement is regained after every half cycle. It is obvious that the maximum value of kinetic and potential energy both are

equal $\left(\text{i.e. } \dfrac{1}{2}kA^2 = \dfrac{1}{2}m\omega^2 A^2\right)$ indicating that the energy exchange is complete. Figure 1.3a shows how the kinetic and potential energy of the harmonic oscillator vary with time, where for simplicity we have set $\phi = 0$ and Fig. 1.3b shows the same variation with displacement.

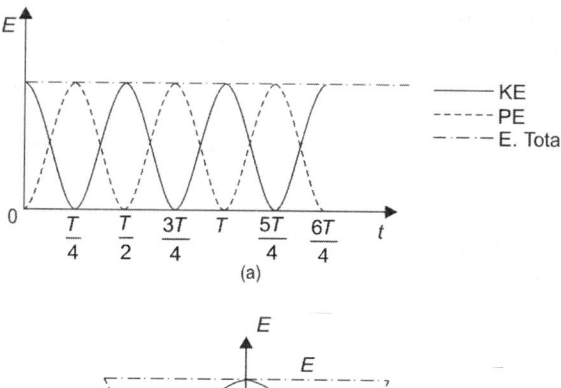

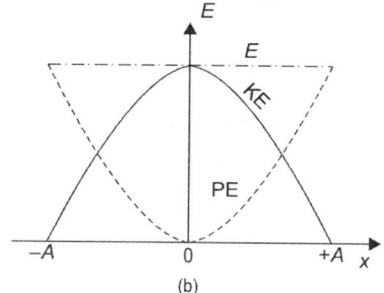

(b)

Fig. 1.3

Equation (1.13) reveals that total mechanical energy of the oscillator is conserved and has the value $\dfrac{1}{2}kA^2$.

Thus, total energy of the particle executing SHM is proportional to the square of the amplitude of motion and force constant. The energy is proportional to the square of the angular frequency. Thus, the amplitude of a SH oscillator of a given total energy E is

$$A = \pm\sqrt{\frac{2E}{k}} \tag{1.14}$$

and maximum particle momentum is

$$p = mv_{max} = mA\omega = m\sqrt{\frac{2E}{k}}\cdot\sqrt{\frac{k}{m}} = \sqrt{2mE} \tag{1.15}$$

Thus, for an oscillator of energy E, the displacement lies between the limits $-\sqrt{\dfrac{2E}{k}}$ and $+\sqrt{\dfrac{2E}{k}}$, while momentum lies between the limits $-\sqrt{2mE}$ and $+\sqrt{2mE}$. These are sometimes referred to as the classical limits for an oscillator of total energy E. In fact, quantum mechanics predicts that an oscillator of total energy E. In fact, quantum mechanics predicts that oscilator of total energy E has a finite probability of being found outside the classical limits. Such a view has found a large confirmation in as much as it is capable of predicting many experimental results which the classical theory fails to do.

1.5 SOME EXAMPLES OF FREE VIBRATIONS

(i) Simple Pendulum

A simple pendulum is a point mass suspended by means of a weightless and inelastic string from a rigid support. Figure 1.4 shows a point mass m suspended by means of a string of length l. Mass when pulled to one side of its equilibrium position and released, the pendulum swings in a vertical plane under the influence of gravity about a horizontal axis passing through the point of suspension O. The motion is periodic and oscillatory and for small values of θ, it can be shown to be simple harmonic.

The forces acting on the mass m in displaced position are mg its weight and tension T in the chord. The radial components of the forces $(T - mg\cos\theta)$ provide the necessary centripetal force to keep the particle moving on a circular arc. The tangenital component $(mg\sin\theta)$ is the restoring force acting on m binding it to return to the equilibrium position. Hence, the restoring force is

$$F = -mg\sin\theta. \tag{1.16}$$

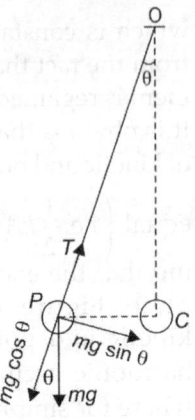

Fig. 1.4

Notice that the restoring force is not proportional to the angular displacement θ but $\sin\theta$. The resulting motion is, therefore, not simple harmonic. However, if the angle θ is small, $\sin\theta$ is very nearly equal to θ in radians[1]. The displacement along the arc is $x = l\theta$ and for small angles, this is nearly straight line motion. Hence, assuming $\sin\theta \approx \theta$, we have

$$F = -mg\theta = -mg\cdot\frac{x}{l} = -\left(\frac{mg}{l}\right)\cdot x \tag{1.17}$$

Hence, for small displacements, the restoring force is proportional to the displacement and oppositely directed, i.e. the motion is simple harmonic with force constant $k = \dfrac{mg}{l}$ and its period of oscillation

$$T = 2\pi\sqrt{\frac{m}{k}} = 2\pi\sqrt{\frac{l}{g}} \tag{1.18}$$

The period is independent of the mass of the suspended particle.

(ii) Compound Pendulum

A compound pendulum is a *rigid body of any shape capable of oscillating in a vertical plane about a horizontal axis passing through it*. Figure 1.5 shows a vertical section of a rigid body free to rotate about point S called the *centre of suspension*. *The distance l between the centre of suspension (S) and centre of gravity (C) of the body is called the length of the pendulum.* The pendulum is given a small angular displacement θ and released. It begins to oscillate about the point S. In displaced position the weight mg of the pendulum acts vertically downwards at C', the new position of the centre of gravity. The pendulum tends to return under the influence of a reactive couple (or torque). The moment of the restoring couple $= -mgl\sin\theta$, the negative sign indicating that the couple is directed opposite to

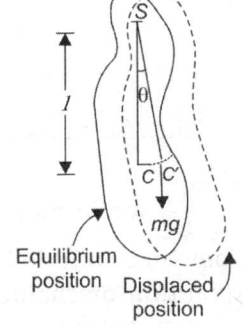

Equilibrium position   Displaced position

Fig. 1.5

the displacement. If I is the moment of inertia of the pendulum about the axis of suspension, this couple is also equal to $I\dfrac{d^2\theta}{dt^2}$ where $\dfrac{d^2\theta}{dt^2}$ is the instantaneous angular acceleration. Thus

$$I\frac{d^2\theta}{dt^2} = -mgl\sin\theta$$

---

1. $\sin\theta = \theta - \theta^3/3! + \theta^5/5! \dots.$ The error in replacing $\sin\theta$ by θ is nearly 1% for $\theta = 0.25$ radians or 15°. For smaller angles the error is even less.

If θ is small, then $\sin\theta \approx \theta$ (θ is expressed in radians) so that is

$$\frac{d^2\theta}{dt^2} = -\frac{mgl}{I}\theta \tag{1.19}$$

Thus, the pendulum executes SHM and its time period is given by

$$T = 2\pi\sqrt{\frac{I}{mgl}} \tag{1.20}$$

It is sometimes more useful and convenient to express the moment of inertia about a parallel axis through C, the centre of mass. If this is expressed as $I_c = mk^2$, where k is the radius of gyration of the pendulum about horizontal axis through C, then using theorem of parallel axes

$$I = I_c + ml^2 = m(k^2 + l^2)$$

$$\therefore \qquad T = 2\pi\sqrt{\frac{m(k^2 + l^2)}{mgl}}$$

$$= 2\pi\sqrt{\frac{\dfrac{k^2}{l} + l}{g}} \tag{1.20a}$$

The time period is the same as that of a simple pendulum of length $L = \dfrac{k^2}{l} + l$. This length is called the *length of the equivalent simple pendulum.*

(iii) Mass Spring System

When a force is applied to a spring to compress or stretch it, the resulting compression or elongation does not bear a simple relationship with the force applied. Figure 1.6 illustrates the force–displacement relationship for a spring. The relationship is, in general, not linear. Only for small displacements, the relationship is linear (portion AB of the curve). The elastic force produced in the linear spring is given by

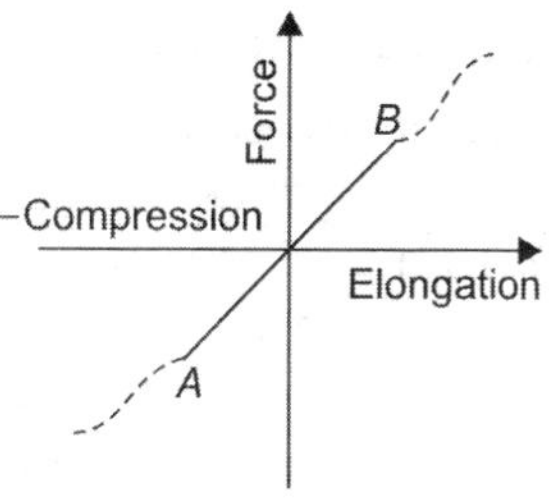

$$F = -kx$$

Fig. 1.6

where x is the change in the length of the spring when force F is applied to it. The constant k is called the *spring constant* or *stiffness constant* and is defined as the force required to produce a unit extension or compression in the spring. The unit of k in SI system is N·m^{-1}. *The constant k is inversely proportional to the unstretched length L of the spring,* i.e.

$$k = \frac{C}{L}.$$

(a) Horizontal oscillations of spring mass system

Consider a massless ideal spring (i.e. the one which obeys Hooke's law) of constant k, one end of which is fixed to a wall and the other end is attached to a body of mass m which is free to move on a frictionless horizontal surface (Fig. 1.7). Figure 1.7a is the position of static equilibrium, the spring being relaxed and no force is acting on it. When the body is pulled to right (Fig. 1.7b) through a small distance x, force exerted by the spring on the body is directed to the left and is given by $F = -kx$. This is the restoring force. Since the restoring force is proportional to the displacement (true only for small displacements) and is opposite in sign to the displacement; the resulting motion is SHM. The body begins to move with a linear acceleration which from Newton's law is given by

$$\frac{d^2x}{dt^2} = -\frac{k}{m}x \qquad (1.21)$$

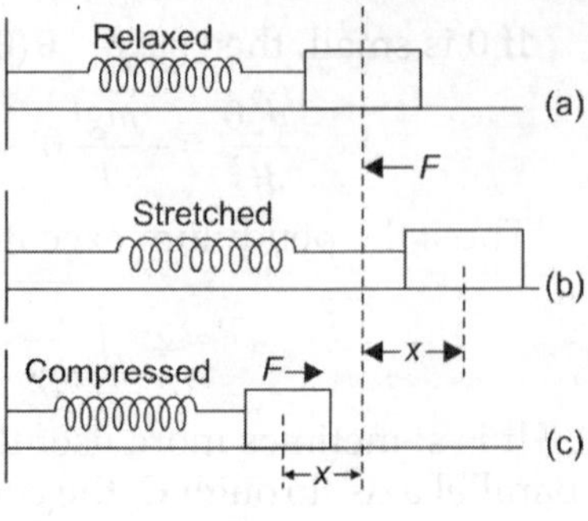

Comparing it with the equation of SHM, i.e.

$$\frac{d^2x}{dt^2} = -\omega^2 x$$

the angular frequency $\omega = \sqrt{\dfrac{k}{m}}$

and time period $\qquad T = \dfrac{2\pi}{\omega} = 2\pi\sqrt{\dfrac{m}{k}} \qquad (1.22)$

and displacement x as a function of time is given by

$$x = A\sin(\omega t + \phi) \qquad (1.23)$$

Notice that angular frequency ω (or period T) is determined for all circumstances by: (a) the inertia factor mass m and (b) the elasticity factor the spring constant k. The other two constants A and ϕ are determined from initial conditions.

Fig. 1.7

(b) Vertical oscillations

Let us now consider the vertical oscillations of a loaded spring. Figure 1.8 illustrates the equilibrium position and force on the spring for two extreme positions of mass m. In this case, the equilibrium state of the loaded spring is the state, when the spring is stretched by a distance l by the force mg where g is the acceleration due to gravity. No net force acts on the body since

$$mg = kl.$$

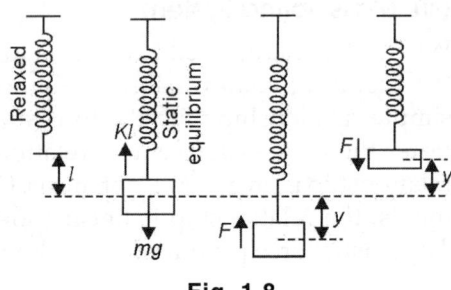

Fig. 1.8

When the body is pulled through a small distance y from the equilibrium position and released, it starts oscillating with SHM. Since the restoring force is given by $F = -ky$

imparting acceleration $\dfrac{d^2y}{dt^2}$ given by

$$\frac{d^2y}{dt^2} = -\frac{k}{m}y \qquad (1.24)$$

which is the equation of SHM, whose period is given by

$$T = 2\pi\sqrt{\frac{m}{k}} \qquad (1.25)$$

which is the same as that for horizontal oscillations. The equation $mg = kl$ determines the spring constant k if m, l and g are known.

(c) Vibrations of a mass suspended by two parallel springs

Figure 1.9a shows a mass m suspended by means of two massless parallel springs. The situation in Fig. 1.9b is identical in which there will be horizontal oscillations. When body is displaced through y and let go, the body oscillates under the influence of both the springs. When body is displaced through y, then change in length of two springs is y, causing the restoring force $F_1 = k_1 y$ and $F_2 = k_2 y$ where k_1 and k_2 are the spring constants of the two springs respectively. Hence, the resultant restoring force on the body is

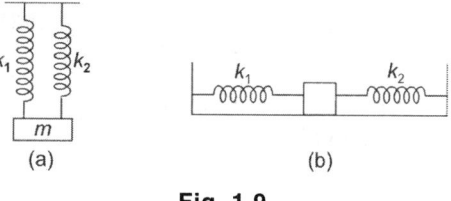

(a)   (b)

Fig. 1.9

$$F = F_1 + F_2 = -(k_1 + k_2)y = -ky$$

where $k = k_1 + k_2$, the combined effective force constant of two springs. Hence, the effective force constant is the sum of the force constants of the individual spring. Hence, time period is

$$T = 2\pi\sqrt{\frac{m}{k}} = 2\pi\sqrt{\frac{m}{(k_1 + k_2)}}.$$

In case of a number of springs of constants $k_1, k_2, k_3, \dots$ are connected in parallel, then effective force constant will be $k = \Sigma_i k_i = k_1 + k_2 + k_3 + \dots + k_i + \dots$ and time period is

$$T = 2\pi\sqrt{\frac{m}{\Sigma_i k_i}}.$$

(d) Vibrations of a mass suspended by two springs in series

Figure 1.10 shows a mass m suspended by two springs of force constant k_1 and k_2 connected in series. The body is pulled a distance y and let go. In stretched position in this case, the stretching of the two springs is different but the restoring force developed are same. This can be proved by the equilibrium of point A of the springs. If the stretching in the two springs be y_1 and y_2 respectively, then total stretch

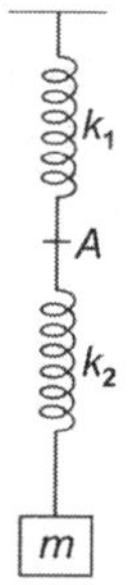

$$y = y_1 + y_2$$

Hence

$$F = -k_1 y_1 \quad \text{and} \quad F = -k_2 y_2$$

or

$$y_1 = -\frac{F}{k_1} \quad \text{and} \quad y_2 = -\frac{F}{k_2}$$

$\therefore$

$$y = y_1 + y_2 = -F\left(\frac{1}{k_1} + \frac{1}{k_2}\right) = -\frac{F}{k}$$

where $\dfrac{1}{k} = \dfrac{1}{k_1} + \dfrac{1}{k_2}$ and k is called the *effective force constant*. Thus in series combination, the two springs will behave like a single spring of constant k given by

$$k = \frac{k_1 k_2}{k_1 + k_2}$$

and

$$F = -ky.$$

Fig. 1.10

On releasing after stretching, the body will execute SHM of time period T given by

$$T = 2\pi\sqrt{\frac{m}{k}} = 2\pi\sqrt{\frac{m(k_1 + k_2)}{k_1 k_2}}$$

In case of a number of springs connected in series having constants $k_1, k_2, k_3, \ldots$ etc., the effective value of constant k is given by

$$\frac{1}{k} = \frac{1}{k_1} + \frac{1}{k_2} + \ldots + \frac{1}{k_n}.$$

(iv) An Ideal L-C Circuit

Not only the mechanical systems show the nature of periodicity, but electrical circuits also possess such properties. L–C electrical circuit form yet another example of the harmonic oscillations which have wide applications in electronics. In general, L–C circuit is always associated with some resistance that causes a damping effect. In our present discussion, we shall assume the circuit resistance to be zero.

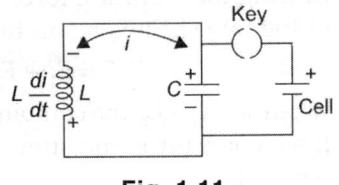

Fig. 1.11

Consider a circuit containing an inductance L and a capacitor of capacity C connected in series. The equilibrium state is the state when the capacitor is uncharged and no current is flowing in the circuit. This state is disturbed by pressing the key, thus, charging the capacitor. Let q be the charge on the capacitor so that $V = q/C$ is the voltage across the capacitor plates. When the key is released, the capacitor starts discharging through the inductor, i.e. the charge changes with time and current $i = \dfrac{dq}{dt}$ is established in the inductor.

In this circuit, the restoring force is due to the force of repulsion between the electrons. This force tries to distribute electrons equally on the capacitor plates so that there is no net charge. Inductance, on the other hand, tries to oppose the redistribution, i.e. it opposes the increase of current. At any instant of time, the voltage across the inductor is

$$V = -L\frac{di}{dt} = -L\frac{d^2q}{dt^2}.$$

The minus sign indicates the voltage opposes the increase of the current. From Kirchhoff's law, this voltage must be equal to the voltage q/C across the capacitor plates giving

$$-L\frac{d^2q}{dt^2} = \frac{q}{C}$$

or
$$\frac{d^2q}{dt^2} = -\omega^2 q \text{ with } \omega = \frac{1}{\sqrt{LC}} \tag{1.27}$$

Thus, the electrical circuit consisting of an inductance L and a capacitance C, the charge oscillates harmonically with an angular frequency $\omega = \dfrac{1}{\sqrt{LC}}$ and period $T = 2\pi\sqrt{LC}$. At any instant of time, the charge q is given by

$$q = q_0 \sin(\omega t + \phi)$$

or
$$i = i_0 \cos(\omega t + \phi)$$

where $i_0 = \omega q_0$ is the maximum value of the current. If V_0 is the applied voltage, then

$$i_0 = \omega q_0 = \frac{1}{\sqrt{LC}} V_0 C = V_0 \sqrt{\frac{C}{L}}.$$

Energy Consideration

An L–C circuit resembles a mass spring system in the sense that each has a characteristic frequency. To understand how an L–C circuit oscillates, let us assume that initially the capacitor carries a charge q and the current in the inductor is zero. At this instant, the electrostatic energy stored in the capacitor is

$$E_e = \frac{1}{2}\frac{q^2}{C}$$

and that in the inductor is zero since $i = 0$ initially. As time passes, the capacitor starts discharging through the inductance and current $i = \dfrac{dq}{dt}$ is established in the inductor. As q decreases, E_e decreases and i increases, so that the energy now appears around inductor as the current is built up. When the capacitor is completely discharged, the magnetic energy

$$E_m = \frac{1}{2} Li^2$$

associated with inductance is maximum because the current is maximum and $E_e = 0$ since $q = 0$. Thus, although at this time $q = 0$, $\dfrac{dq}{dt}$ is not zero; it is in fact maximum. The large current in the inductor starts transporting the charge across the capacitor plates and the capacitor is charged again. It starts discharging again and the current now flows in the opposite direction. Eventually, the current returns to its initial value and the process continues. The energy exchange occurs between the electric field of the capacitance and magnetic field of the inductance. The total energy of the system is conserved, since the system considered here does not contain any resistance, so that there is no dissipation of energy. Thus,

$$E = E_e + E_m$$

$$= \frac{1}{2}\frac{q^2}{C} + \frac{1}{2}Li^2 = \text{constant}$$

and differentiating with respect to time, we have

$$\frac{q}{C}\frac{dq}{dt} + Li\frac{di}{dt} = 0 \quad \text{But} \quad i = \frac{dq}{dt} \text{ and } \frac{di}{dt} = \frac{d^2q}{dt^2}$$

$$\therefore \qquad \frac{q}{C}\frac{dq}{dt} + L\frac{d^2q}{dt^2}\cdot\frac{dq}{dt} = 0$$

or $\qquad\qquad L\dfrac{d^2q}{dt^2} + \dfrac{q}{C} = 0 \qquad\qquad\qquad\qquad\qquad\qquad$ (1.28)

Equation (1.28) is identical to Eq. (1.6) obtained earlier. The comparison indicates that the mass in mechanical system and magnetic field in electrical system play analogous role of inertia. The mass controls the velocity changes for a given force and the magnetic field controls the rate of change of current for a given voltage.

Example 1.1: Prove that in simple harmonic motion, the average PE equals the average KE when average is taken with respect to time over one period of the motion and that each average equals $\frac{1}{4}ka^2$. But when average is taken with respect to position over one cycle, the average PE equals to $\frac{1}{6}ka^2$ and the average KE equals $\frac{1}{3}ka^2$. Explain why the two results are different.

Solution: At any time t the PE is given by

$$u = \frac{1}{2}kx^2 = \frac{1}{2}ka^2 \sin^2(\omega t + \phi)$$

$\therefore$ Its time average is

$$\langle u \rangle = \frac{\int_0^T u\,dt}{\int_0^T dt} = \frac{1}{T}\int_0^T \frac{1}{2}ka^2 \sin^2(\omega t + \phi)\,dt$$

$$= \frac{ka^2}{2T}\int_0^T \left\{\frac{1 - \cos 2(\omega t + \phi)}{2}\right\}dt$$

$$= \frac{ka^2}{4} \quad \because \int_0^T \cos 2(\omega t + \phi)\,dt = 0.$$

Kinetic energy at any time t is given by

$$KE = K = \frac{1}{2}mv^2 = \frac{1}{2}m\omega^2 a^2 \cos^2(\omega t + \phi)$$

$$= \frac{1}{4}ka^2\{1 + \cos 2(\omega t + \phi)\}$$

here $k = m\omega^2$ is *force constant*.

$$\therefore \qquad \langle K \rangle = \frac{\int_0^T K\,dt}{\int_0^T dt} = \frac{1}{T}\int_0^T \frac{1}{4}ka^2\{1 + \cos 2(\omega t + \phi)\}dt$$

or $$\langle K \rangle = \frac{1}{4}ka^2$$

Thus $$\langle u \rangle = \langle K \rangle = \frac{1}{4}ka^2.$$

Position average of PE is

$$\langle u \rangle = \frac{\int_{-a}^{+a} u\,dx}{\int_{-a}^{+a} dx} = \frac{1}{2a}\int_{-a}^{+a} \frac{1}{2}kx^2\,dx$$

$$= \frac{k}{4a}\left[\frac{x^3}{3}\right]_{-a}^{+a} = \frac{ka^2}{6}$$

Position average of KE is

$$\langle K \rangle = \frac{\int_{-a}^{+a} K\,dx}{\int_{-a}^{+a} dx} = \frac{1}{2a}\int_{-a}^{+a} \frac{1}{2}K(a^2 - x^2)\,dx$$

$$= \frac{k}{4a} \left[a^2 x - \frac{x^3}{3} \right]_{-a}^{+a} = \frac{ka^2}{3}.$$

Actually, the position average is determined by the relation

$$\langle u \rangle = \int_{-a}^{+a} u(x) f(x) dx$$

where $f(x)$ is the probability density which is assumed to be uniform $= \frac{1}{2a}$ here, and it is

incorrect in present case. The correct form of probability density is $f(x) = \frac{1}{2\sqrt{a^2 - x^2}}$ which

actually gives the same result.

Example 1.2: A mass M is suspended by a spring of finite mass m. Assuming that the spring is stretched uniformly by loading and its force constant is k. Show that the time

period of oscillations of M is given by $T = 2\pi \sqrt{\dfrac{M + M/3}{k}}$.

Solution: Let equilibrium length of the spring be L_0 and its instantaneous length after it is stretched and let free be l. Then instantaneous restoring force
$$F = -k \, (l - L_0).$$

If instantaneous velocity of mass M be denoted by i, the power delivered by the force F to the system is
$$P = Fi = -k(l - L_0)i$$

The power delivered by the elastic restoring force must obviously be equal to the rate at which the kinetic energy E_k of the system increases. Since mass of the spring is not negligible, the total KE of the system consists of two parts.

(i) The kinetic energy of the suspended mass (k_1). This equals $k_1 = \dfrac{1}{2} M i^2$.

(ii) The kinetic energy of the spring k_2. This is calculated as follows:

Let the point O, the point of suspension be taken as the origin and a line vertically downwards along the length of the spring as Y-axis. Consider a small element of length dy of a spring at a distance y below O (Fig. 1.12). The mass of this element is

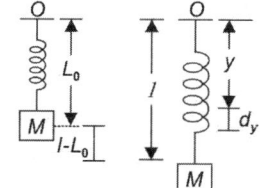

Fig. 1.12

$dm = \dfrac{m}{l} dy$ while its instantaneous velocity is $v = \dfrac{y}{l} i$. (This

is according to the assumption that the extension of the spring is uniform throughout its length). The KE of the element is thus

$$dE_k = \frac{1}{2} \frac{m}{l} dy v^2 = \frac{1}{2} \frac{m}{l} \frac{y^2 i^2}{l^2} dy$$

$$= \frac{1}{2} \frac{m}{l^3} i^2 y^2 dy.$$

Hence, total kinetic energy of the whole spring is

$$\int dE_k = \int_0^l \frac{m i^2 y^2}{2 l^3} dy = \frac{1}{6} m i^2.$$

∴ Total kinetic energy of the system

$$E = \frac{1}{2} M i^2 + \frac{1}{6} m i^2 = \frac{1}{2} \left(M + \frac{m}{3} \right) i^2$$

Hence, instantaneous power

$$P = \frac{dE}{dt} = \frac{1}{2}\left(M + \frac{m}{3}\right) 2 i \frac{di}{dt}$$

$$= \left(M + \frac{m}{3}\right) i \frac{di}{dt}$$

$$\therefore \quad \left(M + \frac{m}{3}\right) i \frac{di}{dt} = -k\,(l - L_0)\,i$$

or

$$\frac{di}{dt} = -\frac{k(l - L_0)}{(M + m/3)}$$

where $(l - L_0)$ = instantaneous displacement of mass M. We notice that acceleration of mass is proportional to its displacement having tendency to restore the particle to its equilibrium position. Thus, the motion of mass M is SHM and period is

$$T = 2\pi \sqrt{\frac{M + m/3}{k}}$$

Example 3: A body is attached to a mass-less horizontal spring of force constant k (Fig. 1.13). The body can roll without slipping on the horizontal surface. Show that under these conditions, the centre of mass of system executes SHM. Calculate its time period.

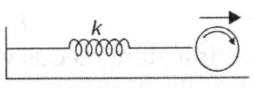

Fig. 1.13

Solution: Let the body has mass m and is pulled from equilibrium position and let go. At any instant let the extension in the spring be x and the linear velocity of the centre of mass of the body be v and angular velocity of body ω.

$$\therefore \qquad v = \omega r$$

where r is radius of body.

Its PE: $\qquad E_p = \frac{1}{2} k x^2$

KE $\qquad E_k = E_{rot} + E_{tr}$

$$= \frac{1}{2} I \omega^2 + \frac{1}{2} m v^2$$

If radius of gyration of the body about the axis of rotation which in present case is horizontal and passing through centre of mass of body be k', then

$$I = m k'^2 \quad \text{and} \quad \omega = \frac{v}{r}$$

$$\therefore \qquad E_k = \frac{1}{2} m k'^2 \frac{v^2}{r^2} + \frac{1}{2} m v^2$$

$$= \frac{1}{2} m k'^2 \left[1 + \frac{k'^2}{r^2} \right]$$

$\therefore$ Total energy of the system

$$E = E_k + E_p = \frac{1}{2} k x^2 + \frac{1}{2} m v^2 \left[1 + \frac{k'^2}{r^2} \right]$$

Since total energy is constant independent of time, hence

$$\frac{dE}{dt} = 0$$

$$\frac{dE}{dt} = k \cdot x \frac{dx}{dt} + mv\left(1 + \frac{k'^2}{r^2}\right)\frac{dv}{dt} = 0$$

$$\therefore \quad \frac{dv}{dt} = -\frac{k}{m\left(1 + \dfrac{k'^2}{r^2}\right)}x.$$

i.e. the acceleration of the centre of mass is proportional to displacement. Hence, motion is SHM and time period:

$$T = 2\pi\sqrt{\frac{m\left(1 + \dfrac{k'^2}{r^2}\right)}{k}}$$

Special Cases:

For ring $\qquad \dfrac{k'^2}{r^2} = 1 \quad \therefore T = 2\pi\sqrt{\dfrac{2m}{k}}$

For disc and cylinder $\quad \dfrac{k'^2}{r^2} = \dfrac{1}{2} \quad \therefore T = 2\pi\sqrt{\dfrac{3m}{k}}$

For solid sphere $\qquad \dfrac{k'^2}{r^2} = \dfrac{2}{5} \quad \therefore T = 2\pi\sqrt{\dfrac{7m}{5k}}$

Example 1.4: Two masses m_1 and m_2 are connected by a massless spring of normal length l and force constant k. A horizontal force F is applied to one of the masses. The masses move on a frictionless floor. Calculate the maximum compression of the spring.

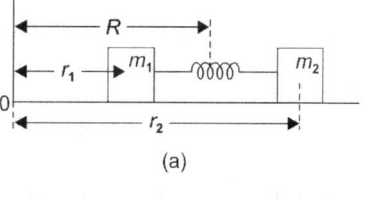

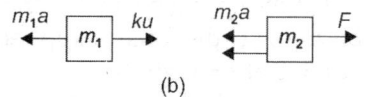

Fig. 1.14

Solution: The centre of mass of the system has acceleration

$$a = \frac{F}{m_1 + m_2}$$

If the position of the masses be denoted by r_1 and r_2, the centre of mass is at

$$R = \frac{m_1 r_1 + m_2 r_2}{m_1 + m_2}$$

with respect to the origin.

With respect to the centre of mass, the position of masses are

$$r_1' = r_1 - R = \frac{m_2(r_1 - r_2)}{m_1 + m_2}$$

and $\qquad r_2' = r_2 - R = \dfrac{m_2(r_2 - r_1)}{m_1 + m_2}$

The instantaneous length of the spring is $r_2 - r_1 = r_2' - r_1'$. If natural length of the spring is l, then stretching is

$$r_2' - r_1' - l = u.$$

In the centre of mass system (accelerated), the force acting on the masses are shown in Fig. 1.14b. The equations of motion are:

$$m_1 \ddot{r}_1' = ku - \frac{m_1 F}{m_1 + m_2} \tag{i}$$

$$m_2 \ddot{r}_2' = -ku - \frac{m_2 F}{m_1 + m_2} + F = -ku - \frac{m_1 F}{m_1 + m_2} \tag{ii}$$

Multiplying (i) by m_2 and (ii) by m_1 and subtracting, we have

$$m_1 m_2 (\ddot{r}_2' - \ddot{r}_1') = -ku(m_1 + m_2) + m_1 F.$$

But $\ddot{r}_2' - \ddot{r}_1' = \ddot{u}$

or $m_1 m_2 \ddot{u} = -ku(m_1 + m_2) + m_1 F$

or $\ddot{u} = -ku\left(\dfrac{1}{m_1} + \dfrac{1}{m_2}\right) + \dfrac{m_1}{m_1 m_2} F$

or $\dfrac{d^2 u}{dt^2} = \dfrac{-ku}{\mu} + \dfrac{F}{m_2}$

where $\dfrac{1}{\mu} = \dfrac{1}{m_1} + \dfrac{1}{m_2}$, and μ is called *reduced mass of the system.*

This equation is identical to the harmonic oscillator equation for a single particle of mass μ excepting for the constant $\dfrac{F}{m_2}$, whose derivatives are zero. The solution can be written as

$$u = A \sin(\omega t + \phi) + \frac{F}{m_2 \omega^2} \quad \text{where } \omega^2 = \frac{k}{\mu}.$$

Let the spring be unstretched at $t = 0$ at this time we also assume the relative velocity to be zero, i.e. the force is applied at $t = 0$

$\therefore u = 0$ at $t = 0$ gives

$$A \sin \phi = -\frac{F}{m_2 \omega^2}$$

and $\dfrac{du}{dt} = 0$ at $t = 0$ gives $A\omega \cos \phi = 0$, i.e. $\phi = \dfrac{\pi}{2}$ and $A = -\dfrac{F}{m_2 \omega^2}$ which gives

$$u = \frac{F}{m_2 \omega^2} \{1 - \cos \omega t\}$$

Hence, the maximum extension is at $\omega t = \pi$

$\therefore$ $u_{max} = \dfrac{2F}{m_2 \omega^2}.$

Example 1.5: A mass m is attached to a spring of spring constant K through a frictionless pulley of radius r and mass M as shown in Fig. 1.15. Determine the frequency of vertical oscillations of mass using: (a) Newton's laws of motion, and (b) energy consideration.

Solution: As shown in Fig. 1.15a in equilibrium the stretching of spring is $r\theta_0$ and hence

$$mg = kr\theta_0 \qquad\qquad (i)$$

If the mass is pulled and let free, then at any time, if stretching is x, then force equation for mass m is

$$m\frac{d^2x}{dt^2} = mg - T' \qquad\qquad (ii)$$

and torque equation for the pulley of mass M is

$$I\frac{d^2\theta}{dt^2} = T'r - kr^2(\theta + \theta_0) \qquad\qquad (iii)$$

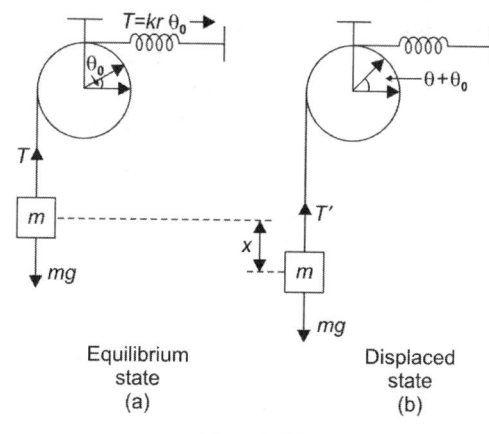

Equilibrium state (a)

Displaced state (b)

Fig. 1.15

Eliminating T' between equations (ii) and (iii) and putting $I = \dfrac{Mr^2}{2}$, we have

$$m\frac{d^2x}{dt^2} + \frac{1}{2}Mr\frac{d^2\theta}{dt^2} = -kr\theta \qquad\qquad (iv)$$

Here we have used $mg = kr\theta_0$. Putting $x = r\theta$ and $\dfrac{d^2x}{dt^2} = r\dfrac{d^2\theta}{dt^2}$, we have

$$\left(m + \frac{M}{2}\right)\frac{d^2x}{dt^2} = -kx$$

and

$$\omega^2 = \frac{2k}{2m + M}$$

or

$$\omega^2 = \sqrt{\frac{2k}{2m + M}}.$$

(b) Energy Consideration

The total energy E of the system is the sum of the kinetic energy of the mass, the rotational kinetic energy of the pulley and the potential energy of the spring.

Translational KE of mass $m = \dfrac{1}{2}m\left(\dfrac{dx}{dt}\right)^2$

Rotational KE of pulley $= \dfrac{1}{2}I\left(\dfrac{d\theta}{dt}\right)^2 = \dfrac{1}{4}Mr^2\left(\dfrac{d\theta}{dt}\right)^2$

$$= \frac{1}{4}M\left(\frac{dx}{dt}\right)^2 \qquad \left(\because \frac{dx}{dt} = r\frac{d\theta}{dt}\right)$$

PE of the spring $= \dfrac{1}{2}kx^2$

∴ Total energy of the system

$$E = \text{KE of mass} + \text{KE of pulley} + \text{PE of spring}$$

$$= \frac{1}{2}\left(m + \frac{M}{2}\right)\left(\frac{dx}{dt}\right)^2 + \frac{1}{2}kx^2$$

Since the total energy of the system remains constant, i.e.

$$\frac{dE}{dt} = 0$$

which gives

$$\left(m + \frac{M}{2}\right)\frac{dx}{dt}\frac{d^2x}{dt^2} + kx\frac{dx}{dt} = 0$$

Since $\dfrac{dx}{dt} \neq 0$ always.

Hence,

$$\frac{d^2x}{dt^2} = -\frac{2k}{2m + M}x$$

giving

$$\omega = \sqrt{\frac{2k}{2m + M}}$$

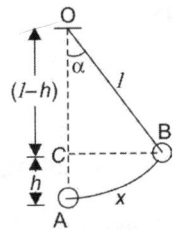

Example 1.6: A simple pendulum consists of a rod of mass m and length l which is pivoted at O and carries a mass M at the other end as shown in Fig. 1.16. Using energy consideration determines the frequency of the pendulum if (a) $m \leq M$, and (b) m is comparable with M.

Fig. 1.16

Solution: Let x be the displacement of the pendulum at any instant of time, where its velocity is $\dfrac{dx}{dt}$ and α is the angle subtended with the vertical.

(a) Mass of rod is not negligible. Then

$$\text{KE} = \text{ME of mass } M + \text{KE of rod}$$

$$= \frac{1}{2}M\left(\frac{dx}{dt}\right)^2 + \frac{1}{2}I\omega^2$$

$$= \frac{1}{2}M\left(\frac{dx}{dt}\right)^2 + \frac{1}{2}\left(\frac{ml^2}{3}\right)\left(\frac{1}{l}\frac{dx}{dt}\right)^{2*}$$

$$= \frac{1}{2}\left(M + \frac{m}{3}\right)\left(\frac{dx}{dt}\right)^2.$$

Since rod is uniform, its weight mg acts through centre of mass.

Hence, PE of the system is

$$\text{PE} = \text{PE of mass} + \text{PE of rod}$$

$$= Mgh + \frac{mgh}{2}$$

But

$$h = l(1 - \cos\alpha) = 2l\sin^2\alpha/2$$

∴

$$\text{PE} = \left(M + \frac{m}{2}\right)g \cdot h$$

$$= \left(M + \frac{m}{2}\right)g 2l\sin^2\alpha/2$$

$^*\ M \cdot I \cdot T = \dfrac{1}{3}ml^2$ and $\omega = \dfrac{v}{r} = \dfrac{1}{l}\dfrac{dx}{dt}$, $r = l$ and $v = \dfrac{dx}{dt}$.

For small value of α and displacement $\sin \alpha = \alpha = \dfrac{x}{l}$

$$\therefore \quad \sin^2 \alpha/2 \approx (\alpha/2)^2 = \left(\dfrac{x}{2l}\right)^2$$

$$\therefore \qquad PE = \left(M + \dfrac{m}{2}\right) g 2l \dfrac{x^2}{4l^2}$$

$$= \dfrac{1}{2}\left(M + \dfrac{m}{2}\right) \dfrac{g}{l} \cdot x^2$$

Hence, total energy

$$E = KE + PE$$

$$= \dfrac{1}{2}\left(M + \dfrac{m}{3}\right)\left(\dfrac{dx}{dt}\right)^2 + \dfrac{1}{2}\left(M + \dfrac{m}{2}\right)\dfrac{g}{l} x^2$$

Since energy is constant, hence setting $\dfrac{dE}{dt} = 0$, we have

$$\left(M + \dfrac{m}{3}\right)\dfrac{dx}{dt}\dfrac{d^2x}{dt^2} + \left(M + \dfrac{m}{2}\right)\dfrac{g}{l} x \dfrac{dx}{dt} = 0$$

or

$$\dfrac{d^2x}{dt^2} = -\dfrac{\left(M + \dfrac{m}{2}\right)}{\left(M + \dfrac{m}{3}\right)}\dfrac{g}{l} x$$

i.e. motion is SHM with angular frequency

$$\omega = \left\{\dfrac{\left(M + \dfrac{m}{2}\right)}{\left(M + \dfrac{m}{3}\right)}\dfrac{g}{l}\right\}^{1/2} = \left\{\dfrac{3}{2}\dfrac{g}{l}\dfrac{(2M + m)}{(3M + m)}\right\}^{1/2}$$

In case $m \ll M$, we can neglect m in comparison to M, we have

$$\omega = \sqrt{\dfrac{g}{l}}$$

Example 1.7: A string of length L is stretched with a tension T between fixed points A and B as shown in Fig. 1.17. A mass m is fixed at a distance a from point A. Determine the frequency of the vertical oscillations of mass m assuming that the tension remains constant for small displacements.

Solution: Fig. 1.17b shows the displaced position of mass. The net restoring force acting on the mass is

$$F = -T \sin \theta_1 - T \sin \theta_2$$

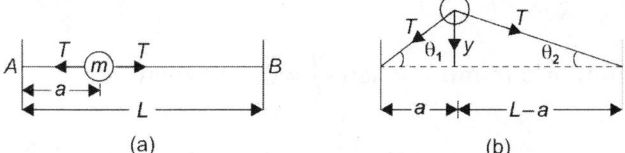

(a)   (b)

Fig. 1.17

If displacement y is small then

$$\sin\theta_1 \simeq \tan\theta_1 = \frac{y}{a} \quad \text{and} \quad \sin\theta_2 \simeq \tan\theta_2 = \frac{y}{L-a}$$

$$\therefore \qquad F = -T\left[\frac{1}{a} + \frac{1}{L-a}\right]y = -\frac{TLy}{a(L-a)} = -ky$$

where $k = \dfrac{TL}{a(L-a)}$ and $\omega = \sqrt{\dfrac{k}{m}} = \sqrt{\dfrac{TL}{a(L-a)m}}$.

Example 1.8: Show that in compound pendulum, there are four colinear points, the time period of the pendulum when suspended from these points is same. Also calculate minimum and maximum time period of pendulum.

Solution: When the body is suspended from the centre of suspension, its period of oscillation is given by

$$T = 2\pi\sqrt{\frac{\dfrac{k^2}{l}+l}{g}}$$

where l is the distance of S from C [Fig. 1.5, Eq. 1.20(a)].

Let the body now be inverted and suspended from point O such that $OC = k^2/l$. This point is called *centre of oscillation*. The new period of oscillation T is then

$$T = 2\pi\sqrt{\frac{\dfrac{k^2}{l}+\dfrac{k^2}{k^2/l}}{g}} = 2\pi\sqrt{\frac{k^2/l+l}{g}} = T$$

Thus, the period of oscillation about S and O are equal i.e. centre of suspension and oscillation are interchangeable.

Similarly, centre of suspension can be at distance l from C on the other side of centre of mass m as shown in Fig 1.18a, corresponding to this point S as point of suspension O' is the centre of oscillation. Hence, there are four colinear points (S, O', O, S') for which time period is same.

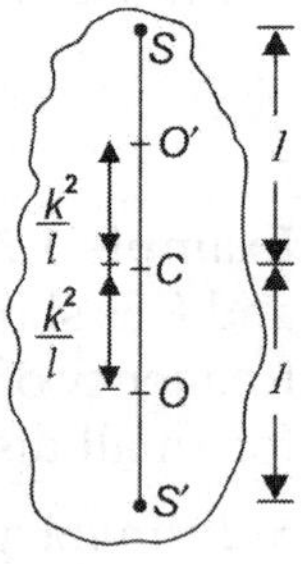

Time period is given by

$$T^2 = \frac{4\pi^2}{g}\left(\frac{k^2}{l}+l\right)$$

differentiating with respect to l, we get

Fig. 1.18a

$$2T\frac{dt}{dl} = \frac{4\pi^2}{g}\left(-\frac{k^2}{l^2}+1\right)$$

T will be maximum or minimum when $\dfrac{dt}{dl} = 0$, i.e. when

$$-\frac{k^2}{l^2}+1 = 0 \qquad \text{or } l = \pm k$$

With this value l, the value of $\dfrac{d^2T}{dl^2}$ is negative. Hence

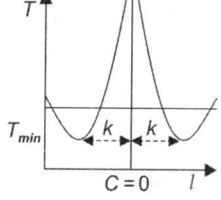

$$T_{min} = 2\pi\sqrt{\dfrac{\dfrac{k^2}{k}+k}{g}} = 2\pi\sqrt{\dfrac{2k}{g}}$$

Fig. 1.18b

If the pendulum is suspended from its centre, i.e., $l = 0$, then $T = \alpha$.

This is the maximum period. This means the pendulum suspended from its centre of mass will not oscillate. The graph between T and l is shown in Fig. 1.18b.

Example 1.9: A massless spring with no mass attached to it hangs from a rigid support. A mass m is now hung on the lower end of the spring. The mass is supported on a platform, so that the spring is relaxed. The supporting platform is suddenly removed. The mass begins to oscillate. The lowest position of the mass during the oscillation is 5 cm below the place where it was existing on the platform. (a) What is the frequency of oscillation? (b) What is the velocity, when the mass is 2.5 cm below the original resting place? Take $g = 10$ ms^{-2}.

Solution: Figure 1.19 shows that the separation between the two extreme positions of the oscillating mass is 5 cm. Therefore, the equilibrium position is 2.5 cm below the supporting platform, i.e. the force mg produces an extension of $y = 2.5$ cm in the spring. If k is spring constant, we have

$$mg = ky$$

or

$$\dfrac{m}{k} = \dfrac{y}{g} = \dfrac{2.5 \times 10^{-2}}{10} = 2.5 \times 10^{-3}$$

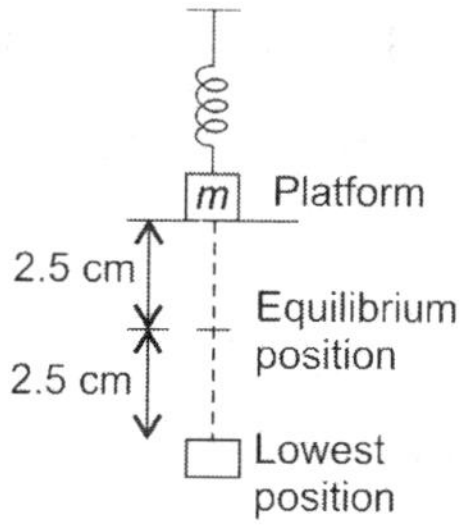

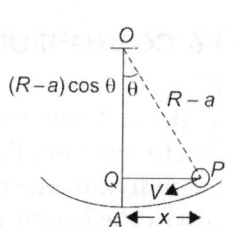

Fig. 1.19

Hence, angular frequency ω of the system is

$$\omega = \sqrt{\dfrac{k}{m}} = \sqrt{\dfrac{1}{2.5 \times 10^{-3}}} = 20 \text{ rad s}^{-1}$$

Hence, frequency $v = \dfrac{\omega}{2\pi} = \dfrac{20}{2\pi} = \dfrac{10}{\pi} = 3.18$ Hz.

When the mass is 2.5 cm below the platform, it is passing through the equilibrium position and has maximum velocity given by

$$V_{max} = A\omega = 2.5 \times 20 = 50 \text{ cm·s}^{-1} = 0.5 \text{ m·s}^{-1}.$$

Example 1.10: A small steel ball is placed a little away from the centre of a concave mirror, whose radius of curvature is R. When the ball is released, it begins to roll and oscillate about the centre of mirror. What is its period of oscillation?

Solution: Let the ball at any instant be at P, with angular displacement θ. Its energy is

(i) Potential energy $u = (R - a) (1 - \cos\theta)$, where a is radius of small sphere as shown in Fig. 1.20, PE is taken zero at A.

Fig. 1.20

(ii) If its velocity is V, its $KE = \dfrac{1}{2}mV^2$, where m is mass of steel ball.

(iii) Its rotational $KE = \dfrac{1}{2}I\omega^2$. For pure rotation $\omega = \dfrac{V}{a}$ and $I = \dfrac{2}{5}ma^2$.

$\therefore$ Total energy $E = m(R-a)(1-\cos\theta)\,g + \dfrac{1}{2}mV^2 + \dfrac{1}{2}\cdot\dfrac{2}{5}ma^2\dfrac{V^2}{a^2}$

or $\qquad\qquad E = mg(R-a)(1-\cos\theta) + \dfrac{1}{2}mV^2 + \dfrac{1}{2}\cdot\dfrac{2}{5}mV^2$

$\qquad\qquad\qquad = mg(R-a)(1-\cos\theta) + \dfrac{1}{2}m\times\dfrac{7}{5}V^2$

Since energy is conserved for motion in gravitational field, we have

$$\frac{dE}{dt} = 0$$

$\therefore \qquad \dfrac{dE}{dt} = mg\,(R-a)\sin\theta\dfrac{d\theta}{dt} + m\dfrac{7}{5}V\dfrac{dv}{dt} = 0$

But $(R-a)\dfrac{d\theta}{dt} = V$ and for small θ. $\sin\theta \approx \theta$ and $\theta = \dfrac{x}{(R-a)}$, where x is arc AP which is practically a straight line for small θ.

$\therefore \qquad \dfrac{V\cdot x\,mg}{R-a} + \dfrac{7}{5}mV\dfrac{dv}{dt} = 0$

or $\qquad\qquad \dfrac{7}{5}\dfrac{dv}{dt} = -\dfrac{gx}{R-a}$

or $\qquad$ acceleration $= \dfrac{dv}{dt} = -\dfrac{5}{7}\dfrac{g}{R-a}x$

i.e. acceleration is $\propto$ displacement x and angular frequency (ω)

$$\omega = \sqrt{\frac{5}{7}\frac{g}{R-a}}$$

and time period is $T = \dfrac{2\pi}{\omega} = 2\pi\sqrt{\dfrac{7}{5}\dfrac{(R-a)}{g}}$

for $\qquad R >> a\ T = 2\pi\sqrt{\dfrac{7}{5}\dfrac{R}{g}}.$

1.6 COMPOSITION OF SIMPLE HARMONIC MOTIONS

When a particle is simultaneously under the influence of two or more simple harmonic motions, it will describe a motion which is the resultant of all these motions. When the constituent vibrations are in a straight line, the resultant is given by the algebraic sum of the individual displacements and when at right angles, the particle describes a closed curve called *Lissajous figure*. The figure depends on the frequency ratio, ratio of amplitudes and phase difference. Let us treat the following cases:

(a) Two Simple Harmonic Vibrations of Same Angular Frequency but Different Amplitudes and Phases

Suppose we have two SHM of equal frequencies but of different amplitudes and phase constants acting on a particle in Y direction. The displacement Y_1 and Y_2 of the two harmonic motions of the same angular frequency ω are given by

$$y_1 = A_1 \sin(\omega t + \phi_1) \qquad (1.29)$$

and $$y_2 = A_2 \sin(\omega t + \phi_2) \qquad (1.29a)$$

where A_1 and A_2 are the amplitudes and ϕ_1 and ϕ_2 are the phase constants of the two motions. The resultant of these two harmonic motions is found using superposition principle. Thus

$$
\begin{aligned}
Y &= y_1 + y_2 \\
&= A_1 \sin(\omega t + \phi_1) + A_2 \sin(\omega t + \phi_2) \\
&= A_1 \sin \omega t \cos \phi_1 + A_1 \cos \omega t \sin \phi_1 \\
&\quad + A_2 \sin \omega t \cos \phi_2 + A_2 \cos \omega t \sin \phi_2
\end{aligned}
$$

or
$$
\begin{aligned}
y &= (A_1 \cos \phi_1 + A_2 \cos \phi_2) \sin \omega t \\
&\quad + (A_1 \sin \phi_1 + A_2 \sin \phi_2) \cos \omega t
\end{aligned}
$$

putting $$A \cos \theta = A_1 \cos \phi_1 + A_2 \cos \phi_2$$

and $$A \sin \theta = A_1 \sin \phi_1 + A_2 \sin \phi_2$$

we have

$$
\begin{aligned}
y &= A \sin \omega t \cos \theta + A \cos \omega t \sin \theta \\
&= A \sin(\omega t + \theta) \qquad (1.30)
\end{aligned}
$$

where $$A^2 = A_1^2 + A_2^2 + 2A_1 A_2 \cos(\phi_1 - \phi_2) \qquad (1.31)$$

and $$\tan \theta = \frac{A_1 \sin \phi_1 + A_2 \sin \phi_2}{A_1 \cos \phi_1 + A_2 \cos \phi_2} \qquad (1.31a)$$

Thus, we observe that the resultant effect of two collinear SHMs of equal frequencies is a SHM of the same frequency but having amplitude and phase constant given by equation (1.31) and (1.31a) respectively.

Let us consider the following special cases:

(i) If $\phi_1 - \phi_2 = 2n\pi$, where $n = 0, 1, 2, \ldots$ etc., i.e. they have phase difference which is even multiple of π then

$$A^2 = (A_1 + A_2)^2$$

or $$A = (A_1 + A_2)$$

(ii) If $\phi_1 - \phi_2 = (2n + 1)\pi$, where $n = 0, 1, 2, \ldots$ etc. the component vibrations are in opposite phase then

$$A_2 = (A_1 - A_2)^2 \text{ or } A = (A_1 - A_2)$$

Obviously, in this case if $A_1 = A_2$, $A = 0$ i.e. the particle will remain at rest.

(iii) If $A_1 = A_2$ and $\phi_1 \neq \phi_2$ then

$$A = 2A_1 \cos \frac{1}{2}(\phi_1 - \phi_2)$$

and $$\tan \theta = \frac{\sin \phi_1 + \sin \phi_2}{\cos \phi_1 + \cos \phi_2} = \tan \frac{1}{2}(\phi_1 + \phi_2)$$

or $$\theta = \frac{1}{2}(\phi_1 + \phi_2)$$

The amplitude will be maximum $A = 2A$, when $\cos\dfrac{1}{2}(\phi_1 - \phi_2) = 1$ or $\phi_1 - \phi_2 = 2n\pi$, where n is an integer. This is the case when the component vibrations are in phase.

The amplitude will be zero when $\cos\dfrac{1}{2}(\phi_1 - \phi_2) = 0$ or $\phi_1 - \phi_2 = (m + 1)$ this is the case when the component vibrations are in opposite phase.

(b) Vector Addition of Amplitudes (Geometrical Method)

The geometrical method is explained in Figs 1.21a and 1.21b, for obtaining the resultant of two SHMs of same frequencies. In Fig. 1.21a, OA is a rotating vector of length A_1—the amplitude of first SHM making an angle $(\omega t + \phi_1)$ with the X-axis at time t, where ω is the angular frequency of the oscillation and ϕ_1 its phase constant. The projection ON_1 of OA, on X-axis is the displacement x_1 of this motion at time t. Similarly, OB is a rotating vector of length A_2 at an angle $(\omega t + \phi_2)$. Its projection ON_2 on the X-axis is the second SHM of same frequency ω, amplitude A_2, and phase constant ϕ_2. The superposition vector sum of these two motions is then represented by the vector OC as defined by the parallelogram law of vector addition. As OA and

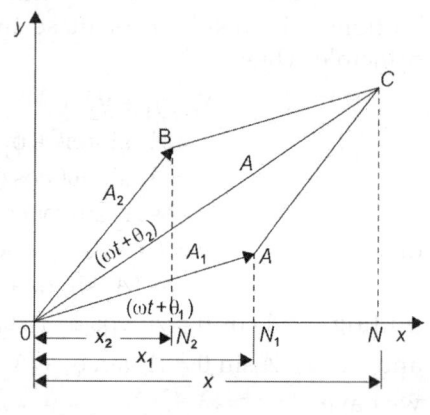

Fig. 1.21a

OB rotate at the same angular frequency, we can imagine the parallelogram $OACB$ rotating at the same angular frequency. The resulting vector OC can be obtained as the vector sum of OA and AC as shown in Fig. 1.21b, vector AC being equal to vector OB. Since angle $\angle AON_1 = \omega t + \phi_1$ and $\angle BON_2 = (\omega t + \phi_2)$, the angle between OA and OB is $(\phi_2 - \phi_1) = \phi$. Thus, we have

$$A^2 = (A_1 + A_2 \cos\phi)^2 + (A_2 \sin\phi)^2$$
$$= A_1^2 + A_2^2 + 2A_1A_2 \cos\phi$$
$$= A_1^2 + A_2^2 + 2A_1A_2 \cos\phi(\phi_2 - \phi_1)$$

which is same as shown in Eq. (1.131).

The resultant phase of the motion is given by $\angle CON$. Let this be $(\omega t + \delta)$, where δ is the phase constant of the resultant motion. It is evident from Fig. 1.18b that

$$\delta = \beta + \phi_1$$
$$\therefore \qquad \tan\delta = \tan(\beta + \phi_1)$$
$$= \frac{\tan\beta + \tan\phi_1}{1 + \tan\beta \tan\phi_1}$$

and $\qquad \tan\beta = \dfrac{A_2 \sin(\phi_2 - \phi_1)}{A_1 + A_2 \cos(\phi_2 - \phi_1)}$

Substituting for $\tan\beta$ and simplifying, we have

$$\tan\delta = \frac{A_1 \sin\phi_1 + A_2 \sin\phi_2}{A_1 \cos\phi_1 + A_2 \cos\phi_2}$$

which is the equation obtained earlier.

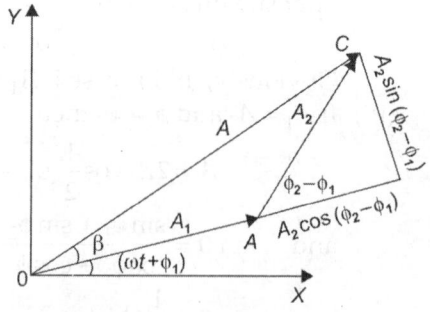

Fig. 1.21b

1.7 SUPERPOSITION OF MANY SIMPLE HARMONIC MOTIONS

The method of combining two harmonic oscillations outlined in previous section can readily be extended to an arbitrarily large number of oscillations. The general case in which the amplitude, frequencies and initial phase of the component oscillatons are all different is of no great importance, as it does not find applications in physics. Two situations in particular are of great interest and have wide applications. These are as follows:

(i) Superposition of large number of harmonic oscillations, all of the same frequency and amplitude and with equal successive initial phase difference.

(ii) Superposition of number of harmonic oscillation, all of the same amplitude and initial phase difference and with equal successive frequency difference.

The former finds application in the analysis of multisource interference effects in Optics (such as multibeam interferometer and diffraction), while the latter have special relevance to the problem of wave groups of packets. We shall deal with these two cases separately.

(i) Successive Phase Difference

Let there be n component vibrations of the same amplitude a and frequency ω but with a progressive phase difference ϕ. The different components then are

$$y_1 = a \sin \omega t$$
$$y_2 = a \sin (\omega t + \phi)$$
$$y_3 = a \sin (\omega t + 2\phi)$$
$$\vdots$$
$$y_n = a \sin \{\omega t (n-1)\phi\}$$

These functions can be written in exponential form like

$$y_1 = ae^{i\omega t}$$
$$y_2 = ae^{i(\omega t + \phi)}$$
$$y_3 = ae^{i(\omega t + 2\phi)}$$
$$\vdots$$
$$y_n = ae^{i\{\omega t + (n-1)\phi\}}.$$

The sine terms of these functions are the representation of the given functions. In fact, in such representation, real as well as imaginary parts, both represent the SHM functions. Thus, the superposition of all these gives the complex function $y(t)$, whose sine term gives the resultant of all these functions. Thus

$$y(t) = a\left[e^{i\omega t} + e^{i(\omega t + \phi)} + e^{i(\omega t + 2\phi)} + \ldots + e^{\{i\omega t + (n-1)\phi\}}\right]$$

$$= ae^{i\omega t}\left[1 + e^{i\phi} + e^{2i\phi} + \ldots + e^{i(n-1)\phi}\right]$$

the expression within the bracket is a geometrical series and sum[2] is given by

$$S_n = \left[1 + e^{i\phi} + e^{2i\phi} + \ldots + e^{i(n-1)\phi}\right] = \frac{e^{in\phi} - 1}{e^{i\phi} - 1}$$

---

2. $S_n = 1 + e^{i\phi} + e^{2i\phi} + \ldots + e^{i(n-1)\phi}$ put $= e^{i\phi} = A$

∴ $S_n = 1 + A + A^2 + \ldots + A^{(n-1)}$

∴ $AS_n = A + A^2 + A^3 + \ldots + A^n$

∴ $(A-1)S_n = A^n - 1$ or $S_n = \dfrac{A^n - 1}{A - 1} = \dfrac{e^{in\phi} - 1}{e^{i\phi} - 1}$

$$\therefore \qquad y(t) = ae^{i\omega t}\frac{e^{in\phi}-1}{e^{i\phi}-1}$$

$$= ae^{i\omega t}\cdot \frac{e^{\frac{in\phi}{2}}\left(e^{\frac{in\phi}{2}}-e^{-\frac{in\phi}{2}}\right)}{e^{\frac{i\phi}{2}}\left(e^{\frac{i\phi}{2}}-e^{-\frac{i\phi}{2}}\right)}$$

$$= a\frac{\sin\left(\dfrac{n\phi}{2}\right)}{\sin\dfrac{\phi}{2}}e^{i\omega t}\cdot e^{i(n-1)\frac{\phi}{2}}$$

$$= a\frac{\sin\left(\dfrac{n\phi}{2}\right)}{\sin\dfrac{\phi}{2}}e^{i\left\{\omega t+(n-1)\frac{\phi}{2}\right\}} \qquad\qquad (1.32)$$

Recovering the sine terms from the complex function, one gets the resultant motion

$$y = a\frac{\sin\left(\dfrac{n\phi}{2}\right)}{\sin\dfrac{\phi}{2}}\sin\left\{\omega t+(n-1)\frac{\phi}{2}\right\} \qquad \left(\sin\frac{\phi}{2}\approx\frac{\phi}{2},\phi\text{ being small}\right)$$

Now, amplitude of the motion is

$$A = a\frac{\sin\left(\dfrac{n\phi}{2}\right)}{\sin\dfrac{\phi}{2}} \qquad\qquad (1.33)$$

This expression is used in the analysis of the diffraction pattern by a plane optical diffraction grating. When n become very large and ϕ become very small, then:

$$\sin\frac{\phi}{2}\to\frac{\phi}{2}$$

and amplitude A becomes

$$A = a\frac{\sin\dfrac{n\phi}{2}}{\sin\dfrac{\phi}{2}} = a\frac{\sin\dfrac{n\phi}{2}}{\dfrac{\phi}{2}}$$

$$= an\frac{\sin\dfrac{n\phi}{2}}{\left(\dfrac{n\phi}{2}\right)}$$

and in limiting case when $n\phi\to 0$. The amplitude is

$$A = an \qquad\qquad (1.34)$$

From equation (1.32) it is apparent that the resultant phase $\theta = (n-1)\dfrac{\phi}{2}$.

The same result can be obtained vectorially by applying graphical method. The lines separating the displacements to be combined become the sides of an incomplete regular polygon and the line closing is the resultant displacement A and having the phase θ [Fig. 1.22(a)].

When the amplitude a and the successive phase difference ϕ is infinitely small, i.e., both na and $n\phi$ are finite, then the vibration polygon coincides with its circumscribing circle. The resultant displacement R is represented by chord OA of the circular arc OBA and resultant phase by $\angle AOD$ [Fig. 1.22(b)].

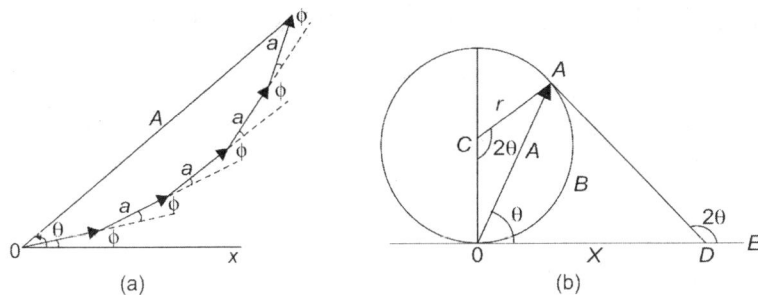

Fig. 1.22

From Fig. 1.22b geometrically

$$\angle ADE = n\phi \quad \text{and} \quad \angle AOD = \theta$$

$$A = OA = 2r \sin \theta = 2r \sin \left(\frac{n\phi}{2} \right)$$

where r is the radius of circumscribing circle.
Also

$$na = \text{Arc } OBA = 2r\theta$$

$$r = \frac{na}{2\theta}$$

Hence,

$$A = \frac{2na}{2\theta} \sin \theta = na \frac{\sin \theta}{\theta}$$

$$A = na \frac{\sin \dfrac{n\phi}{2}}{\dfrac{n\phi}{2}}$$

This is the result obtained analytically.

(ii) Oscillations of Equal Amplitudes, Equal Phase Constant and Equal Successive Frequency Differences

Consider a superposition of N different harmonic oscillations having equal amplitudes a, equal phase constants (assumed zero, for simplicity) and angular frequencies distributed uniformly between the lowest frequency ω_1 and the highest frequency ω_2. Such situation finds applications in propagation of pulse and wave packets.

Here we shall only obtain the resultant motion corresponding to the above superposition, which is given by

$$\begin{aligned} Y = {} & a \sin \omega_1 t + a \sin (\omega_1 + \delta\omega) t \\ & + a \sin (\omega_1 + 2\delta\omega) t + \dots + \dots + a \sin (\omega_2) \end{aligned} \tag{1.35}$$

where

$$\delta\omega = \frac{\omega_2 - \omega_1}{N-1} = \frac{\Delta\omega}{N-1} \tag{1.36}$$

where $\Delta\omega = (\omega_2 - \omega_1)$ is called the *band width* and $\delta\omega$ is the frequency spacing between neighbouring components. As we have done previously the superposition is the imaginary part of the function $f(t)$ where

$$f(t) = a\left[e^{i\omega_1 t} + e^{i(\omega_1 + \delta\omega)t} + e^{i(\omega_1 + 2\delta\omega)t} + \dots + e^{i\{\omega_1 + (N-1)\delta\omega\}t} \right]$$

$$= ae^{i\omega_1 t}\left[1 + p + p^2 + p^3 + \dots + p^{N-1} \right]$$

$$= ae^{i\omega_1 t} S$$

where

$$p = e^{i\delta\omega t} \quad \text{and} \quad S = 1 + p + p^2 + p^3 + \dots + p^{N-1}$$

$$\because \qquad S = 1 + p + p^2 + p^3 + \dots + p^{N-1}$$

$$\therefore \qquad ps = p + p^2 + p^3 + \dots + p^{N-1} + p^N$$

$$\therefore \qquad (p-1)S = p^N - 1$$

or

$$S = \frac{p^N - 1}{p - 1} = \frac{e^{iN\delta\omega t} - 1}{e^{i\delta\omega t} - 1}$$

$$= \frac{e^{\frac{iN\delta\omega t}{2}}\left[e^{\frac{iN\delta\omega t}{2}} - e^{\frac{-iN\delta\omega t}{2}} \right]}{e^{\frac{iN\delta\omega t}{2}}\left[e^{\frac{i\delta\omega t}{2}} - e^{\frac{-i\delta\omega t}{2}} \right]}$$

or

$$S = \exp\left\{ \frac{i(N-1)}{2}\delta\omega t \right\} \frac{\sin\left(\frac{1}{2}N\delta\omega t\right)}{\sin\left(\frac{1}{2}\delta\omega t\right)}$$

$$\therefore \qquad f(t) = a \exp i\left\{ \omega_1 t + \frac{1}{2}(N-1)\delta\omega t \right\} \frac{\sin\left(\frac{1}{2}N\delta\omega t\right)}{\sin\left(\frac{1}{2}\delta\omega t\right)}$$

But

$$\omega_1 + \frac{1}{2}(N-1)\delta\omega = \omega_1 + \frac{1}{2}(N-1)\frac{(\omega_2 - \omega_1)}{N-1}$$

$$= \frac{\omega_1 + \omega_2}{2} = \omega_a$$

where ω_a is the average of two extreme frequencies then we have

$$f(t) = ae^{i\omega_a t} \frac{\sin\left(\frac{1}{2}N\delta\omega t\right)}{\sin\left(\frac{1}{2}\delta\omega t\right)} \tag{1.37}$$

Now, y is the sin function of $f(t)$. Hence

$$y = a \frac{\sin\left(\frac{1}{2}N\delta\omega t\right)}{\sin\left(\frac{1}{2}\delta\omega t\right)} \sin \omega_a t$$

$$= A_m \sin \omega_a t \tag{1.38}$$

where A_m is the modulation amplitude.

$$A_m = a \frac{\sin\left(\frac{1}{2}N\delta\omega t\right)}{\sin\left(\frac{1}{2}\delta\omega t\right)} \tag{1.39}$$

Since A_m is time dependent, the resulting oscillation is not harmonic. This reduces to its familiar form of beats when there are just two terms. Setting $N = 2$ in Eq. (1.39), we have

$$A_m = a \frac{\sin\left(\delta\omega t\right)}{\sin\left(\frac{1}{2}\delta\omega t\right)}$$

$$= 2a\cos\left(\frac{1}{2}\delta\omega t\right) \tag{1.40}$$

Thus

$$y = 2a \cos\left(\frac{\omega_2 - \omega_1}{2}\right) t \sin\left(\frac{\omega_1 + \omega_2}{2}\right) t \tag{1.40a}$$

which is the equation of beats.

This result of Eq. (1.40a) is illustrated in Fig 1.23. It represents a sinusoidal oscillations at the average frequency $(\omega_1 + \omega_2)/2$ having a displacement amplitude of $2a$ which modulates; that it varies between $2a$ and zero under the influence of the cosine term of a much lower frequency equal to half the difference $(\omega_2 - \omega_1)/2$ between the original frequencies.

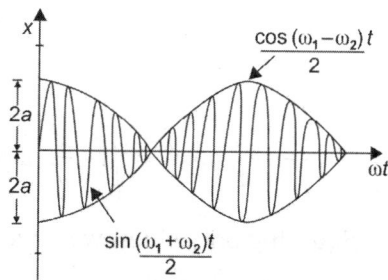

Fig. 1.23

When ω_1 and ω_2 are almost equal, the sine term has a frequency very close to both (ω_1 and ω_2) which the cosine envelope modulates the amplitude at a frequency $(\omega_2 - \omega_1)/2$ which is very slow.

Acoustically, this growth and decay of the amplitude is registered as beats of strong reinforcement when two sounds of almost equal frequency are heard. The frequency of 'beats' is $\left(\frac{\omega_2 - \omega_1}{2\pi}\right)$ the difference between the separate frequencies (not half the difference), because the maximum amplitude of $2a$ occurs twice in every period associated with the frequency $(\omega_2 - \omega_1)/2$. We will discuss again with coupling of two oscillators.

1.8 SUPERPOSITIONING OF TWO PERPENDICULAR HARMONIC OSCILLATIONS (LISSAJOUS FIGURES)

Suppose a particle is subjected simultaneously to two perpendicular harmonic motions, one along the X-axis and the other along the Y-axis. The particle will describe a close curve called *Lissajous figure*. The shape of the curve depends on the frequency ratio of the two harmonic motions, their amplitues, and phase difference between them.

(i) Oscillations Having Equal Frequencies

Let a particle be subjected to two perpendicular harmonic oscillations of equal frequency, which are mutually perpendicular. Let a_1 and a_2 respectively be the amplitudes of the x and y oscillations. For simplicity, let us assume that the phase constant of X-oscillation is zero and that of the Y-oscillation is ϕ, so that ϕ is the phase difference between them. In doing so, there is no loss of generality. Thus, the two rectangular SHM can be written as

$$x = a_1 \sin \omega t \tag{i}$$

and
$$y = a_2 \sin (\omega t + \phi) \tag{ii}$$

where x and y are displacements along two mutually perpendicular directions. In order to get the information of the path of the resultant vibration, we eliminate t between the above equations. We have

$$\sin \omega t = \frac{x}{a_1} \text{ and } \cos \omega t = \sqrt{1 - \frac{x^2}{a_1^2}}$$

Now

$$\frac{y}{a_2} = \sin \omega t \cos \phi + \cos \omega t \sin \phi$$

putting the values of $\sin \omega t$ and $\cos \omega t$, we have

$$\frac{y}{a_2} = \frac{x}{a_1} \cos \phi + \sqrt{1 - \frac{x^2}{a_1^2}} \sin \phi$$

or
$$\frac{y}{a_2} - \frac{x}{a_1} \cos \phi = \sqrt{1 - \frac{x^2}{a_1^2}} \sin \phi$$

Squaring both sides, we have

$$\left(\frac{y}{a_2} - \frac{x}{a_1} \cos \phi \right)^2 = \left(1 - \frac{x^2}{a_1^2} \right) \sin^2 \phi$$

or
$$\frac{y^2}{a_2^2} - \frac{2xy}{a_1 a_2} \cos \phi + \frac{x^2}{a_1^2} \cos^2 \phi = \sin^2 \phi - \frac{x^2}{a_1^2} \sin^2 \phi$$

or
$$\frac{y^2}{a_2^2} - \frac{2xy}{a_1 a_2} \cos \phi + \frac{x^2}{a_1^2} = \sin \phi \tag{1.41}$$

This is the equation of an ellipse inclined to the axes of coordi-nates and the ellipse may be inscribed in a rectangle of sides $2a_1$ and $2a_2$.

Let us examine the following particular cases:

(i) When $\phi = 0$, i.e. when the component vibrations are in phase. Equation (1.41) becomes

$$\left(\frac{y^2}{a_2^2} - \frac{2xy}{a_1 a_2} + \frac{x^2}{a_1^2} \right) = 0$$

or

$$\pm \left(\frac{y}{a_2} - \frac{x}{a_1} \right) = 0$$

or

$$y = \frac{a_2}{a_1} x \qquad (1.42)$$

This equation represents a pair of coincident straight lines lying in the first and third quadrants with slope $\dfrac{a_2}{a_1}$, i.e. equal to the ratio of the amplitudes of the motions.

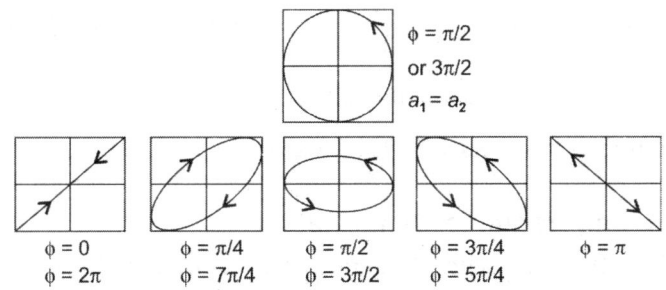

Fig. 1.24

(ii) When phase difference $\phi = \dfrac{\pi}{4}$

$$\sin \frac{\pi}{4} = \frac{1}{\sqrt{2}} \quad \text{and} \quad \cos \frac{\pi}{4} = \frac{1}{\sqrt{2}}$$

and equation 1.41 reduces to

$$\frac{x^2}{a_1^2} + \frac{y^2}{a_2^2} - \frac{\sqrt{2}xy}{a_1 a_2} = \frac{1}{2} \qquad (1.43)$$

This equation represents an oblique ellipse shown in Fig. 1.24 described in clockwise direction.

(iii) When phase difference $\phi = \dfrac{\pi}{2}$ the general equation reduces to

$$\frac{x^2}{a_a^2} + \frac{y^2}{a_2^2} = 1 \qquad (1.44)$$

This is the equation of symmetrical ellipse with its semi axis a_1 and a_2 coinciding with the X- and Y-axis respectively.

If in addition to $\phi = \dfrac{\pi}{2}$, the amplitudes of two SHM are equal, then general equation reduces to

$$x^2 + y^2 = a^2 \qquad (1.44a)$$

which is the equation of a circle. Thus, a uniform circular motion may be considered to be constituted of two similar SHM at right angles to each other and having phase difference $\dfrac{\pi}{2}$.

(iv) When phase difference $\phi = \pi$, the general equation becomes

$$\frac{x^2}{a_1^2} + \frac{y^2}{a_2^2} + \frac{2xy}{a_1 a_2} = 0$$

or $\qquad \pm\left(\dfrac{x}{a_1} + \dfrac{y}{a_2}\right) = 0$ $\hfill$ (1.45)

This represents a pair of straight lines lying in second and fourth quadrant with negative slope of magnitude $\dfrac{a_2}{a_1}$, i.e., equal to the ratio of the amplitudes of two motions.

The figure for phase difference $\dfrac{5\pi}{4}, \dfrac{3\pi}{2}, \dfrac{7\pi}{4}$ and 2π are respectively same as for phase differences $\dfrac{3\pi}{4}, \dfrac{\pi}{2}, \dfrac{\pi}{2}$ and zero. With the difference that formers are traced clockwise whereas these are traced anticlockwise (Fig. 1.24).

(ii) Oscillations Having Frequency Ratio 1:2

Let us first consider the case when the frequence ω_2 of the Y oscillation is twice the frequency ω_1 of the X oscillation, i.e. $\omega_1 = \omega$ and $\omega_2 = 2\omega$. The two SHM are given by

$$x = a_1 \sin \omega t \hfill (1.46)$$
$$y = a_2 \sin (2\omega t + \phi) \hfill (1.46a)$$

where a_1 and a_2 are their respective amplitudes and ϕ is the phase difference between them. The Lissajous figure can be obtained by eliminating t between the above two equations. From Eq. (1.46), we have

$$\sin \omega t = \frac{x}{a_1} \quad \text{and} \quad \cos \omega t = \sqrt{1 - \frac{x^2}{a_1^2}}$$

and Eq. (1.46a) gives

$$\frac{y}{a_2} = \sin 2\omega t \cos\phi + \cos 2\omega t \sin\phi$$
$$= 2\sin \omega t \cos \omega t \cos\phi + (1 - 2\sin^2 \omega t) \sin\phi$$

Substituting the value of $\sin \omega t$ and $\cos \omega t$, we have

$$\frac{y}{a_2} = 2\frac{x}{a_1}\sqrt{1 - \frac{x^2}{a_1^2}}\cos\phi + \left(1 - 2\frac{x^2}{a_1^2}\right)\sin\phi$$

Rearranging the terms, we have

$$\left\{\frac{y}{a_2} - \left(1 - \frac{2x^2}{a_1^2}\right)\sin\phi\right\} = \frac{2x}{a_1}\sqrt{1 - \frac{x^2}{a_1^2}}\cos\phi$$

Squaring, we have

$$\left\{\frac{y}{a_2} - \left(1 - \frac{2x^2}{a_1^2}\right)\sin\phi\right\}^2 = \frac{4x^2}{a_1^2}\left(1 - \frac{x^2}{a_1^2}\right)\cos^2\phi$$

or

$$\left\{\left(\frac{y}{a_2} - \sin\phi\right) + \frac{2x^2\sin\phi}{a_1^2}\right\}^2 = \frac{4x^2}{a_1^2}\cos^2\phi - \frac{4x^4}{a_1^4}\cos^2\phi$$

or

$$\left(\frac{y}{a_2} - \sin\phi\right)^2 + \frac{4x^2}{a_1^2}\sin\phi\left(\frac{y}{a_2} - \sin\phi\right) + \frac{4x^4\sin^2\phi}{a_1^4} = \frac{4x^2}{a_1^2}\cos^2\phi - \frac{4x^4}{a_1^4}\cos^2\phi$$

or

$$\left(\frac{y}{a_2} - \sin\phi\right)^2 + \frac{4x^2}{a_1^2}\frac{y}{a_2}\sin\phi - \frac{4x^2}{a_1^2}\sin^2\phi + \frac{4x^4}{a_1^4}\sin^2\phi = \frac{4x^2}{a_1^4}\cos^2\phi$$

or

$$\left(\frac{y}{a_2} - \sin\phi\right)^2 + \frac{4x^2}{a_1^2}\frac{y}{a_2}\sin\phi = \frac{4x^2}{a_1^2} - \frac{4x^4}{a_1^4}$$

or

$$\left(\frac{y}{a_2} - \sin\phi\right)^2 + \frac{4x^2}{a_1^2}\left\{\frac{x^2}{a_1^2} + \frac{y}{a_2}\sin\phi - 1\right\} = 0 \tag{1.47}$$

This is an equation of the fourth degree, which in general, represents a closed curve having two loops. For a given value of ϕ, the curve corresponding to Eq. (1.47) can be traced using the knowledge of co-ordinate geometry.

Let us examine particular cases one by one.

(i) When $\phi = 0$ or π, $\sin\phi = 0$ and the equation reduces to

$$\frac{y^2}{a_2^2} + \frac{4x^2}{a_1^2}\left\{\frac{x^2}{a_1^2} - 1\right\} = 0 \tag{1.48}$$

This equation represents the figure of eight (8).

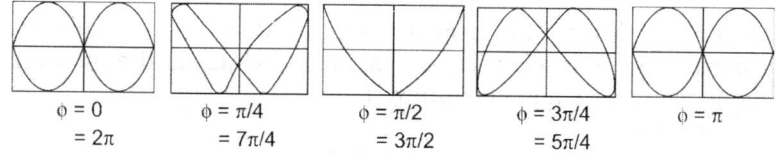

$\phi = 0$	$\phi = \pi/4$	$\phi = \pi/2$	$\phi = 3\pi/4$	$\phi = \pi$
$= 2\pi$	$= 7\pi/4$	$= 3\pi/2$	$= 5\pi/4$	

Fig. 1.25

(ii) When $\phi = \dfrac{\pi}{2}$, $\sin\phi = +1$, we get

$$\left(\frac{y}{a_2} - 1\right)^2 + \frac{4x^2}{a_1^2}\left\{\frac{x^2}{a_1^2} + \frac{y}{a_2} - 1\right\} = 0$$

or

$$\left(\frac{y}{a_2} - 1\right)^2 + \left(\frac{2x^2}{a_1^2}\right)^2 + \frac{4x^2}{a_1^2}\left(\frac{y}{a_2} - 1\right) = 0$$

or

$$\left\{\left(\frac{y}{a_2} - 1\right) + \frac{2x^2}{a_1^2}\right\}^2 = 0 \tag{1.49}$$

This represents two coincident parabolas and each of which has the equation

$$\left(\frac{y}{a_2} - 1\right) + \frac{2x^2}{a_1^2} = 0$$

or

$$x^2 = -\frac{a_1^2}{2}\left(\frac{y}{a_2} - 1\right)$$

$$= -\frac{a_1^2}{2a_2}(y - a_2)$$

The different figures obtained depending on the values of the phase difference are given in Fig. 1.25.

A study of Fig. 1.25 reveals some very interesting features of Lissajous figures, when the frequencies of the two perpendicular oscillations are in commensurate ratio.

(a) The resultant curve is inscribed in rectangle of sides $2a_1$ and $2a_2$ where a_1 and a_2 are the amplitudes of the component oscillations.

(b) The resulting motion is periodic since the curve returns to itself.

(c) The sides of the rectangle are tangential to the curve at number of points and the ratio of the number of these tangential points along the X-axis to those along Y-axis is the inverse of the ratio of the corresponding frequencies.

Example 1.11: Two vibrations along the same line are described by equations $x_1 = 0.03 \cos 10\pi t$, $x_2 = 0.03 \cos 12\pi t$, where x_1 and x_2 are measured in meters and t in seconds obtain the equation describing the resultant motion and hence find the beat period.

Solution: Using the superposition principle, the resultant motion is given by

$$x = x_1 + x_2 = 0.03\,(\cos 10\pi t + \cos 12\pi t)$$

Using the identity $\cos\alpha + \cos\beta = 2\cos\dfrac{(\alpha + \beta)}{2}\cos\dfrac{(\alpha - \beta)}{2}$, we have

$$x = 0.06 \cos \pi t \cos 11\pi t$$

and this is of the form

$$x = a \cos \omega_a t$$

with amplitude $a = 0.06 \cos \pi t$. It is clear that amplitude is not constant but varies with time and the oscillation frequency $\omega_a = 11\pi$ rad/s which is the average of the two component frequencies. Now a is maximum if

$$\cos \pi t = \pm 1$$

or $\qquad\qquad \pi t = n\pi \qquad$ where $n = 0, 1, 2, 3, \dots$

i.e. $\qquad\qquad t = 0, 1, 2, \dots$

Hence the beat period is 1 sec, i.e. the time interval between two consecutive maximum values of (amplitude)^2.

Example 1.12: A particle is simultaneously subjected to three simple harmonic motions, all of the same frequency and in the same directions. If the amplitudes are 0.5 mm, 0.4 mm and 0.3 mm respectively and the phase difference between the first and second is 45°, between the second and third is 30°, find the amplitude, the resultant displacement, and its phase relative to the first motion of amplitude 0.5 mm.

Solution: In this case let three motions be represented by equations

$$y_1 = 0.5 \sin \omega t$$
$$y_1 = 0.4 \sin (\omega t + 45°)$$

and $\qquad\qquad y_3 = 0.3 \sin (\omega t + 75°)$

The resultant of these three due to superposition is given by

$$y = y_1 + y_2 + y_3 = 0.5 \sin \omega t + 0.4 \sin (\omega t + 45°) + 0.3 \sin (\omega t + 75°)$$
$$= (0.5 + 0.4 \cos 45° + 0.3 \cos 75°) \sin \omega t + (0.4 \sin 45° + 0.3 \sin 75°) \cos \omega t$$
$$= A \sin \omega t$$

where
$$A = \sqrt{(0.86)^2 + (0.573)^2} = 1.03 \text{ mm}$$

and
$$\tan \theta = \frac{0.573}{0.86} = 0.667$$

or
$$\theta = \tan^{-1} (0.667) = 33.67°$$

∴
$$y = 1.03 \sin (\omega t + 33.67°)$$

Hence the resulting amplitude is 1.03 mm and the resulting phase with initial motion is 33.67°.

Example 1.13: Two tuning forks A and B of nearly equal frequencies are employed in an optical experiment to produce Lissajous figures. On slightly loading fork A it is observed that the cycle of change of figure slows down from 10 to 20 sec. If the frequency of fork B is 256 Hz determine the frequency of fork A before and after the loading.

Solution: Let v_1 be the frequency of fork A and t_1 be the time for a complete cycle of a Lissajous figure before the fork is loaded. Then if v_2 is the frequency of fork B, we have

$$v_1 = v_2 \pm \frac{1}{t_1}$$

$$= 256 \pm \frac{1}{10} = 256.1 \text{ Hz or } 255.9 \text{ Hz}$$

Thus, the frequency v_1 of fork A is either 255.9 Hz or 256.1 Hz. We know that on loading the frequency of fork decreases. If v is 255.9 Hz then, on loading it v_1 decreases hence, the difference $v_2 - v_1$ must increase, which would decrease the value of t_c the time for a complete cycle of change of figures with the loaded fork. But t_c is observed to increase from 10 s to 20 s. Hence the frequency of fork A cannot be 255.9 Hz, it must be 256.1 Hz before loading. After loading the frequency of fork A must become

$$v_1' = 256 + \frac{1}{20} = 256.05 \text{ Hz}$$

Example 1.14: A particle is simultaneously subjected to two SHM in the same direction each of frequency 5 Hz. If the amplitudes are 0.005 m and 0.002 m respectively and phase difference between them is 45°, find the amplitude of the resultant displacement and its phase relative to first component. Write down the expression for the resultant as function of time.

Solution: Let phase constant of the first component be zero then the phase component of the second is 45°. Since the amplitude of the two components respectively are $A_1 = 0.005$ m and $A_2 = 0.002$ m, then the amplitude (A) of the resultant motion is given by

$$A^2 = A_1^2 + A_2^2 + 2A_1 A_2 \cos \phi$$

Substituting the values we have

$$A_2 = (0.005)^2 + (0.002)^2 + 2(0.005)(0.002) \cos 45$$

$$= 25 \times 10^{-6} + 4 \times 10^{-6} + \frac{20}{\sqrt{2}} \times 10^{-6} = 43.14 \times 10^{-6}$$

or
$$A = 6.57 \times 10^{-3} \text{ m} = 0.00657 \text{ m}$$

and resultant phase is given by

$$\tan\theta = \frac{A_2\sin\phi}{A_1 + A_2\cos\phi}$$

or

$$\tan\theta = \frac{0.002\sin 45}{0.005 + 0.002\cos 45} = 0.2205$$

or

$$\theta = \tan^{-1}(0.2205) = 12.5° = \frac{\pi}{14.4}$$

Since frequency of each vibration $v = 5$ Hz

∴ Angular frequency $\omega = 2\pi v = 10\pi$ rad/s.

Using these values, the resultant motion has the equation

$$y = 6.57\times 10^{-3}\sin\left(10\pi t + \frac{\pi}{14.4}\right)$$

Here y is in metres and t in seconds.

Example 1.15: The pulley shown in Fig. 1.26 has a moment of inertia I about its axis and mass m. Find the time period of vertical oscillations of its centre of mass. The spring has spring constant k and the string does not slip over the pulley.

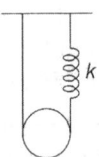

Fig. 1.26

Solution: Let us first find the equilibrium position. For rotational equilibrium of the pulley, the tension in the two strings should be equal. Only then the torque on the pulley will be zero. Let this tension be T. The extension of the spring will be $y = \dfrac{T}{k}$ as the tension in the spring will be the same as the tension in the string. For transnational equilibrium of the pulley

$$2T = mg \quad\text{or}\quad 2ky = mg \quad\text{or}\quad y = \frac{mg}{2k}$$

The spring is extended by distance $\dfrac{mg}{k}$ when the pulley is in equilibrium.

Now suppose, the the centre of mass of the pulley goes down further by a distance X. The total increase in the length of the string plus the spring is $2x$ (x of the left of the pulley and x on the right). As the string has a constant length, the extension of the spring is $2x$. Hence, the energy of the system is

$$U = \frac{1}{2}I\omega^2 + \frac{1}{2}mv^2 - mgx + \frac{1}{2}k\left\{\frac{mg}{2k} + 2x\right\}^2$$

$$= \frac{1}{2}\left\{\frac{I}{r^2} + m\right\}v^2 + \frac{m^2 g^2}{8k} + 2kx^2$$

As the system is conservative $\dfrac{du}{dt} = 0$ gives

$$0 = \left\{\frac{I}{r^2} + m\right\}v\frac{dv}{dt} + 4kx\frac{dx}{dt}$$

∵

$$\frac{dx}{dt} = v$$

We have

$$\frac{dv}{dt} = -\frac{4k}{\dfrac{I}{r^2} + m}x$$

or acceleration is proportional to displacement, i.e. motion of the centre of mass of the pulley is SHM with

$$\omega^2 = \frac{4k}{\dfrac{I}{r^2} + m}$$

and time period

$$T = 2\pi\sqrt{\dfrac{\dfrac{I}{r^2} + m}{4k}}.$$

REVIEW QUESTIONS AND PROBLEMS

1. Show that whatever be the nature of the potential function $u = f(x)$, small oscillations about an equilibrium position will always be SH.

2. (i) Prove that in SHM, the average PE equals average KE when the average is taken with respect to time over one period of motion and that each average $\frac{1}{4}kA^2$.

   (ii) Prove that when average is taken with respect to position over one cycle, the average PE equals $\frac{1}{6}kA^2$ and average KE equal to $\frac{1}{3}kA^2$.

   (iii) Explain physically why two results are different.

3. A particle is subjected simultaneously to N simple harmonic motions of the same frequency. If the amplitude of each oscillations is A_0 and ϕ is phase difference between successive oscillations, show that the amplitude of the resultant oscillation is given by

$$A = A_0 \frac{\sin\left(\dfrac{N\phi}{2}\right)}{\sin\left(\dfrac{\phi}{2}\right)}$$

4. State the principle of superposition and prove that it holds only for linear differential equations.

5. Two collinear simple harmonic motions acting simultaneously on a particle are given by

$$x_1 = A_1\cos\omega t \quad \text{and} \quad x_2 = A_2\cos(\omega t + \phi)$$

Show that the resultant motion of the particle is simple harmonic. Also obtain the expression for the amplitude and phase constant of the resultant motion in terms of A_1 and A_2 and ϕ.

6. A particle is subjected simultaneously to N simple harmonic oscillation having frequency distributed uniformly between v_1 and v_2. If the amplitude of each oscillation is A_0, initial phase of each is zero and δv is frequency difference between

successive components, show that the resultant displacement of the particle is given by

$$x(t) = A_0 \frac{\sin(\pi N \delta v t)}{\sin(\pi \delta v t)} \cos[\pi(v_1 + v_2)t].$$

7. What do you understand by periodic and simple harmonic motion? What is the criteria for the motion to be simple harmonic?

8. Deduce the expression for the velocity and acceleration of a particle executing, simple harmonic motion and discuss diagrammatically their phase relationship.

9. Using the rotating vector representation obtain the resultant motion of a particle subjected simultaneously to two SHM in same direction having equal amplitudes and equal frequencies and different in phase by $\frac{\pi}{4}$.

10. Trace graphically or analytically the motion of a particle that is subject to two perpendicular simple harmonic motion of equal frequencies, different amplitude and phase differing by (i) zero (ii) $\frac{\pi}{2}$.

11. What are Lissajous figures? A particle is subjected to two SHM of slightly different frequencies. Explain how the shape of the curve traced out by the particle changes with time and show that the frequency of repetition of a particular curve equals to the difference of the frequencies of the two component oscillations.

12. A pendulum clock gives correct time at the equator, will it gain time or loose as it is taken to the poles?

13. The force acting on a particle moving along X-axis is $F = -K(x - v_0 t)$ where K is a positive constant. An observer moving at a constant velocity v_0 along the X-axis looks at the particle. What kind of motion does he find for the particle?

14. A block of known mass is suspended from fixed support through a light spring. Can you find the time period of vertical oscillation only by measuring the extension of the spring when the block is in equilibrium.

15. A hollow sphere filled with water is used as the bob of a pendulum. Assume that the equation for SH pendulum is valid with the distance between the point of suspension and centre of mass of the bob acting as the effective length of the pendulum. If water slowly leaks out of the bob, how will the time period vary?

16. A particle of mass m is attached to three springs, A, B and C of equal force constant K as shown in Fig. 1.27. If the particle is pushed against the spring C and released, find the time period of oscillation.

$$\left[Ans.\ 2\pi \sqrt{\frac{m}{2k}} \right]$$

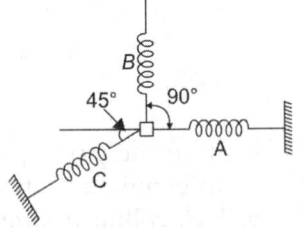

Fig. 1.27

17. The block of mass m_1 shown in Fig. 1.28 is fastened to the spring and block of mass m_2 is placed against it. (a) Find the compression of the spring in the equilibrium position. (b) The blocks are pushed; further distance $\left(\frac{2}{k}\right)(m_1 + m_2)$ against the spring and released. Find the position where the two blocks separate. (c) What is the common speed of blocks at the time of separation?

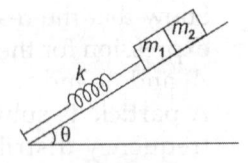

Fig. 1.28

[*Ans.* (a) $\dfrac{(m_1 + m_2)}{k}\, g \sin\theta$; (b) When the spring acquires its natural length;

(c) $\sqrt{\dfrac{3}{2k}(m_1 + m_2)\, g \sin\theta}$].

18. A uniform plate of mass M stays horizontally and symmetrically on two wheels rotating in opposite direction (Fig. 1.29). The separation between the wheels is L. The coefficient of friction between each wheel is L. Find the time period of oscillation of the plate if it is slightly displaced along its length and released.

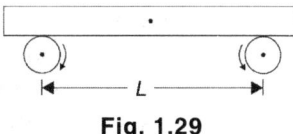

Fig. 1.29

$$\left[Ans.\ 2\pi\sqrt{\dfrac{L}{2\mu g}} \right]$$

19. Assume that a tunnel is dug across the earth passing through its centre. Find the time a particle takes to cover the length of tunnel if (a) it is projected into the tunnel with a speed of $\sqrt{gR}$, (b) it is released from height R above the tunnel, (c) it is thrown vertically upwards along the length of tunnel with a speed of $\sqrt{gR}$.

$$\left[Ans.\ \dfrac{\pi}{2}\sqrt{\dfrac{R}{g}}\ \text{in each case} \right]$$

20. Calculate the frequency of oscillation of an L–C circuit in which the inductance and capacitance have the respective values of 10 milihenry and 1 $\mu\mu$F. Calculate the energy of the system if the maximum voltage across the capacitor is 1 volt.

$$\left[Ans.\ \dfrac{10^7}{2\pi}\,\text{Hz}, 0.5 \times 10^{-12}\ \text{J} \right]$$

21. How long after the start of the motion, a harmonically oscillating point will be out of the equilibrium position by half the amplitude? Oscillation period is 24 s and the initial phase is zero.   [*Ans.* 2 sec.]

22. A mass moves under a potential $V(x) = v_0 \cos\left(\dfrac{x}{x_0}\right)$ where v_0 and x_0 are constants. Find the position of stable equilibrium. (b) Show that the frequency of small vibrations about the equilibrium position is the same as it would be if the same mass was vibrating on a spring of stiffness $\dfrac{v_0}{x_0^2}$.   [*Ans.* $x = 0$]

23. A particle is subjected simultaneously to two SHM of the same frequency and in the same direction. If their amplitudes are 5 mm and 3 mm respectively and the phase of the second component relative to the first is 30°, find the amplitude of the resultant and its phase relative to the first component.   [*Ans.* $a = 7.74$ mm and $\theta = 11.4°$]

24. Two vibrations at right angles to each other are described by the equations

$x = 3 \cos 5\pi t$ and $y = 2 \cos\left(5\pi t + \dfrac{\pi}{3}\right)$ where x and y are expressed in centimetre and t in seconds. Construct the curve for the combined motion.

[*Ans.* It is an ellipse whose axes are inclined to coordinate axis.]

25. A particle is simultaneously subjected to three simple harmonic motions, all of the same frequency and in the same direction. If the amplitudes are 1.0 cm, 0.5 cm and

0.25 cm respectively and the phase of the second relative to the first is 30° and that of the third relative to the second is 60°. Find the amplitudes of the resultant displacement and its phase relative to the first component.   [*Ans.* 1.52 cm, 19.23°]

26. What is a potential well? Show that the motion of a particle in a potential well is always simple harmonic for small oscillations, whatever may be the nature of potential function.

27. A body when suspended from a spring and made to oscillate has time period T_1 and when suspended from another spring and made to oscillate has time period T_2. Show that when the same body will be suspended from both the springs together and made it oscillate, time period will be τ such that

$$\tau = \frac{T_1 T_2}{\sqrt{T_1^2 + T_2^2}}.$$

28. The displacement of a moving particle is $x = 4 \sin (4t) + 3 \cos(4t)$ where x is in cm and t in seconds. Show that its motion is SHM; calculate its period and maximum velocity and maximum acceleration.

[*Ans.* $T = 11/2$, $V_{max} = 20$ cm·s^{-1}, $A_{max} = 80$ cm·s^{-2})

29. (a) What is potential well?
   If potential energy $u(x)$ is function of position x as $u(x) = a \cos \alpha x$, show that the position of stable equilibrium are given by

$$x = (2n + 1) \frac{\pi}{\alpha}.$$

   (b) The motion of an oscillation is governed by the relation

$$\frac{1}{2} mv^2 + \frac{1}{2} kx^2 = E.$$

   Obtain the relation for its displacement in terms of the constant given in the above relation.

30. The displacement of a simple harmonic oscillator is represen-ted by $x = a \sin \omega t$. If the displacement x and velocity are plotted on rectangular axis then prove that
   (i) Locus of (X, Y) points will be an ellipse, and
   (ii) This ellipse represents a path of constant energy.

31. What is a simple harmonic oscillator? Derive differential equation of simple harmonic oscillator and obtain expression for displacement, velocity and time period.

32. A body suspended from a spring is oscillating. Compare the kinetic and potential energies of the body with its total energy when its displacement is half of its amplitude.

SHORT ANSWER QUESTIONS

1. (a) Can a motion be periodic but not oscillatory. (b) Can a motion be oscillatory but not SH? If your answers are yes, give an example and if not, explain why?
   Ans: (a) Yes. Uniform circular motion. (b) Yes. When a ball is thrown from a height on a perfectly elastic plane surface, the motion is oscillatory but not SH as resorting force $F = mg$ = counts and not $F\alpha(-x)$.

2. A body is executing SHM of amplitude A and of period T. What is (a) the distance moved, (b) the displacement of body in time T?

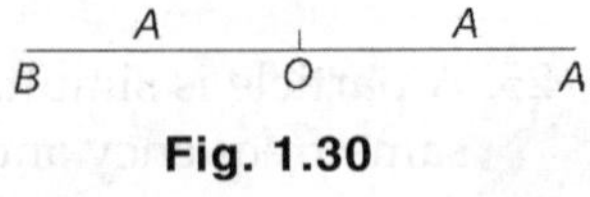

Fig. 1.30

Ans: In one time period particle executing SHM goes from O to A, A to O, O to B and B to A. Thus, distance moved is $4A$. Whereas its position is same as at $t = 0$, hence displacement is zero.

3. Can a body have acceleration without having velocity?

Ans: In SHM at extreme position velocity is zero but acceleration is maximum, i.e. $\omega^2 A$.

In SHM at mean position, acceleration is zero $(x = 0)$ and velocity is maximum, i.e. ωA. At other positions, the particle has both acceleration as well as velocity.

4. A simple pendulum L is suspended from the ceiling of a cart which is sliding without friction on an inclined plane of inclination θ. What will be the time period of the pendulum?

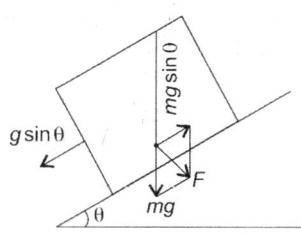

Fig. 1.31

Ans: Since trolly is moving with acceleration $g \sin \theta$ down the slope, hence pendulum bob is under the influence of force (i) mg vertically downwards, and (ii) a pseudo force mg $\sin \theta$ at an angle of $90° + \theta$ with vertical (i.e. with mg).

Hence, net force on bob is

$$F^2 = m^2 g^2 + m^2 g^2 \sin^2\theta + 2m^2 g^2 \sin\theta \cdot \cos(90 + \theta)$$
$$= m^2 g^2 + m^2 g^2 \sin^2\theta - 2m^2 g^2 \sin^2\theta$$
$$= m^2 g^2 (1 - \sin^2\theta) = m^2 g^2 \cos^2\theta$$

or $\qquad F = mg \cos\theta$

∴ Effective value of acceleration due to gravity is $g \cos \theta$. Hence, time period is

$$T = 2\pi \sqrt{\frac{L}{g \cos\theta}}.$$

5. The displacement of a particle of mass 100 g from the mean position is given by (x is in metres and t in seconds)

$$Y = 0.05 \sin 4\pi (5t + 0.4)$$

What is the force constant and total energy y of the particle?

Ans: Comparing equation with equation of SHM

$$Y = a \sin(\omega t + \phi)$$

We have amplitude $a = 0.05$ m, $\omega = 20 \pi$ rad/s, $\phi = 1.6\pi$

But $\omega^2 = \dfrac{k}{m}$ or $k = m\omega^2 = 10^{-1} \times (20\pi)^2 = 400$ N/m

Total energy $= \dfrac{1}{2} ka^2 = \dfrac{1}{2} \times 400\,(.05)^2 = 0.5$ J

6. How are each of the following parameters of a SHM affected by doubling the amplitude (i) period, (ii) maximum velocity, (iii) maximum acceleration, and (iv) total energy.

Ans: (i) Period—Time period of a SHM is given by

$$T = \frac{2\pi}{\omega} = 2\pi \sqrt{\frac{m}{k}}.$$

where m is mass of oscillator and k force constant, since time period is independent of amplitude, hence it remains unchanged.

(ii) Maximum velocity: $V_{max} = \omega a = \sqrt{\dfrac{k}{m}}\, a$, when a is doubled, V_{max} is doubled.

(iii) Maximum acceleration $f_{max} = \omega^2 a = \dfrac{k}{m} \cdot a$, when a is doubled, f_{max} is also doubled.

(iv) Total energy: $E = \dfrac{1}{2} ka^2$, when a is doubled energy becomes 4 times.

7. Determine whether the following quantities can be in the same direction for SHM: (a) displacement and velocity, (b) velocity and acceleration, (c) acceleration and displacement.

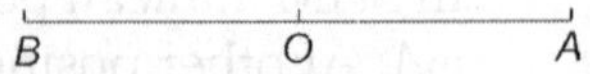

Fig. 1.32

Ans: (a) Yes, when particle goes from O to A, the velocity and displacement are in same direction.

(b) Yes, when particle moves from A to O, the velocity and acceleration are both in same direction towards O.

(c) No, displacement and acceleration are always in opposite direction in SHM.

8. A particle is subjected to two SHM in the same direction having equal amplitudes and frequencies. If the resultant amplitude is equal to the amplitude of individual motion, what is the phase difference between them?

Ans: When two SH motions are given to a particle simultaneously, the resulting amplitude due to superposition is given by
$$A^2 = a^2 + b^2 + 2ab \cos\phi$$
where a and b are the amplitudes of the two motions and ϕ is the phase difference between them. Given is
$$a = b = A$$
$\therefore$ $\qquad A^2 = A^2 + A^2 + 2A^2 \cos\phi = 2A^2 (1 + \cos\phi)$

or $\qquad A^2 = 4A^2 \cos^2 \phi/2$

or $\qquad \cos^2\phi/2 = 1/4$ or $\cos\phi = \pm 1/2$

or $\qquad \dfrac{\phi}{2} = 60°$ or $\phi = 120°$ or $\dfrac{2\pi}{3}$.

9. The force acting on a particle moving along x-axis is $F = -k (x - v_0 t)$, where k is a positive constant

An observer moving with constant velocity v_0 along X-axis looks at particle. What kind of motion does he find executed by the particle?

Ans: Since the frames of motion of particle and that of observer are inertial, hence, the observer observes particle executing SHM because for him $F\alpha\ (-x)$.

10. Two bodies A and B of equal masses are suspended from the springs of spring constants k_1 and k_2 respectively. If the bodies oscillate such that their maximum velocities are equal. Find the ratio of their amplitudes.

Ans: Maximum velocity is given by
$$V_{max} = a\omega = a\sqrt{\dfrac{k}{m}}$$

$\therefore$ $\qquad V_1 = a_1\sqrt{\dfrac{k_1}{m}}$ and $V_2 = a_2\sqrt{\dfrac{k_2}{m}}$

$\therefore$ $\qquad \dfrac{V_1}{V_2} = \sqrt{\dfrac{k_1}{k_2}}$

11. An object of mass 0.2 kg executes SHM along the X-axis with frequency of $\left(\dfrac{25}{\pi}\right)$ Hz. At the point $x = 0.04$ mm, the object has KE $= 0.5$ J and PE $= 0.4$J. Find the amplitude of the motion.

    Ans: For a particle executing SHM, when the displacement is x, its KE and PE is given by

    $$KE = \frac{1}{2}k(a^2 - x^2) \text{ and } PE = \frac{1}{2}kx^2$$

    $$k = m\omega^2 = m(2\pi v)^2$$

    Substituting values, we have

    $$0.5 = \frac{1}{2}k(a^2 - x^2)$$

    $$0.4 = \frac{1}{2}kx^2$$

    Adding

    $$0.9 = \frac{1}{2}ka^2$$

    $$k = m\left(2\pi \cdot \frac{25}{\pi}\right)^2 = m(50)^2 = 0.2 \times 25 \times 10^2 \qquad (\because k = m\omega^2)$$

    $$= 5 \times 10^2 = 500 \text{ N/m}$$

    $\therefore \qquad a^2 = \dfrac{2 \times 0.9}{500}$ or $a = 0.6$ m

12. A mass m attached to a spring oscillates with a period of 2 s. If the mass is increased by 2 kg, the period increases by 1 s. Assume Hooke's law is obeying, then calculate initial mass m.

    Ans: Time period of oscillation of a SHM is given by

    $$T = 2\pi\sqrt{\frac{m}{k}}$$

    $\therefore \qquad 2 = 2\pi\sqrt{\dfrac{m}{k}}$

    When m is increased by 2 kg, T increases by 1 s.

    $\therefore \qquad 3 = 2\pi\sqrt{\dfrac{m+2}{k}}$

    dividing, we get $\quad \dfrac{2}{3} = \sqrt{\dfrac{m}{m+2}}$ or $\dfrac{4}{9} = \dfrac{m}{m+2}$

    or $\qquad m = 1.6$ kg

13. A particle is executing SHM of amplitude 2 cm. At extreme position, the force is 4N. How much force will be at the mid-point between the mean and extreme position?

    Ans: When displacement is x, force is given by
    $$F = -kx$$

When $x = 2$ cm, the force is $4N$

$\therefore$ $4 = -k \times 2 \times 10^{-2}$

When $x = 1$ cm, the force is

$F = -k \times 1 \times 10^{-2}$

$\therefore$ $\dfrac{F}{4} = \dfrac{1}{2}$ or $F = 2N$

14. In case of a SHM, (a) what fraction of total energy is kinetic and what fraction is potential, when displacement is one-half of the amplitude? (b) at what displacement, the kinetic and potential energies are equal?

Ans: (a) $E_k = \dfrac{1}{2}k(a^2 - x^2)$. Total energy $E = \dfrac{1}{2}ka^2$

$\therefore$ $\dfrac{E_k}{E} = \dfrac{(a^2 - x^2)}{a^2}$.   When $x = \dfrac{a}{2}$

$\dfrac{E_k}{E} = \dfrac{\left(a^2 - \dfrac{a^2}{4}\right)}{a^2} = \dfrac{3}{4}$

$E_p = \dfrac{1}{2}kx^2$

$\therefore$ $\dfrac{E_p}{E} = \dfrac{\dfrac{1}{2}kx^2}{\dfrac{1}{2}ka^2} = \dfrac{\left(\dfrac{a}{2}\right)^2}{a^2} = \dfrac{1}{4}$

(b) $E_p = E_u$ i.e. $a^2 - x^2 = x^2$ or $x = \pm\dfrac{a}{\sqrt{2}}$.

15. Two particles execute SHM of same amplitude and frequency along the same straight line. They cross each other while going in opposite direction each time, their displacement is half of their amplitude. Find the phase difference between them.

Fig. 1.33

Ans: For SHM displacement

$x = a \sin \omega t$

For first particle $x = a/2$ gives

$\dfrac{a}{2} = a \sin \omega t$ or $\sin \omega t = \dfrac{1}{2} = \sin\dfrac{\pi}{6}$

or $\omega t = \pi/6$

For second particle $\dfrac{a}{2} = a \sin(\omega t + \phi)$

or $\sin(\omega t + \phi) = \dfrac{1}{2}$ or $\omega t + \phi = \dfrac{5\pi}{6}$

or $\phi = \dfrac{5\pi}{6} - \dfrac{\pi}{6} = \dfrac{4\pi}{6} = \dfrac{2\pi}{3}$.

16. The displacement of a particle executing periodic motion is given by

$$Y = 4\cos^2\left(\dfrac{1}{2}t\right)\sin 1000t$$

Find the independent components of simple harmonic motion of which it is superposition.

Ans: $y = 4\cos^2\left(\dfrac{1}{2}t\right)\sin 1000\,t = 2\,(1+\cos t)\sin 1000\,t$

or $\quad y = 2\sin 1000\,t + \sin 1001\,t + \sin 999\,t$

Hence, it is superposition of 3 simple harmonic motions of frequencies 1000 Hz, 1001 Hz and 999 Hz.

17. A block is resting on a platform which is moving vertically with SHM of period 1.0 s. At what amplitude of motion block and platform will separate?

Ans: The block and platform will separate when downwards acceleration of platform becomes greater than g.

In limiting case:

$$\omega^2 A = g \quad \text{or} \quad A = \frac{g}{\omega^2} \qquad \text{But } T = \frac{2\pi}{\omega} \quad \text{or} \quad \omega = \frac{2\pi}{T}$$

$$\therefore \qquad A = \frac{g \cdot T}{4\pi^2} = \frac{1}{4} = 0.25 \text{ m} \qquad \because T = 1\text{ s and } \pi^2 = g$$

18. A block rests on a horizontal slab which is connected to a spring and is executing SHM of frequency 2 Hz. The co-efficient of static friction between block and slab is 0.5. How large can be the amplitude, so that block does not slip on the slab?

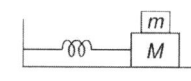

Fig. 1.34

Ans: For no slip of block

$$mA < f$$

where f is force of friction.

In limiting case $f = \mu mg$.

$\therefore mA < \mu mg \quad$ or $\quad A < \mu g$

In SHM, maximum acceleration $A = \omega^2 a$.

When a is amplitude and A is acceleration

$$\omega = \frac{2\pi}{T}$$

$\therefore \qquad A = \mu g \quad$ or $\quad \omega^2 a = \mu g \quad$ or $\quad a = \dfrac{\mu g}{\omega^2}$

or $\qquad a = \dfrac{\mu g}{\left(\dfrac{2\pi}{T}\right)^2} = \dfrac{\mu g \cdot T^2}{4\pi^2} = \dfrac{0.5 \times g}{4 \times \pi^2 \times (2)^2}$

$$= \frac{0.5}{16} \text{ m} = 3.1 \text{ cm}$$

$\because \qquad g = \pi^2 \text{ and } T = \dfrac{1}{v} = \dfrac{1}{2}\text{ s.}$

19. Two simple pendulums of length 1 m and 16 m respectively are both given small displacement in the same direction and at the same instant. They will be again in phase after the shorter pendulum completed n oscillations. Calculate n.

Ans: The time period of the two pendulums are in the ratio

$$\frac{T_1}{T_2} = \frac{1}{4}$$

They will again be in same phase when one completes (of higher frequency) one vibration more than the other one (of lower frequency). Then

$$nT_1 = (n-1) T_2 \quad \text{But} \quad \frac{T_1}{T_2} = \frac{1}{4}$$

∴ $$n = 4(n-1) \quad \text{or} \quad 3n = 4 \quad \text{or} \quad n = \frac{4}{3}$$

20. A horizontal spring block system (force constant k) executes SHM with amplitude A. When the block of mass M is passing through its equilibrium position, an object of mass m is put on it and the two move together. Find the new amplitude and frequency of vibration.

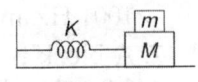

Fig. 1.35

Ans: Initial frequency is:

$$v = \frac{1}{2\pi} \sqrt{\frac{k}{M}}$$

and final frequency is

$$v' = \frac{1}{2\pi} \sqrt{\frac{k}{M+m}}$$

or $$\frac{v}{v'} = \sqrt{\frac{M}{M+m}}$$

If V is the velocity when block is passing through mean position V' after object is placed on it then conservation of momentum

$$V' = \frac{Mv}{M+m}$$

∴ New amplitude A is as given by

$$\frac{1}{2} kA^2 = \frac{1}{2} (M+m) V'^2 = \frac{1}{2} (M+m) \frac{M^2 V^2}{(M+m)^2} = \frac{1}{2} \frac{M^2 V^2}{M+m}$$

∴ $$A = \sqrt{\frac{M^2 V^2}{K(M+m)}} = \sqrt{\frac{M}{M+m}} \, a, \quad \text{where} \quad a = \sqrt{\frac{MV^2}{k}}$$

a is the initial amplitude of system without object.

21. Two blocks each of mass m are connected by a spring of stiffness constant k. A third identical block C and mass m moves with velocity V towards blocks A and B. If the collision is elastic, then calculate the maximum compression of the spring and the kinetic energy of the blocks A and B at this compression.

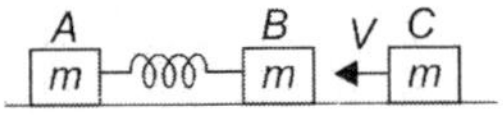

Ans: After collision B will have velocity V and at the time of maximum compression, A and B will have same velocity $V/2$. If x is maximum compression, then conservation of energy gives

$$\frac{1}{2} kx^2 + \frac{1}{2} 2m \left(\frac{V}{2}\right)^2 = \frac{1}{2} mV^2$$

or $$x = V \sqrt{\frac{m}{2k}}$$

and KE of the blocks is $\dfrac{1}{2} 2m \left(\dfrac{V}{2}\right)^2 = \dfrac{mV^2}{4}$

22. A particle is executing simple harmonic motion in a straight line *ABCD*. Its KE at *B* and *C* is half the maximum KE. If period of oscillation of particle is *T* and amplitude *a*, then find the time taken by the particle to go from *O* to *B*.

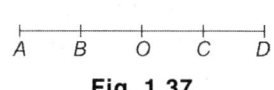

Fig. 1.37

Ans: If *x* is displacement when KE is half the maximum KE, then

$$\dfrac{1}{2} \cdot \dfrac{1}{2} ka^2 = \dfrac{1}{2} k(a^2 - x^2) \quad \text{or} \quad x = \pm \dfrac{a}{\sqrt{2}}.$$

Equation of motion is $x = a \sin \omega t$.
When $x = a/\sqrt{2}$

$$a \sin \omega t = \dfrac{1}{\sqrt{2}} \quad \text{or} \quad \sin \omega t = \dfrac{1}{\sqrt{2}} = \sin \dfrac{\pi}{4}$$

or $\qquad t = \dfrac{\pi}{4\omega} = \dfrac{2\pi}{8\omega} = \dfrac{T}{8} s$

23. A block of mass *M* falls through a height *h* on the pan of negligible mass resting on a spring of force constant *k* placed vertically. After *M* sticks to the pan, the pan performs simple harmonic motion. Find the angular frequency and amplitude of oscillation.

Ans: When mass sticks to the pan, let its compression be *x*, then
$$Mg = kx$$

Energy due to compression is $= \dfrac{1}{2} kx^2$

Initial energy $= Mgh$ is also transferred to the system on impact.
If *A* is the amplitude of motion, then

$$\dfrac{1}{2} kA^2 = \dfrac{1}{2} kx^2 + Mgh$$

or $\qquad kA^2 = k \cdot \dfrac{M^2 g^2}{k^2} + 2Mgh = \dfrac{M^2 g^2}{k} + 2Mgh$

or $\qquad A = \sqrt{\left(\dfrac{Mg}{k}\right)^2 + \dfrac{2Mgh}{k}}.$

24. Figure 1.38 shows two particles *P* and *Q* executing SHM of same frequency between lines *A* and *B*. Their positions and direction of motion at a given time are shown. Find the phase difference between these motions. If both arrows were in the same way what will be the phase difference?

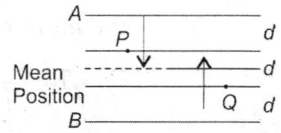

Fig. 1.38

Ans: The two particles reach opposite ends simultaneously or are crossing mean position simultaneously in opposite direction. Hence, phase difference is 180°. If they both move in same direction in given position, phase difference is

$$\phi = 2\sin^{-1}\left(\dfrac{1}{3}\right)$$

25. A ring of radius r is suspended from a point on its circumference. Determine its angular frequency.

    Ans: The system can be treated as a compound pendulum. The angular frequency of compound pendulum is given by

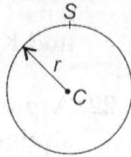

Fig. 1.39

$$\omega = \sqrt{\frac{g}{L}}$$

where L is length of equivalent simple pendulum

$$L = \frac{K^2}{l} + l$$

Here l is the distance between centre of mass of body C and point of suspension S. K is the radius of gyration about an axis passing through centre of mass and perpendicular to the plane of ring.

In case of ring $l = r$ and $K = r$

$$\therefore \qquad \omega = \sqrt{\frac{g}{\dfrac{K^2}{l} + l}} = \sqrt{\frac{g}{\dfrac{r^2}{r} + r}} = \sqrt{\frac{g}{2r}}$$

26. In an arrangement shown in Fig. 1.40, the sleeve P of mass m is fixed between two identical springs whose stiffness constant is k. The sleeve can slide without friction over a horizontal bar AB. The arrangement rotates with a constant angular speed ω about a vertical axis passing through the middle of the bar. Find the period of oscillations of the sleeves. At what value of ω will there be no oscillations of the system?

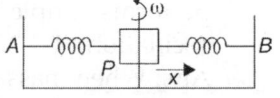

Fig. 1.40

    Ans: When sleeve moves a distance x, net force on it is

$$F = 2kx - m\omega^2 x$$

$m\omega^2 x$ is the centrifugal force.

Hence, equation of motion is

$$\frac{md^2x}{dt^2} = -F = -[2kx - m\omega^2 x]$$

or

$$\frac{d^2x}{dt^2} = -\left[\frac{2k}{m} - \omega^2\right]x$$

Hence, the motion is SHM and its time period is

$$T = \frac{2\pi}{\sqrt{\dfrac{2k}{m} - \omega^2}}$$

The sleeve will not perform oscillation when

$$\frac{2k}{m} \leq \omega^2$$

27. Two tuning forks produce Lissajous figures of the shape of straight line, ellipse, and finally to the straight line is 2s. If one of the tuning forks has a frequency of 100 Hz, find the possible frequencies of the other.

Ans: The Lissajous fitgures are straight line and ellipse, hence the frequencies of two forks are in the ratio 1:1.

As the cycle of figures is traced in 2s, the difference in frequencies must be $\dfrac{1}{2}$ Hz.

Hence, other fork has possible frequencies

$$\nu = 100 \pm \frac{1}{2} \quad \text{i.e.} \quad 100.5 \text{ Hz} \quad \text{or} \quad 99.5 \text{ Hz.}$$

28. The Lissajous figures in case of two tuning forks is a parabola. Prove that their frequencies are in the ratio 1:2.

    Ans: Let us draw co-ordinate axes across the given parabola. Let the fork vibrating along Y-axis has frequency ν_1 and the one vibrating along X-axis has frequency ν_2. Then, if parabola cuts Y-axis P_y times and X-axis P_x times, then

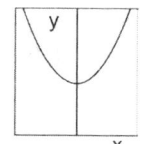

Fig. 1.41

$$\frac{\nu_1}{\nu_2} = \frac{P_x}{P_y}$$

In diagram, $P_x = 2$ and $P_y = 1$

$\therefore \qquad \dfrac{\nu_1}{\nu_2} = \dfrac{2}{1}$

Hence the result.

29. Two tuning forks produce Lissajous figures. The figure changes in form from a parabola to a figure of eight and again to same parabola. The whole cycle occupying 6s. If the frequency of one fork is 100, find the possible frequencies of the other.

    Ans: Since the figure changes in form from a parabola to figure of eight and again to a parabola, the frequency ratio must be 2:1. Since frequency of one fork is 100 Hz, hence, the frequency of the other must be 50 Hz or 100 Hz.

    Case 1: When the unknown frequency ν is lower than the known frequency 100 Hz: Since cycle of figures completes in 6s, the unknown fork of lower frequency n will make $6n$ vibrations in this time, while the known fork will make $2(6n) \pm 1$ vibrations during the same time. Then, we have

    $2(6n) \pm 1 = 6 \times 100$

    or $\qquad 12n \pm 1 = 600$

    or $\qquad 12n = 600 \mp 1$

    or $\qquad n = 50 \pm 1/12$

    Case II: When the unknown frequency n is higher than the known frequency 100: The known fork of lower frequency 100 will make 600 vibrations during the interval 6s after which the cycle of figures is repeated. Clearly, in this time unknown fork will make $2(600) \pm 1$ vibrations and this should be equal to $6n$ where n is frequency of the unknown fork. Hence,

    $6n = 1200 \pm 1$

    or $\qquad n = 200 \pm 1/6$

    Thus, possible frequencies of unknown fork are $\left(50 \pm \dfrac{1}{12}\right)$ and $\left(200 \pm \dfrac{1}{6}\right)$ Hz.

30. A particle is simultaneously subjected to two simple harmonic motions in the same direction each of frequency 5 Hz. If the amplitudes are 0.005 m and 0.002 m

respectively and phase difference between them is 45°. Find the amplitude of the result and displacement and its phase relative to the first component. Write down the expression for the resultant displacement as a function of time.

Ans: Resultant amplitude is given by

$$A^2 = a_1^2 + a_2^2 + 2a_1a_2 \cos\phi$$

$$= (0.005)^2 + (0.002)^2 + 2(0.005)(0.002)\cos 45°$$

$$= 43.14 \times 10^{-6} \text{ m}^2$$

or $\quad A = 6.57 \times 10^{-3}$ m

Phase constant of resultant is

$$\tan \delta = \frac{0.002 \sin 45°}{0.005 + 0.002 \cos 45} = 0.2205$$

or $\quad \delta = 12.5° = \dfrac{\pi}{14.4}$

Since frequency is 5 Hz. $\quad \therefore \omega = 2\pi f = 10\pi$ rad s^{-1}

Hence, equation of motion is

$$x = 6.57 \times 10^{-3} \cos\left(10\pi t + \frac{\pi}{14.4}\right)$$

where x-displacement is in metres and t in seconds.

MULTIPLE CHOICE QUESTIONS

1. A particle moves such that the displacement $x = a \cos \omega t$, the distance covered by the particle during the interval $t = 0$ to $t = \dfrac{3\pi}{2\omega}$ is ...

   (a) $\dfrac{a}{2}$ (b) $\dfrac{3a}{2}$ (c) $2a$ (d) $3a$

2. The ratio of the time taken in displacing a particle form $x = 0$ to $x = \dfrac{a}{2}$ and $x = \dfrac{a}{2}$ to $x = a$, when the motion is according to the equation $x = a \sin\dfrac{t}{T}$ will be

   (a) 1:1   (b) 1:2   (c) 2:1   (d) $1:\sqrt{2}$

3. The length of path of a particle performing SHM is A and the time period is T, then its maximum velocity will be

   (a) $\dfrac{2A\pi}{T}$ (b) $\dfrac{A\pi}{T}$ (c) $\dfrac{A\pi}{2T}$ (d) $\dfrac{\pi A^2}{T}$

4. Motion of a particle is SHM. When its displacements are 4 cm and 5 cm from the mean position, its velocity are 10 cm/s and 8 cm/s respectively. The period of the particle will be (in sec):

   (a) π (b) 2π (c) $\dfrac{\pi}{2}$ (d) 1.5π

5. The maximum acceleration and maximum velocity of a particle performing simple harmonic motion are α and β respectively. The amplitudes of motion is

   (a) $\dfrac{\beta^2}{\alpha}$ (b) $\dfrac{\alpha^2}{\beta}$ (c) $\alpha\beta$ (d) $\dfrac{1}{\alpha\beta}$

6. The frequency of change of kinetic energy into potential energy of an oscillation of frequency n is

(a) $\dfrac{n}{2}$ (b) n (c) $2n$ (d) $4n$

7. A particle of mass m is in one dimensional potential field and its potential energy is given by the following equation $U(x) = U_0(1 - \cos \alpha x)$, where U_0 and α are constants. The period of the particle for small oscillations near the equilibrium position is

(a) $2\pi \sqrt{\dfrac{m\alpha^2}{U_0}}$ (b) $2\pi \sqrt{m\alpha^2 U_0}$ (c) $2\pi \sqrt{\dfrac{m}{\alpha^2 U_0}}$ (d) $2\pi \sqrt{\dfrac{\alpha^2 U_0}{m}}$

8. A solid sphere of radius R is attached to one end of a string to form a pendulum. The distance between the pivot and the centre of sphere is L. T_0 is the period of simple pendulum of length, L, the period of given pendulum will be

(a) $T_0 \sqrt{\dfrac{2R^2}{5L^2}}$ (b) $T_0 \sqrt{\dfrac{1 + 2R^2}{5L^2}}$ (c) $T_0 \sqrt{1 + \dfrac{2R^2}{5L^2}}$ (d) T_0

9. After suspending a long thin rod of length 1 m from its one end, it is made to oscillate in a vertical plane. The time period is found to be T. If a 20 cm long lower part of the rod is broken, the time period will be

(a) $\dfrac{4T}{5}$ (b) $\dfrac{2T}{5}$ (c) $\sqrt{\dfrac{5}{4}}\,T$ (d) $T \cdot \sqrt{\dfrac{2}{5}}$

10. A metre scale of mass M is pivoted at its centre and two identical light springs are attached to its end as shown in Fig. 1.42. Spring constant for each spring is K. If one end of the scale is pressed downwards and left the period of oscillation will be

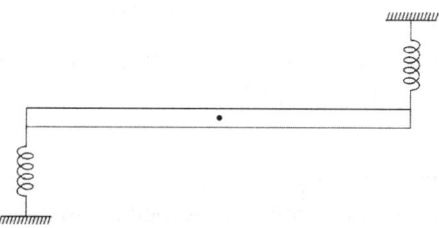

Fig. 1.42

(a) $2\pi \sqrt{\dfrac{m}{6k}}$ (b) $2\pi \sqrt{\dfrac{m}{2k}}$ (c) $2\pi \sqrt{\dfrac{m}{k}}$ (d) $2\pi \sqrt{\dfrac{2m}{k}}$

11. A cylinder of mass M and radius R is resting on a horizontal platform (parallel to XY plane) in such a way that its axis is stationary along Y-axis and it is free to rotate about its own axis. Platform is moved in the X-direction with $X = A \cos \omega t$. The cylinder does not slip on the platform. Maximum torque on it during the motion is

(a) $M\omega^2 AR$ (b) $\dfrac{MR^2\omega^2}{A}$ (c) $\dfrac{M\omega \cdot A}{R}$ (d) $\dfrac{M\omega \cdot R}{A}$

12. In a SHM at time $t = 1$ sec and $t = 2$ sec, the displacement are x and $\sqrt{3}x$ respectively. The period of motion is

(a) 3 sec   (b) 6 sec   (c) 12 sec   (d) 16 sec

13. A horizontal spring is connected to a mass M. It executes SHM. When mass M passes through its mean position, object of mass m is put on it and the two move together. The ratio of amplitude in two cases is

(a) $\left(1+\dfrac{m}{M}\right)^{1/2}$ (b) $\left(1+\dfrac{m}{M}\right)$ (c) $\left(\dfrac{M}{M+m}\right)$ (d) $\left(\dfrac{M}{M+m}\right)^{1/2}$

14. A cubical body of mass 0.1 kg and side 0.1 m floats on water. After pressing, it is released, then it starts oscillating. If $g = 9.86 \text{ m/s}^2$, the time period will be
    (a) 0.2 sec   (b) 1.0 sec   (c) 2.0 sec   (d) 5 sec

15. Which of the following expressions does not represent SHM
    (a) $A \cos \omega t$ (b) $A \sin 2\omega t$ (c) $A \sin \omega t + B \cos \omega t$ (d) $A \sin^2 \omega t$

16. Two SHM are given by $y_1 = a\sin\left\{\left(\dfrac{\pi}{2}\right)t + \phi\right\}$ and $y_2 = a\sin\left\{\left(\dfrac{2\pi}{3}\right)t + \phi\right\}$. The phase difference between these two after 1 sec is

    (a) π (b) $\dfrac{\pi}{2}$ (c) $\dfrac{\pi}{4}$ (d) $\dfrac{\pi}{6}$

17. A particle is subjected to two mutually perpendicular SHM such that $x = 2 \sin \omega t$ and $y = 2 \sin\left(\omega t + \dfrac{\pi}{2}\right)$. The path of the particle will be

    (a) an ellipse   (b) a straight line   (c) a parabola   (d) a circle

18. A particle is executing SHM when its displacement from mean position is equal to half the amplitude of motion, the ratio of KE and PE is
    (a) 3:1   (b) 2:1   (c) 1:1   (d) 1:2

19. A body is placed on a piston, which is executing SHM in vertical plane. Its period is 1 sec. Maximum amplitude of the piston so that body leaves the piston is (in metres)
    (a) 0.25   (b) 2.5   (c) 0.025   (d) 25

20. Figure 1.43 shows curves for a particular type of motion. For this type of motion, the curve that represents displacement, velocity and acceleration respectively are

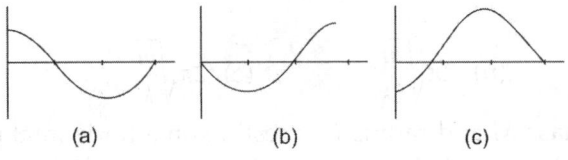

(a)   (b)   (c)

Fig. 1.43

(a) a, b, c (b) c, b, a (c) a, c, b (d) c, a, b

21. A spring of stiffness constant K has normal length l. It is broken in two parts of length l_1 and l_2 such that $l_1 = nl_2$, where n is an integer. Then the stiffness constant K_1 of part of length l_1 is

    (a) $\dfrac{K(n+1)}{n}$ (b) $K(n+1)$ (c) $\dfrac{Kl}{l_2}$ (d) $\dfrac{Kl_1}{l}$

22. A particle executes simple harmonic motion of period T and amplitude a. The time taken by the particle to go from $x = \dfrac{\sqrt{3}a}{2}$ to $x = a$ is

(a) $\dfrac{T}{6}$ (b) $\dfrac{T}{8}$ (c) $\dfrac{T}{9}$ (d) $\dfrac{T}{12}$

23. Which of the following quantities are always positive in simple harmonic motion

(a) $F \cdot a$ (b) $V \cdot r$ (c) $a \cdot r$ (d) $F \cdot r$

where F is force, a is acceleration, V is velocity and r is displacement.

24. Two masses m_1 and m_2 are suspended together by a massless spring of constant K. When the masses are in equilibrium, mass m_1 is removed without disturbing the system. The angular frequency and amplitude of motion respectively are

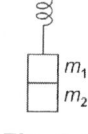

Fig. 1.44

(a) $\omega = \sqrt{\dfrac{K}{m_2}}, A = \dfrac{m_1 g}{K}$

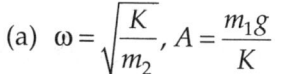

(b) $\omega = \sqrt{\dfrac{K}{m_1}}, A = \dfrac{m_1 g}{K}$

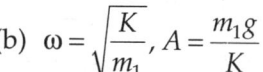

(c) $\omega = \sqrt{\dfrac{K}{m_1 + m_2}}, A = \dfrac{m_1 g}{K}$ (d) $\omega = \sqrt{\dfrac{K}{m_2}}, A = \dfrac{(m_1 m_2)}{K}$

25. The time period of a particle in simple harmonic motion is equal to the time between consecutive appearances of the particle at a particular point in its motion. This point is

(a) the mean position   (b) its extreme position

(c) between the mean position and the positive extreme

(d) between the mean position and the negative extreme

26. A particle moves on X-axis according to the equation $x = x_0 \sin^2 \omega t$. The motion is SHM

(a) with amplitude x_0 (b) with amplitude $2x_0$

(c) with time period $\dfrac{2\pi}{\omega}$ (d) with time period $\dfrac{\pi}{\omega}$

27. A particle executes SHM of period $T = \dfrac{2\pi}{\omega}$. Its velocity is $\sqrt{3}\,b\omega$. When its displacement from its mean position is b, its amplitude is

(a) b (b) $2b$ (c) $4b$ (d) $8b$

28. A body is suspended by a spring. The extension in spring is x when the body is in equilibrium and time period of its vertical oscillation is T. Now, the spring is broken into two equal parts, the body is suspended by both the springs. The extension of both springs is (x') and time period is T', when

(a) $T' = \dfrac{T}{2}$ and $x' = \dfrac{x}{2}$ (b) $T' = \dfrac{T}{4}$ and $x' = \dfrac{x}{4}$

(c) $T' = \dfrac{T}{2}$ and $x' = \dfrac{x}{4}$ (d) $T' = \dfrac{T}{4}$ and $x' = \dfrac{x}{2}$

29. Maximum acceleration and maximum velocity of a particle executing SHM are α and β respectively. Its amplitude and time period respectively are

(a) $\dfrac{\beta^2}{\alpha}, 2\pi \dfrac{\beta}{\alpha}$ (b) $\dfrac{\alpha^2}{\beta}, 2\pi \dfrac{\alpha}{\beta}$ (c) $\alpha\beta, 2\pi\alpha\beta$ (d) $\dfrac{1}{\alpha\beta}, \dfrac{2\pi}{\alpha\beta}$

30. In Fig. 1.45 time displacement curve for a particle executing simple harmonic motion is shown. The acceleration of the particle at $t = 0.6$ s in cm/s^2 is

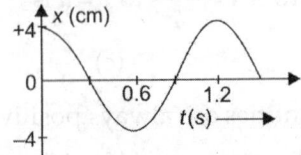

Fig. 1.45

   (a) 60   (b) 90   (c) 110   (d) 120

31. A spring when pulled with force F_1, its length is l, and when pulled with force F_2, its length becomes $2l$. The unstretched length of spring is

   (a) $\dfrac{l(F_2 - 2F_1)}{F_2 - F_1}$ (b) $\dfrac{l(2F_2 - F_1)}{F_2 - F_1}$ (c) $\dfrac{l(F_2 - F_1)}{2F_2 - F_1}$ (d) $\dfrac{l(F_2 - F_1)}{F_2 - 2F_1}$

32. In question 30, if mass of particle is 0.1 kg, then at $t = 0.9$ s, the kinetic energy of the particle in jouls is

   (a) 22×10^{-1} J   (b) 2.2×10^{-2} J   (c) 2.2×10^{-3} J   (d) 2.2×10^{-4} J

33. When a mass m is suspended by a spring, it extends. If mass is slightly pulled downwards and let go, it oscillates in vertical plane. When the mass passes through the mean position at that time, the force on mass due to spring is

   (a) zero   (b) mg upwards
   (c) mg downwards   (d) 2 mg downwards

34. Two masses m and $3m$ are attached to the two ends of a massless spring with force constant K. If $m = 100$ g and $K = 0.3$ N·m^{-1}, then the natural angular frequency of oscillation in rad/s is

   (a) 4   (b) 3   (c) 2   (d) 1

35. A spring block system undergoes SHM on a smooth horizontal surface. The block is now given some positive charge and a uniform horizontal electric field to the right is switched on. As a result

   (a) the time period of oscillation will increase
   (b) the time period of oscillation will decrease
   (c) the time period of oscillation will remain unaffected
   (d) the mean position will shift to the right

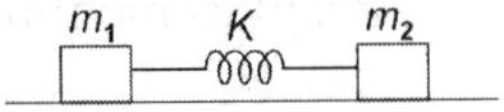

36. In adjacent figure $m_1 = 1.0$ kg, $m_2 = 3.0$ kg and $K = 250$ N·m^{-1}, then the ratio of kinetic energies as the two blocks E_1/E_2 is

Fig. 1.46

   (a) 3:1   (b) 2:1   (c) 1:3   (d) 1:2

37. In Question No. 36, the frequency of oscillation in Hz is

   (a) 1   (b) 2   (c) 3   (d) 4

38. A particle executing SHM has time-displacement curve as shown in Fig. 1.47. The velocity displacement curve is

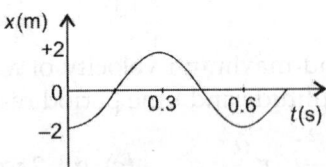

Fig. 1.47

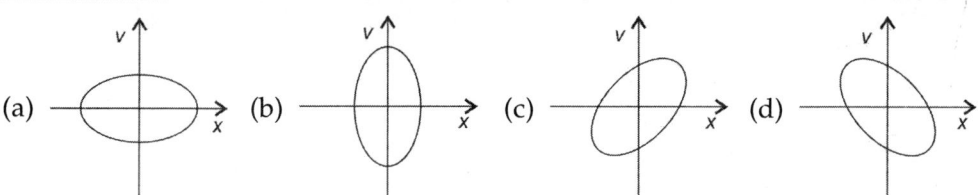

(a)   (b)   (c)   (d)

39. The displacement equations for two particles executing SHM are

$$y = 2\,A\,\sin\,(\omega t + \phi) \quad \text{and} \quad y = A\,\sin\,(2\omega t + \phi)$$

They acquire maximum acceleration:
(a) at the same time and of same magnitude
(b) at the same time and of different magnitude
(c) at different times but of same magnitude
(d) at different times and of different magnitudes

40. A particle executes SHM of time period T and amplitude a. Its speed, when its displacement is $a/2$ is

(a) $\dfrac{\sqrt{3}\,\pi a}{T}$ (b) $\dfrac{\pi a}{T}$ (c) $\dfrac{\sqrt{3}}{2}\dfrac{\pi A}{T}$ (d) $\dfrac{3\pi A}{T}$

ANSWERS

1. (d)   2. (b)   3. (b)   4. (a)   5. (a)   6. (d)   7. (c)   8. (c)   9. (c)   10. (a) 11. (a)
12. (c) 13. (a)   14. (a) 15. (d) 16. (d) 17. (d) 18. (a) 19. (a) 20. (a) 21. (a) 22. (d)
23. (a) 24. (c)   25. (b) 26. (d) 27. (b) 28. (c) 29. (a) 30. (c) 31. (a) 32. (c) 33. (b)
34. (c) 35. (c, d) 36. (a) 37. (c) 38. (b) 39. (d) 40. (a)

2

Damped Harmonic Oscillations

2.1 INTRODUCTION

In the preceding chapter we studied entirely free oscillations of undamped physical system. In that case we observed that total energy of a harmonic oscillator remains constant. Once started the oscillations continue forever with a constant amplitude and a constant frequency. Such oscillations which persist indefinitely without loss of amplitude are called *free* or *undamped oscillations*. However, observations on the free oscillations of a real physical system reveal that the energy of the oscillator gradually decreases with time and the oscillator eventually comes to rest. For example, the amplitude of mass spring system in air decreases with time and it ultimately stops. Similarly, the oscillations of simple pendulum die away with the passage of time. It is because in actual physical system, the friction (or damping) is always present and this resist the motion.

The presence of resistance to motion means that damping or frictional forces act on the system which oppose the motion of the system. Thus, system has to do the work against this force leading to dissipation of energy. When a body moves through a medium such as air, water, etc. its energy is dissipated due to friction and appears in the form of heat either in the system itself or in surrounding medium or both. The energy of an oscillator may also decrease due to radiation. The oscillating body imparts periodic motion to the particles of the medium in which it is oscillating, thus producing waves. For example, a tuning fork produces sound waves in the medium in which it is vibrating resulting in the decrease of its energy. All surrounding bodies are subject to dissipate forces, otherwise there would be no loss of energy by the body and consequently, no emission of sound energy occur. Thus, sound waves are produced by radiation from mechanical oscillatory system. We shall see later on that the electromagnetic waves are produced by radiations from oscillating electric and magnetic fields.

The effects of radiation by an oscillating system and of friction present in the system is that the amplitude of oscillations gradually diminishes with time. The reduction of amplitude (or energy) of an oscillator is called *damping* and the oscillations are said to be *damped*.

2.2 DAMPING FORCES

In real systems the damping is a complex phenomenon involving various types of forces, such as viscous forces, coulomb, frictional forces, and structural deforming forces. In general, it is very difficult to predict the magnitude of the damping forces; we have to rely on experience and experiment to make a reasonably good estimate. It is a common practice to approximate the damping of a system by an equivalent viscous damping, for the simple reason that viscous damping is the most convenient to handle mathematically. Thus, according to this approximation, the magnitude of the viscous force to be used in particular problem is chosen to be the one that would produce the same rate of energy dissipation as actual damping forces. This usually provides good estimate.

The damping force of a fluid (liquid or gas) to a moving object is a function of the velocity of the object. The damping force that depends on velocity is referred to as *viscous damping force*. The magnitude of this force is given by equation

$$F = \beta_1 v + \beta_2 v^2 \tag{2.1}$$

where v is the magnitude of the velocity of the oscillator. The direction of the resistive force is opposite to that of the velocity. If v is small compared to the ratio β_1/β_2, the damping force will be proportional to the first power of velocity. Thus, for small velocities

$$F = -\beta v \tag{2.1a}$$

where β is called the *viscous damping co-efficient* and represents damping force per unit velocity. Actually the viscous force is a retarding force. Since the velocity of the most oscillating systems is usually small, the damping force exerted by the fluid in contact with the system is likely to be viscous, which are much smaller than inertial and elastic forces in the system.

The inclusion of damping forces complicates the analysis considerably. But in actual systems, the damping forces are usually small and can often be ignored. In situations, where they are not negligibly small, the viscous damping model is the most convenient mathematically. Here we shall use this model, under simplifying assumptions, that the velocity of the moving parts of the system is small, so that damping force is the linear function of velocity.

2.3 DAMPED OSCILLATIONS OF A SYSTEM HAVING ONE DEGREE OF FREEDOM

We shall now investigate the effect of damping on the harmonic oscillations of a simple system having one degree of freedom. One such system is shown in Fig. 2.1. When the system is displaced from its equilibrium state and released, it begins to move. The forces acting on the system are:

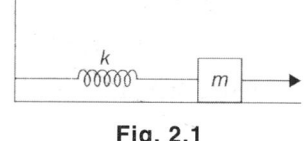

Fig. 2.1

(i) A restoring force $-kx$, where k is the spring constant, i.e. the co-efficient of the restoring force, and x is the displacement.

(ii) The damping force $-\beta\dfrac{dx}{dt}$, where β, is the co-efficient of damping force and $\dfrac{dx}{dt}$ is the velocity of the moving part of the system.

Using Netwon's law of motion, we have, these forces must balance with Newton's force $m\dfrac{d^2x}{dt^2}$, where m is the mass of the oscillator and $\dfrac{d^2x}{dt^2}$, its instantaneous acceleration. Since restoring force and the damping act in a direction opposite to Newton's force, we have

$$m\frac{d^2x}{dt^2} = -kx - \beta\frac{dx}{dt}.$$

It should be kept in mind that this equation holds only for small displacements and small velocities. Rearranging the terms, the equation becomes

$$\frac{d^2x}{dt^2} + 2b\frac{dx}{dt} + \omega_0^2 x = 0 \tag{2.2}$$

with $2b = \beta/m$ and $\omega_0^2 = k/m$.

Here, ω_0 is the natural frequency of oscillator, i.e. the frequency in absence of the damping.

The dimensions of $b = \dfrac{\beta}{m} = \dfrac{\text{Force}}{\text{Velocity} \times \text{Mass}} = \dfrac{MLT^{-2}}{LT^{-1}M} = T^{-1}$ are the same as those of frequency.

It is apparent from equation (2.2) that the damping is characterised by the quantity $2b$, having the dimensions of frequency and the ω_0, the natural frequency of the oscillator. Equation (2.2) is called the *differential equation of the damped oscillator*. In order to study how the displacement of such oscillator varies with time, we have to solve Eq. (2.2).

General solution: To solve Eq. (2.2), we make use of the exponential functions. Let us assume the solution is of the form

$$x = A\, e^{\alpha t}$$

and solve for α. Here constants A and α are arbitrary and have yet to be determined.

Differentiating above Eq., we have

$$\frac{dx}{dt} = \alpha\, A\, e^{\alpha t} \qquad \text{and,} \qquad \frac{d^2 x}{dt^2} = \alpha^2 A\, e^{\alpha t}$$

and substituting these values in Eq. (2.2), we have:

$$\alpha^2 A e^{\alpha t} + 2b\alpha\, A e^{\alpha t} + \omega_0^2\, A e^{\alpha t} = 0$$

or,

$$\alpha^2 + 2b\alpha + \omega_0^2 = 0$$

which gives

$$\alpha = \frac{-2b \pm \sqrt{4b^2 - 4\omega_0^2}}{2}$$

i.e., the two roots of the equations are

$$\alpha_1 = -b + \sqrt{b^2 - \omega_0^2} \quad \text{and} \quad \alpha_2^2 = -b - \sqrt{b^2 - \omega_0^2}$$

Hence, general solution can be written as

$$x = c_1 e^{\alpha_1 t} + c_2 e^{\alpha_2 t} \tag{2.3}$$

Here c_1 and c_2 are arbitrary constants to be determined from the initial conditions. Putting values of α_1 and α_2 in Eq. (2.3), we have

$$x = c_1 e^{\left(-b + \sqrt{(b^2 - \omega_0^2)}\, t\right)^2 t} + c_2 e^{\left(-b - \sqrt{(b^2 - \omega_0^2)}\, t\right) t}$$

$$= e^{-bt} \left\{ c_1 e^{\sqrt{(b^2 - \omega_0^2)}\, t} + c_2 e^{-\sqrt{(b^2 - \omega_0^2)}\, t} \right\} \tag{2.4}$$

The nature of the motion depends on the character of the roots α_1 and α_2. The roots may be real or complex depending on whether $b > \omega_0$ or $b < \omega_0$ respectively. In fact, three different kinds of motions are possible depending on whether $b > \omega_0$, $b = \omega_0$ or $b < \omega_0$. Each condition describes a particular kind of behaviour of the system. We shall now treat each case separately.

Case (i): $b > \omega_0$: Case of Overdamping

In this case, the damping term b dominates the stiffness term ω_0 and the term $(b^2 - \omega_0^2)^{1/2}$ is a real quantity, i.e.

$$\sqrt{(b^2 - \omega_0^2)} = q \text{ (say)}$$

So, displacement is given by

$$x = e^{-bt}(c_1 e^{qt} + c_2 e^{-qt}) \qquad (2.5)$$

and velocity is given by

$$\frac{dx}{dt} = e^{-bt}\{c_1(-b+q)e^{qt} - c_2(b+q)e^{-qt}\} \qquad (2.6)$$

These equations describe the behaviour of heavily damped oscillator such as the motion of a pendulum in a viscous medium such as dense oil.

The constants c_1 and c_2 are determined from the initial conditions. Let us assume that oscillator is at its equilibrium position, $x = 0$ at $t = 0$. At this instant, it is given a kick so that it has a finite velocity, say v_0 at this time, i.e. at $t = 0$:

$$x = 0 \quad \text{and} \quad \frac{dx}{dt} = v_0$$

Equations (2.5) and (2.6) give

$$c_1 + c_2 = 0$$

and $\qquad c_1(-b+q) - c_2(b+q) = v_0$

giving $\qquad c_1 = -c_2 = \dfrac{v_0}{2q} \qquad (2.7)$

Thus, under the above given initial conditions Eq. (2.5) and (2.6) become

$$x = \frac{v_0}{2q} e^{-bt}(e^{qt} - e^{-qt})$$

$$= \frac{v_0}{q} e^{-bt} \sin h(\dot{q}t) \qquad (2.8)$$

and $\qquad \dfrac{dx}{dt} = v_0 e^{-bt}\left\{\cos h(qt) - \dfrac{b}{q}\sin h(qt)\right\} \qquad (2.8a)$

Figure 2.2 illustrates the behaviour of a heavily damped system when it is disturbed from equilibrium by a sudden impulse at $t = 0$. For small values of time t, the term e^{-bt} is very nearly unity, the displacement increases with time since $\sin h(qt)$ increases as t increases. Very soon, however, the term e^{-bt} starts contributing and the displacement decays exponentially with time and ultimately becomes zero. The turning point occurs at a time $t = t_0$, when $\dfrac{dx}{dt} = 0$. From equation (2.8a), it is apparent that it happens at time $t = t_0$ satisfying

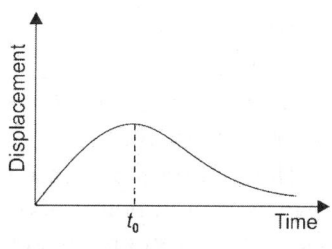

Fig. 2.2

$$\tan h(qt_0) = \frac{q}{b}$$

Thus, displacement increases until time $t = t_0$ after which it slowly returns to zero. Since displacement x never becomes negative, there is no oscillation at all. Such motion is called *dead beat*. We come across such a motion in case of *dead beat galvanometer*.

Case (ii): $b = \omega_0$: Case of Critical Damping

In this case, the two values of α, i.e. α_1 and α_2 are equal, i.e.

$$\alpha_1 = \alpha_2 = -b$$

and the general solution in this case is given by

$$x = (c_1 + c_2)\, t e^{-bt} \tag{2.9}$$

and

$$\frac{dx}{dt} = \{c_2 - b\,(c_1 + c_2)\,t\}\, e^{-bt} \tag{2.9a}$$

The constant c_1 and c_2 are determined from initial conditions. If at, $t = 0$, $x = 0$ and $\frac{dx}{dt} = v_0$, we have

$$c_1 = 0$$
$$c_2 = v_0$$

thus under these conditions, the displacement and velocity are given by the equations

$$x = v_0\, t\, e^{-bt} \tag{2.10}$$

$$\frac{dx}{dt} = v_0\,(1 - bt)\, e^{-bt} \tag{2.10a}$$

Figure 2.3 gives the plot x against time (t) with $b = \omega_0$, when this oscillator is given sudden impulse in equilibrium position. For small values of t, the term e^{-bt} is very nearly unity and the displacement increases nearly linearly with time t. After some time e^{-bt} starts changing and the displacement decays exponentially with time, eventually becoming zero. The turning point occurs at time t_0. When $\frac{dx}{dt} = 0$ given by

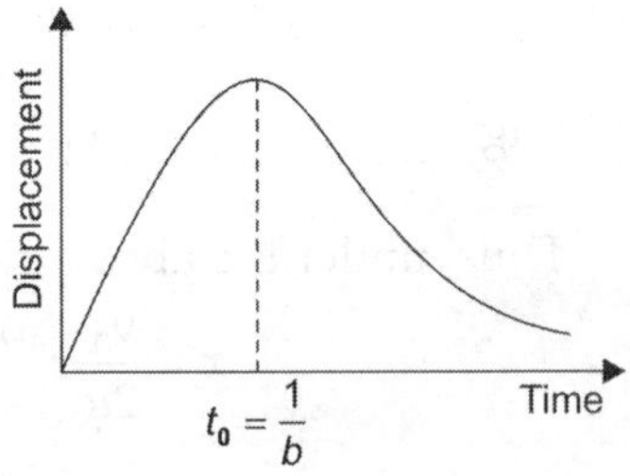

Fig. 2.3

$$1 - bt_0 = 0 \quad \text{or} \quad t_0 = \frac{1}{b} = \frac{1}{\omega_0}$$

The displacement increases from time $t = 0$ to time $t = t_0 = \frac{1}{b}$ after which, it decays to zero. A comparison of Eqs (2.8 and 2.10) reveals that the decay rate is much faster for $b = \omega_0$ than for $b > \omega_0$. In both cases, there are no oscillations as the displacement never becomes negative.

The motion represented by equation (2.10) is called *critically damped motion*. This is made use of in pointer-type of galvanometers where the pointer moves immediately to the correct position and stays there without oscillating.

Case (iii): $b < \omega_0$: Under Damped Oscillator or Oscillatory Case

Where $b < \omega_0$, the damping is small and this gives the most important kind of behaviour called *oscillatory damped harmonic motion*. In this case, the expression $\sqrt{b^2 - \omega_0^2}$ in exponential is an imaginary quantity.

Writing

$$\sqrt{b^2 - \omega_0^2} = \sqrt{-1}\,\sqrt{(\omega_0^2 - b^2)} = i\omega$$

where, $\omega = \sqrt{(\omega_0^2 - b^2)}$ is a real positive quantity and the Eq. (2.4) can be written as

$$x = e^{-bt}\,\{c_1\, e^{i\omega t} + c_2\, e^{-i\omega t}\} \tag{2.11}$$

This equation can be written as

$$x = ce^{-bt} \sin(\omega t + \phi) \tag{2.12}$$

where, c and ϕ are arbitrary constants and are determined from initial conditions. The velocity of oscillator is obtained by differentiating this equation with respect to time.

$$\frac{dx}{dt} = ce^{-bt}\{\omega\cos(\omega t + \phi) - b\sin(\omega t + \phi)\} \tag{2.12a}$$

The equation (2.12) shows that the motion is oscillatory but oscillations are not simple harmonic as its amplitude is not constant but is function of time i.e.

$$A = c\,e^{-bt} \tag{2.13}$$

which decreases exponentially with time. The motion is not even periodic since it never repeats itself; each swing being of smaller amplitude than the preceding one. However, if b is very small compared to ω_0, the amplitude will remain fairly constant over a large number of oscillations of the harmonic term $\sin(\omega t + \phi)$ in which case, the motion is nearly periodic and simple harmonic. The angular frequency of the oscillation ω is given by

$$\omega = \omega_0 \left(1 - \frac{b^2}{\omega_0^2}\right)^{1/2} \tag{2.13a}$$

which is less than the natural angular frequency of free undamped oscillations. Strictly speaking, we are in fact not justified in using the terms 'amplitude' and 'frequency' for a motion, which is not periodic. But when the damping is small, the motion is nearly periodic.

To understand the behaviour of a weakly damped oscillator, let us choose the initial conditions that at $t = 0$, $x = 0$ and $\frac{dx}{dt} = v_0$. These conditions give

$$0 = c \sin \phi$$

and

$$v_0 = c\{\omega \cos \phi - b \sin \phi\}$$

Since $c \neq 0$, hence $\phi = 0$, and $c = \dfrac{v_0}{\omega}$.

Using these values of c and ϕ, we have

$$x = \frac{v_0}{\omega} e^{-bt} \sin \omega t$$

$$= A(t) \sin \omega t \tag{2.14}$$

with $A(t) = \dfrac{v_0}{\omega} e^{-bt} = A_0 e^{-bt}$, A_0 being the value of $A(t)$ for $b = 0$

and

$$\frac{dx}{dt} = v_0 e^{-bt} \left\{\cos \omega t - \frac{b}{\omega} \sin \omega t\right\} \tag{2.14a}$$

Figure 2.4 shows the behaviour of a weakly damped oscillator. It is a graph of x versus t of the motion described by equation (2.14). The constant A_0 is the value of $A(t) = \dfrac{v_0}{\omega} e^{-bt}$ in absence of damping ($b = 0$), i.e. $A_0 = \dfrac{v_0}{\omega}$. Since the maximum values of $\sin(\omega t)$ are $+1$ and -1 alternatively, the displacement-time curve of oscillation is bounded by the curver $A_0 e^{-bt}$ and $-A_0 e^{-bt}$ (exponential curves).

Even though the amplitude decreases exponentially with time, the weakly damped oscillator executes some sort of oscillatory motion. The motion does not repeat itself and is, therefore, not periodic in the usual sense of the term. However, it still has time period $t = \dfrac{2\pi}{\omega}$ which is the time-interval between two alternate zeroes of displacement. The time interval between two successive zeroes of displacement is $T/2$. This is

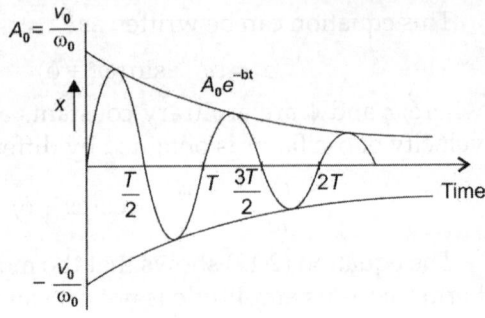

Fig. 2.4

also the time-interval between maximum and next minimum value of the displacement, but the maxima and minima are not exactly halfway between the zeroes which is obvious from equation (2.14a). At a maximum or a minimum displacement, the velocity is zero, giving

$$\cos \omega t - \frac{b}{\omega} \sin \omega t = 0$$

or

$$\tan \omega t = \frac{\omega}{b}$$

Those values of t which satisfy this equation are the instants at which displacement x is either a positive maximum or a negative maximum. In the case when $b \ll \omega_0$. So that,

$$\frac{\omega_0}{b} \to \infty$$

$$\omega t \to \frac{\pi}{2}, \frac{3\pi}{2}, \frac{5\pi}{2}, \ldots$$

The first maxima of displacement x occurs at a time $t = t_1$ given by

$$\omega t_1 \to \frac{\pi}{2} \quad \text{or} \quad t_1 = \frac{\pi}{2\omega} = \frac{T}{4}$$

i.e. maximum is exactly midway between the two zeroes of displacement x. Thus, only in case of negligibly small damping, are the maxima and minima halfway between the zeroes of displacement as in the case of simple harmonic motion.

Effect of damping: The effect of damping is two-fold.

(a) The amplitude of oscillation decreases exponentially with time as

$$A(t) = A_0 \, e^{-bt}$$

where A_0 is the amplitude in the absence of damping.

(b) The angular frequency ω of damped oscillator is less than ω_0, the frequency of the undamped oscillators and the relation between the two is

$$\omega = \sqrt{\omega_0^2 - b^2} = \omega_0 \left(1 - \frac{b^2}{\omega_0^2} \right)^{1/2}.$$

2.4 ENERGY OF A WEAKLY DAMPED OSCILLATOR

Now, we will see how the average energy of a weakly damped oscillator varies with time. In case of weak damping ($b < \omega_0$), the displacement and velocity of the oscillator are represented by Eqs (2.14 and 2.14a).

If m is the mass of the oscillator, its instantaneous translational kinetic energy

$$KE = \frac{1}{2} m \left(\frac{dx}{dt}\right)^2$$

$$= \frac{1}{2} m \, v_0^2 \, e^{-2bt} \left\{\cos \omega t - \frac{b}{\omega} \sin \omega t\right\}^2$$

$$= \frac{1}{2} m \, v_0^2 \, e^{-2bt} \left\{\cos^2 \omega t + \frac{b^2}{\omega^2} \sin^2 \omega t - \frac{2b}{\omega} \sin \omega t \cos \omega t\right\}$$

The instantaneous PE is

$$PE = \frac{1}{2} k x^2 \qquad\qquad \text{where, } k = m\omega_0^2$$

$$PE = \frac{1}{2} m\omega_0^2 \, \frac{v_0^2}{\omega^2} \, e^{-2bt} \sin^2 \omega t$$

$$= \frac{1}{2} m v_0^2 \, \frac{(\omega^2 + b^2)}{\omega^2} \, e^{-2bt} \sin^2 \omega t \qquad [\because \omega_0^2 = \omega^2 + b^2]$$

$\therefore$ Total Energy E = KE + PE

or

$$E = \frac{1}{2} m v_0^2 \, e^{-2bt} \left[(\cos^2 \omega t + \sin^2 \omega t) + \frac{2b^2}{\omega^2} \sin^2 \omega t - \frac{b}{\omega} \sin 2\omega t\right]$$

$$= \frac{1}{2} m v_0^2 \, e^{-2bt} \left[\langle 1\rangle + \frac{2b^2}{\omega^2} \sin^2 \omega t - \frac{b}{\omega} \sin 2\omega t\right]$$

Hence, average value of energy for one time period T will be

$$\langle E\rangle = \frac{1}{2} m v_0^2 \, e^{-2bt} \left[\langle 1\rangle + \frac{2b^2}{\omega^2} \langle\sin^2 \omega t\rangle - \frac{b}{\omega} \langle\sin 2\omega t\rangle\right]$$

Here, we have assumed that e^{-2bt} is fairly constant during one period T of the oscillations.

We know that $\langle\sin^2 \omega t\rangle = \frac{1}{2}, \langle\sin 2\omega t\rangle = 0$ and $\langle 1\rangle 1$.

$\therefore$

$$\langle E\rangle = \frac{1}{2} m v_0^2 \, e^{-2bt} \left(1 + \frac{b^2}{\omega^2}\right) = \frac{1}{2} m \frac{v_0^2}{\omega^2} \omega_0^2 \, e^{-2bt} = \frac{1}{2} m \, A_0^2 \, \omega_0^2 \, e^{-2bt}$$

$$\langle E\rangle = E_0 \, e^{-2bt} \qquad\qquad\qquad (2.15)$$

where $E_0 = \frac{1}{2} m A_0^2 \omega_0^2$ is the total energy of an undamped oscillator. We observe that the energy of a weakly damped oscillator diminishes exponentially with time. The decay of the energy with time is plotted in Fig. 2.5. The average power dissipated during one time period is given by

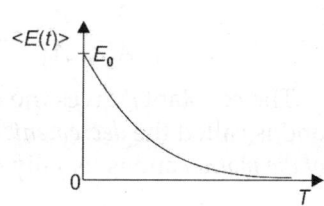

Fig. 2.5

$$\langle P(t)\rangle = \text{rate of loss of energy}$$

$$= \frac{d}{dt} \langle E(t)\rangle = \frac{d}{dt} (E_0 \, e^{-2bt})$$

$$= -2b \cdot E_0 \, e^{-2bt} \quad \text{(–ve sign shows loss of energy)}$$

or $\qquad \langle P(t)\rangle = 2b\langle E(t)\rangle \qquad\qquad\qquad (2.15a)$

2.5 LOGARITHMIC DECREMENT

This method measures the rate at which the amplitude decreases with time. In case of a weakly damped oscillator, initially at rest, at its equilibrium position, when it is given an impulse, the future motion of the oscillator is described by equation

$$x = A_0 e^{-bt} \sin \omega t$$
$$= A(t) \sin \omega t$$

or
$$x = A(t) \sin \frac{2\pi t}{T}$$

where $A(t) = A_0 e^{-bt}$, A_0 being the amplitude in absence of damping ($b = 0$) and $T = \dfrac{2\pi}{\omega}$ is the time period of oscillations. Figure 2.6 shows the displacement-time curve of the motion.

At time $t = \dfrac{T}{4}$, the displacement is given by

$$x = A_0 e^{-bt} \sin\left(\frac{2\pi}{T} \cdot \frac{T}{4}\right) = A_0 e^{-\frac{bT}{4}}$$

and attains the first maximum value. Let us call this A_1. Hence amplitude A_1 is given by

$$A_1 = A_0 e^{-\frac{bT}{4}}$$

At $t = \dfrac{3T}{4}$ the displacement again becomes maximum (this time it is negative). The amplitude A_2 (Fig. 2.6) is given by

$$A_2 = A_0 e^{-\frac{3bT}{4}}$$

The next maximum of displacement x occurs at $t = \dfrac{5T}{4}$ given by

$$A_3 = A_0 e^{-\frac{5bT}{4}}$$

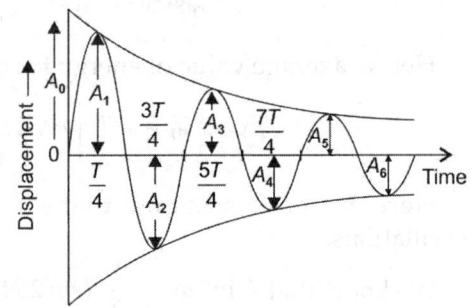

Fig. 2.6

and so on.

We observe that

$$\frac{A_1}{A_2} = \frac{A_2}{A_3} = \ldots \frac{A_{n-1}}{A_n} = e^{\frac{bT}{2}} = d \text{ (constant)}.$$

The constant d gives the ratio of two successive amplitudes of the damped oscillations and is called the *decrement*. The logarithm of decrement d is called *logariothmic decrement of the motion* and is usually denoted by the symbol λ.

$$\therefore \qquad \lambda = \log_e d = \frac{bT}{2} \qquad\qquad (2.16)$$

Logarithmic decrement is the natural logarithm of the ratio of two successive amplitudes that are separated by half a period $T/2$ (larger amplitude to the smaller one). Thus,

$$\frac{A_1}{A_n} = \frac{A_1}{A_2} \cdot \frac{A_2}{A_3} \cdot \frac{A_3}{A_u} \dots \frac{A_{n-1}}{A_n}$$

$$= d^{n-1} = e^{(n-1)\lambda}$$

or
$$\lambda = \frac{1}{(n-1)} \log_e \left(\frac{A_1}{A_n} \right).$$

Thus, logarithmic decrement λ of the motion can be measured by observing several successive amplitudes on both sides of the equilibrium position. For example, if $n = 11$ (eleven successive throws)

$$\lambda = \frac{1}{10} \log_e \left(\frac{A_1}{A_{11}} \right) = \frac{2.303}{10} \log_{10} \left(\frac{A_1}{A_2} \right) \tag{2.17}$$

This method of describing damping by parameter, λ is used in making the necessary correction in the first deflection of the first throw θ_1, of the ballistic galvanometer, when a certain quantity of charge is passed through its coil. The relation between the correct throw θ_0, i.e. the throw if damping were absent and the first throw θ_1, is

$$\theta_1 = \theta_0 \, e^{-\frac{bT}{4}}$$

where T is the period of oscillation of the coil. Thus,

$$\theta_0 = \theta_1 \, e^{-\frac{bT}{4}} = \theta_1 \, e^{\frac{\lambda}{2}}.$$

If $\lambda << 1$ (as is usually the case), we have, $e^{\frac{\lambda}{2}} = 1 + \frac{\lambda}{2}$

$\therefore$
$$\theta_0 = \theta_1 \left(1 + \frac{\lambda}{2} \right) \tag{2.18}$$

Thus knowing the logarithmic decrement (λ) for a given galvanometer, the first throw (θ) can easily be corrected for damping.

2.6 RELAXATION TIME

Another way of expressing the damping effect on the motion is in terms of relaxation time (τ). Relaxation time is defined as the time taken by the amplitude to decrease to $(1/e)^{th}$ of its original value. This time is also called *modulus of decay*. We know that amplitude at any time is given by

$$A(t) = A_0^{-bt}$$

Since after time $t = \frac{1}{\tau}$ (relaxation time) the amplitude $A(t) = \frac{A_0}{e}$. Hence,

$$A_0 \, e^{-1} = A_0 \, e^{-bt} \quad \text{or} \quad \tau = \frac{1}{b}.$$

The relaxation time in fact is the measure of rapidity by which the motion is damped out by friction. Greater the value of b, smaller is the relaxation time. In many problems in Physics involving decay of energy, the magnitude of dissipative process is estimated by measuring the relaxation time. The relaxation time and the logarithmic decrement of the motion are obviously related to each other.

$$\lambda = \frac{bT}{2} = \frac{T}{2\tau} \tag{2.19}$$

i.e. logarithmic decrement is the ratio between half the period of oscillation and relaxation time. Thus, λ is a measure of the fraction of the decrease in amplitude which occurs in one half cycle.

2.7 QUALITY FACTOR OR Q-VALUE

The third method of expressing damping in an oscillatory system measures the rate at which energy decays. In order to do so, we define a parameter called the *quality factor* or *Q-value*. If ω is the angular frequency of the damped oscillations the quality factor

where
$$Q = \frac{\omega}{2b}, \quad \omega = \omega_0 \left(1 - \frac{b^2}{\omega_0^2}\right)^{1/2} \tag{2.20}$$

The Q-value, in fact measures the rate of decay of energy of the system. Equation (2.15a) gives the average rate of loss of energy, i.e.

$$\frac{d}{dt}\langle E(t)\rangle = \langle p(t)\rangle = 2b\langle E(t)\rangle$$

Hence, average energy dissipated in time period $T\left(=\dfrac{2\pi}{\omega}\right)$ is

$$2bT\langle E(t)\rangle = \frac{2\pi}{\omega} 2b\langle e(t)\rangle$$

$$= \frac{2\pi}{Q}\langle E(t)\rangle = \frac{2\pi}{Q} \times (\text{average stored energy})$$

or
$$Q = 2\pi \times \frac{\text{Average energy stored in one period}}{\text{Average energy lost in one period}}$$

Thus the quality factor of a damped harmonic oscillator may be defined as 2π times the ratio between average energy stored and average energy lost per period.

We are mainly concerned with situation in which $b \ll \omega_0$. Under this condition, we can approximately write $\omega \approx \omega_0$ and the quality factor

$$Q = \frac{\omega_0}{2b} \tag{2.20a}$$

and is constant of the damped system. Q-value is a pure number. If Q value is large compared to unity means rate of dissipation of energy of the oscillating system is small. The lower the damping, the higher the value of Q as $b \to 0$ $Q \to \infty$. In such cases the motion of oscillator is given nearly (2.12).

$$x = c\,e^{-bt} \sin(\omega t + \phi)$$

$$= c\,e^{-\frac{\omega_0 t}{2Q}} \sin(\omega_0 t + \phi). \tag{2.21}$$

and average energy of oscillator is given by equation (2.15)

$$\langle E(t)\rangle = E_0\,e^{-2bt} = E_0\,e^{\frac{-\omega_0 t}{Q}} \tag{2.22}$$

It may be noted that Q is closely related to the number of oscillations (or cycles) over which the energy fall to $(1/e)$ of its original value E_0. From equation (2.22), it is clear that energy falls to $(1/e)$ in time $t = \Gamma$, such that

$$\Gamma = \frac{Q}{\omega_0} = \frac{T_0}{2\pi} Q \tag{2.23}$$

where T_0 is the period of oscillation in absence of damping. During this time Γ, the number of n complete oscillations executed by oscillator is obtained from equation (2.21). Thus,

$$n = \frac{\omega_0}{2\pi} \Gamma = \frac{Q}{2\pi} \tag{2.24}$$

Thus, the average energy falls to $\left(\dfrac{1}{e}\right)$ of its original value in $\dfrac{Q}{2\pi}$ cycles of free oscillations. The energy decay time Γ and amplitude decay time τ are related as:

$$\Gamma = \frac{\tau}{2} \tag{2.25}$$

Thus Γ is the mean decay time of damped oscillations. Now, in terms of ω_0 and Q, the differential equation for damped harmonic oscillator becomes

$$\frac{d^2x}{dt^2} + \frac{\omega_0}{Q}\frac{dx}{dt} + \omega_0^2 x = 0 \tag{2.26}$$

In many physical cases, this form of equation is found to be more convenient in analysing the effect of damping—both mechanical and non-mechanical.

Relation between the Logarithmic decrement (λ), *relaxation time* (τ) *and quality factor* (Q): In case $b << \omega_0$, the three parameters are given by

$$\lambda = \frac{bT_0}{2}$$

$$\tau = \frac{1}{b}$$

$$Q = \frac{\omega_0}{2b}$$

In terms of Q, the parameters λ and τ are given by

$$\lambda = \frac{\pi}{2Q}$$

$$\tau = \frac{2Q}{\omega_0}$$

Smaller the damping, the larger are Q and τ, indicating that it takes a longer time for the oscillations to damp out. Also the smaller the damping, the smaller is λ indicating that the reduction in the amplitude in a half cycle is smaller. It may be remarked that these conclusions are independent of the way the system is set into motion (i.e., initial conditions).

2.8 EXAMPLES OF DAMPED OSCILLATIONS

(a) Motion of Simple Pendulum with Viscous Drag

When a body moves in a fluid, it is acted upon by a viscous drag. The effect of the drag is to decrease the velocity of the body. It is found that this viscous drag is proportional to the velocity of the body when it is below the critical velocity. Thus, when the simple pendulum oscillates in air and it is acted upon the viscous forces which are proportional to its velocity in addition to the restoring force. Hence, under the action of viscous forces, the equation of motion for the simple pendulum takes the form

$$m\frac{d^2x}{dt^2} = -\frac{m}{l}gx - \beta\frac{dx}{dt}$$

where β is called *damping constant*, or

$$\frac{d^2x}{dt^2} + \frac{\beta}{m}\frac{dx}{dt} + \frac{g}{l}x = 0$$

or

$$\frac{d^2x}{dt^2} + 2b\frac{dx}{dt} + \omega_0^2 x = 0 \qquad (2.27)$$

where $2b = \dfrac{\beta}{m}$ and $\omega_0^2 = \dfrac{g}{l}$.

This equation is identical to the equation for damped harmonic oscillator. Since the viscous forces are small, solution can be written as

$$x = Ae^{-bt}\sin(\omega t + \phi) \qquad (2.28)$$

where constants A and ϕ are determined from initial condition and $\omega^2 = \omega_0^2 - b^2$.

Thus, the simple pendulum will oscillate with decreasing amplitude and time period now becomes

$$T = \frac{2\pi}{\omega} = \frac{2\pi}{\sqrt{(\omega_0^2 - b^2)}} = \frac{2\pi}{\omega_0}\left(1 - \frac{b^2}{\omega_0^2}\right)^{-\frac{1}{2}}$$

and since $b << \omega_0$

$$T = T_0\left(1 + \frac{b^2}{2\omega_0^2}\right) \qquad (2.28)$$

where T_0 is the time period of pendulum without viscous drag. Since, $2b = \dfrac{1}{\tau}$, hence

$$T = T_0\left(1 + \frac{1}{8\tau^2\omega_0^2}\right) \qquad (2.28a)$$

This equation shows that time period is increased.

(b) Moving Coil Galvanometer

A moving coil galvanometer consists of a current carrying coil (rectangular or circular) on an axis in magnetic field. The magnetic field is provided by a permanent magnet, so shaped that the moving coil experiences the same magnitude of the field at all orientations. The steady current to be measured produces a torque which is proportional to the current. The coil rotates under the electromagnetic torque and comes to an equilibrium position where the turning torque is balanced by the restoring torque due to the stiffness of the suspension. If the deflection of the coil is θ, then this torque due to stiffness of the suspension is given by $c\,\theta$, where

$$c = \frac{\pi\eta r^4}{2l}$$

and is called the *twisting couple* per unit twist, where l is the length of the suspension fiber and r its radius and η is the modulus of rigidity of its material.

The damping of the moving parts in galvanometer apart from external agency (like shunt, etc.) arises due to the following two causes:

(i) *Mechanical damping*: The damping due to the viscosity of air. The damping force is approximately proportional to the angular velocity of the system but is usually negligibly small.

(ii) *Electromagnetic damping*: When suspended coil of a galvanometer rotates in a strong magnetic field, it is resisted in open circuit by induced currents in neighbouring conductors. The open circuit damping couple is proportional to angular velocity $\left(\dfrac{d\theta}{dt}\right)$.

According to the law of electromagnetic induction, this is represented by $-\alpha\dfrac{d\theta}{dt}$, where α is the damping co-efficient. When the circuit is closed, there is an additional damping provided by the induced currents in the coil. The closed circuit damping is universally proportional to the total resistance of the circuit and is given by $-\dfrac{\gamma}{R}\cdot\dfrac{d\theta}{dt}$, where r involves the area of the coil and magnetic flux, etc. Denoting the angular displacement from its new equilibrium position by θ, we get

$$I\frac{d^2\theta}{dt^2} = -c\theta - \alpha\cdot\frac{d\theta}{dt} - \frac{\gamma}{R}\frac{d\theta}{dt}$$

where I is the moment of inertia of the vibrating system. Re-arranging the terms, we have

$$\frac{d^2\theta}{dt^2} + \frac{1}{I}\left(a + \frac{\gamma}{R}\right)\frac{d\theta}{dt} + \frac{c}{I}\theta = 0$$

or
$$\frac{d^2\theta}{dt^2} + 2b\frac{d\theta}{dt} + \omega_0^2\,\theta = 0 \qquad (2.29)$$

where $\omega_0 = \sqrt{\dfrac{c}{I}}$ and $2b = \dfrac{1}{I}\left(a + \dfrac{\gamma}{R}\right)$.

Equation (2.29) is similar to Eq. (2.2) and the general solution can be written as

$$\theta = e^{-bt}\left\{c_1\,e^{\sqrt{(b^2 - \omega_0^2)}\,t} + c_2\,e^{-\sqrt{(b^2 - \omega_0^2)}\,t}\right\} \qquad (2.30)$$

where constants c_1 and c_2 are determined from the initial conditions.

Case (i): Dead Beat Motion

If the damping is high such that $b^2 \gg \omega_0^2$, we have two real roots for the equation of angular displacement. The Eq. (2.30) cannot be further simplified. In this case, oscillatory motion is not possible. This type of motion is called *dead beat* as it decreases exponentially.

Here we see that electromagnetic damping $\left(\dfrac{\gamma}{R}\right)$ depends on resistance. Higher the resistance in series of the galvanometer, the smaller is the damping. In a dead beat galvanometer R is very small, at most equal to the resistance of the coil. When a current is passed in the coil, it slowly deflects to a new equilibrium position without any oscillation.

Case (ii): Criticaly Damped Motion

When $b^2 = \omega_0^2 = \dfrac{c}{I}$, the motion is said to be *critically damped*. In this case, the moving system after displacement comes to rest in minimum time and direction of motion never changes.

This condition in galvanometer is realised by introducing a resistance R which is necessary to just stop the oscillation of the coil. This condition corresponds to what, is called *critical damping* and R is called *critical damping resistance*. Under these conditions of critical damping, the coil smoothly and quickly approaches its new equilibrium positions when a steady current is passed through it. Such a behaviour is highly advantageous in electrical meters (ammeters, voltmeters, etc.), where one would like to take a steady reading immediately after the meter is connected to the circuit.

Case (iii): Ballistic or Under Damped Galvanometer $b^2 < \omega_0^2$

Sometimes, instead of measuring a steady current, it is necessary to measure a transient current (or charge). The current produced in the coil, while the charge is passing, flows only for a short time. The time period of oscillations of coil must be greater than the time for which the charge flows in the coil. In other words, we should make the moving system such that the current, due to flow of charge will have ceased before the moving part has had time to move appreciably from its position of rest and it is deflected only after the charge has passed through the coil. A galvanometer having a long period of mechanical oscillations of its coil will meet this requirement. Such a galvanometer is called *ballistic galvanometer*. Since time period T is given by

$$T = 2\pi\sqrt{\frac{I}{c}}$$

means a ballistic galvanometer should have a large moment of inertia I. If I is large, then $2b = \dfrac{1}{I}\left(\alpha + \dfrac{\gamma}{R}\right)$ is small and hence, the motion of the coil is oscillatory. Thus, for galvanometer to be ballistic, its time period should be long and must have low damping. The damping is further reduced by making resistance R much larger than the critical resistance and γ small. To make γ small, the coil is wound on a non-conducting frame like paper, plastic, etc.

The logarithmic decrement of the galvanometer is measured by observing the successive throws and the first throw is corrected by damping as explained earlier.

(iii) The Series LCR Circuit

Another important example of damped harmonic oscillations arises in case of LCR circuit shown in Fig. 2.7. When $R = 0$, the oscillations of the circuit are undamped with angular frequency

$\omega_0 = \dfrac{1}{\sqrt{LC}}$ (see Chapter 1). Here, we shall see that resistance R plays the part of a resistive or dissipative force analogous to that of friction or viscosity in the case of mechanical oscillations.

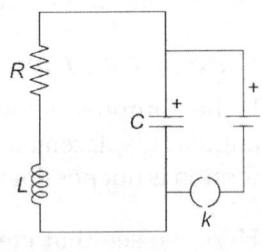

Fig. 2.7

On pressing key k, the capacitor gets charged by the battery. When the key is released, the battery is thrown out of circuit and the capacitor begins to discharge through the inductance and resistance. Let at any instant t, the charge on the capacitor be q and the current in circuit is i, so that, V_0 the voltage across the capacitor $= q/c$. The induced e.m.f. across the inductance is $V_L = L\dfrac{di}{dt}\left(= L\dfrac{d^2q}{dt^2}\right)$ and potential difference across the resistance $V_R = Ri\left(= R\dfrac{dq}{dt}\right)$. Since there is no external e.m.f. (battery is out of the circuit), we have

$$\frac{q}{c} = -L\frac{d^2q}{dt^2} - R\frac{dq}{dt}$$

or

$$L\frac{d^2q}{dt^2} + R\frac{dq}{dt} + \frac{q}{c} = 0$$

or

$$\frac{d^2q}{dt^2} + \frac{R}{L}\frac{dq}{dt} + \frac{q}{LC} = 0$$

or

$$\frac{d^2q}{dt^2} + 2b\frac{dq}{dt} + \omega_0^2 q = 0 \tag{2.31}$$

where $2b = \dfrac{R}{L}$ and $\omega_0^2 = \dfrac{I}{LC}$. The equation is of the same form as that of a damped mechanical oscillator Eq. (2.2) except that the variable x is the charge in this case. It is interesting to note that the inductance is analogous to mass m, the resistance to the viscous damping factor β and the inverse of the capacitance of the stiffness constant K.

When the damping is large, i.e. when $b > \omega_0$ or $R > 2\sqrt{\dfrac{L}{c}}$, the charge decays gradually until the capacitor is discharged. The charge is non-oscillatory or dead beat.

When the damping is critical, i.e. $R = 2\sqrt{\dfrac{L}{c}}$, the discharge became just non-oscillatory and dies out quickly.

When the damping is less than critical, i.e. $R < 2\sqrt{\dfrac{L}{c}}$, an interesting phenomenon takes place, the charge begins to oscillate and the electrical system exhibits damped harmonic oscillations.

The charge on the capacitor repeatedly becomes positive and negative eventually decaying to zero.

The variation of charge with time is given by

$$q = q_0 e^{-bt} \sin(\omega t + \phi)$$

where

$$\omega = \sqrt{\frac{1}{LC} - \frac{R^2}{4L^2}} \quad \text{and} \quad b = \frac{R}{2L}$$

ω gives the angular frequency of the oscillations. Since, $\omega_0 = \dfrac{1}{\sqrt{LC}}$ is the natural frequency in absence of damping (in this case resistance), it is evident that the damping in this case is due to finite value of resistance R, and the energy dissipated appears in the form of heat.

The quality factor of the circuit is

$$Q = \frac{\omega}{2b} = \frac{\omega L}{R}.$$

The Q-value is large (i.e., damping is small) if R is small. In case R is small $\omega \approx \omega_0$, so that

$$Q = \frac{L\omega_0}{R} = \frac{1}{R}\sqrt{\frac{L}{C}}.$$

For purely inductive circuit, i.e., $R = 0$, the quality factor Q will be infinite.

Example 2.1: For vertical spring mass system (Fig. 2.8), take mass $m = 0.1$ kg and spring constant $k = 10$ N/m. The motion of mass is resisted by a force proportional to the velocity. The constant of proportionality $\beta = 0.872$ kg/s. Calculate (a) the time period of damped oscillations, (b) time taken to reach the lowest point, and (c) the maximum extension of the spring. Assuming that the mass was released when the spring was of its natural length.

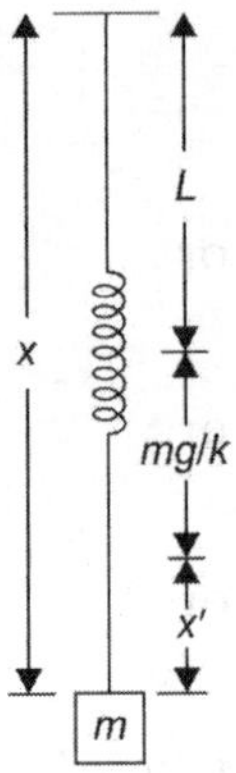

Solution: The forces acting on mass m are: (1) The weight mg always downwards, (2) The spring force $k(x - L)$ in a direction opposite to the direction of the velocity of m, (3) The damping force βv in a direction opposite to the direction of velocity of m.

If we define $x' = x - (L + mg/k)$, the equation of motion is

$$m\frac{d^2x'}{dt^2} + \beta\frac{dx'}{dt} + kx' = 0$$

or

$$\frac{d^2x'}{dt^2} + 2b\frac{dx'}{dt} + \omega_0^2 x' = 0$$

where

$$2b = \frac{\beta}{m} \text{ and } \omega_0^2 = \frac{k}{m}.$$

The solution of this equation can be written as

$$x' = Ae^{-bt}\sin(\omega t + \phi).$$

(a)

$$\omega_0 = \sqrt{\frac{k}{m}} = \sqrt{\frac{10}{0.1}} = 10 \text{ rad/s}$$

$$b = \frac{\beta}{2m} = \frac{0.872}{2 \times .1} = 4.36 \text{ per sec}$$

and

$$\omega = \sqrt{\omega_0^2 - b^2} = \sqrt{100 - (4.36)^2} = 9 \text{ per sec}$$

Hence, time period $T = \dfrac{2\pi}{\omega} = \dfrac{2\pi}{9} = 0.6981$ sec

(b) The velocity

$$v = \frac{dx'}{dt} = Ae^{-bt}\{-b \cdot \sin(\omega t + \phi) + \omega\cos(\omega t + \phi)\}$$

and $t = 0, x' = -mg/k$ and $v = 0$, hence

$$A\sin\phi = mg/k \qquad \text{(i)}$$

and

$$A\{-b\sin\phi + \omega\cos\phi\} = 0 \qquad \text{(ii)}$$

this gives, $\tan\phi = \dfrac{\omega}{b}$ and $A = \dfrac{mg}{k}\sqrt{1 + \dfrac{b^2}{\omega^2}}$

Fig. 2.8

$$\therefore \qquad A = \frac{0.1 \times 10}{10} \sqrt{1 + \left(\frac{4.36}{9}\right)^2} = 0.109 \text{ m}$$

and

$$\tan \phi = \frac{\omega}{b} = \frac{9}{4.36} = 2.0642$$

or

$$\phi = 64.15° = 1.1196 \text{ rad}$$

At the lowest point, the velocity is zero, i.e.

$$\tan(\omega t + \phi) = \frac{\omega}{b} = \tan \phi$$

or

$$\omega t = n\pi \text{ or } t = \frac{n\pi}{\omega}.$$

Hence the time taken to reach the lowest point is

$$t = \frac{\pi}{\omega} = \frac{3.14}{9} = 0.349 \text{ sec.}$$

It may be seen that the equilibrium point ($x' = 0$) is reached whenever $\sin(\omega t + \phi)$ vanishes, i.e. at

$$\omega t + \phi = n\pi$$

or

$$t = \frac{n\pi - \phi}{\omega} = \frac{n\pi - 1.1196}{9}$$

i.e. at $t = 0.2246$ sec, 0.5736 sec ..., etc.

The maximum displacement occurs at $t = 0.349$ sec, which is earlier than the mid-point $\left(t = \frac{6.2246 + 0.5736}{2} = 0.39915 \right)$ between two zeroes of displacement.

(c) The maximum extension of this spring is obtained on substituting the values into the expression for x'

$$x'_{max} = (0.109) e^{-(4.36 \times 0.349)} \sin(\pi + 1.11196)$$
$$= 0.0214 \text{ m}$$

which is less than 25% of the amplitude in the absence of drag.

Example 2.2: A massless spring, suspended from a rigid support, carries a flat disc of mass 100 gm at its lower end. It is observed that the system oscillates with a frequency of 10 Hz and the amplitude of the damped oscillations reduces to half its undamped value in one minute. Calculate

(a) The resistive force constant
(b) Relaxation time of the system
(c) its quality factor, and
(d) the force constant of the spring.

Solution: The amplitude of the damped oscillator at an instant t is given by

$$A = A_0 e^{-bt}$$

since $\dfrac{A}{A_0} = \dfrac{1}{2}$ when $t = 1$ minute or 60 sec.

$$\therefore \qquad e^{-bt} = e^{-60b} = \frac{1}{2}$$

$\therefore$ $\qquad\qquad 60b = 2.303 \log 2 = 0.693$

$\therefore$ $\qquad\qquad b = \dfrac{0.693}{60} = 0.01155$

But $2b = \dfrac{\beta}{m}$ or $\beta = 2bm$, where β is resistive force constant

$\therefore$ $\qquad\qquad \beta = 2bm = \dfrac{2 \times 0.693}{60} \times 0.1 = 0.0023 \text{ Ns} \cdot \text{m}^{-1}.$

(b) The relaxation time

$$\tau = \frac{1}{b} = \frac{60}{0.693} = 86.58 \text{ s}$$

(c) The quality factor $Q = \dfrac{\omega}{2b}$

where $\omega = 2\pi v = 2\pi \times 10 = 20\pi \text{ rad/s}$, v is the frequency of the damped oscillations which is given to be 10 Hz.

$\therefore$ $\qquad\qquad Q = \dfrac{20\pi}{2 \times 0.01155} = 2720$

(d) The force constant k of the spring is given by

$$\omega_0 = \sqrt{\frac{k}{m}}$$

or $\qquad\qquad k = m\omega_0^2 = m(\omega^2 + b^2)$

$$= 0.1[(20\pi)^2 + (.01155)^2]$$

$$= 394.8 \text{ N·m}^{-1}$$

Example 2.3: The viscous forces on sphere of radius a moving with velocity V in a medium of co-efficient of viscosity η is $-6\pi\eta av$. Determine the effects of air viscosity on the amplitude and period of simple pendulum consisting of an aluminium bob of radius 0.5 cm suspended by means of a 1 m long thread. Take the density of aluminium at 2.65 g·cm^{-3} and the co-efficient of viscosity of air at room temperature (20°C) as 1.78×10^{-4} g·cm^{-1}s^{-1}.

Solution: When the deflection of the bob from vertical is θ, the linear velocity of the bob is given by

$$v = l\frac{d\theta}{dt}.$$

The resistive viscous force is then

$$F = -6\pi\eta av = -6\pi\eta al\frac{d\theta}{dt}.$$

For small angular displacements the restoring gravitational force

$$f = mg\theta$$

Hence the equation of motion of bob is

$$lm\frac{d^2\theta}{dt^2} = -6\pi\eta al\left(\frac{d\theta}{dt}\right) - mg\theta$$

or $$\frac{d^2\theta}{dt^2} + 2b\frac{d\theta}{dt} + \omega_0^2\theta = 0$$

where $$2b = \frac{6\pi\eta al}{ml} = \frac{6\pi\eta a}{\frac{4}{3}\pi a^3\rho} = \frac{9\eta}{2a^2\rho}$$

where ρ is density of material of bob.

$$2b = \frac{9}{2} \times \frac{1.78 \times 10^{-4}}{(0.5)^2 \times 2.65} = 12.08 \times 10^{-4}\ \text{s}^{-1}$$

$$\omega_0^2 = \frac{g}{l} = \frac{9.8}{1} = 9.8$$

The amplitude of oscillations of the pendulum decrease, as

$$A = A_0\, e^{-bt} = A_0\, e^{-0.00604}$$

To have an idea, let us calculate the time in which amplitude reduces by 10%, i.e.

$$0.9 = e^{-0.000604}$$

or $$t = 174\ \text{sec.} = 2.90\ \text{min}$$

The angular frequency ω becomes

$$\omega = \sqrt{\omega_0^2 - b^2} = \sqrt{(9.8) - (6.04 \times 10^{-4})^2}$$
$$\approx \sqrt{9.8} = \omega_0$$

It is thus seen that the frequency of a pendulum is not appreciably influenced by viscosity; although the amplitude is found to reduce appreciably with time.

Example 2.4: A massless spring of spring constant 10 N·m^{-1} is suspended from a rigid support and carries a mass of 0.1 kg at its lower end. The system is subjected to resistive force $-\beta v$ where β is a constant and v is the velocity. It is observed that the system performs damped oscillatory motion and its energy decays to $1/e$ of its initial value in 50 s:

(a) What is the value of β?
(b) What is the Q value of the oscillator?
(c) Show that the fractional change in the frequency of the damped oscillator is $\approx (8Q^2)^{-1}$. What is the percentage change in frequency due to damping? What conclusion do you draw from it?

Solution: (a) The decay of energy of the damped oscillator is given by

$$E(t) = E_0\, e^{-2bt}$$

where E_0 is the initial energy and $2b = \dfrac{\beta}{m}$, where β the constant of resistive force. It is given that when $t = 50$ s

$$\frac{E(t)}{E_0} = \frac{1}{e}$$

or $$e^{-50(2b)} = \frac{1}{e} = e^{-1}$$

or $$2b = \frac{1}{50} = 2 \times 10^{-2}\ \text{sec}$$

Hence, $\beta = m(2b) = 0.1 \times 2 \times 10^{-2} = 2 \times 10^{-3}\ \text{kgs}^{-1}$

(b) The angular frequency ω, in the absence of the damping is

$$\omega_0 = \sqrt{\frac{k}{m}} = \sqrt{\frac{10}{0.1}} = 10 \text{ rad/s}$$

The angular frequency of the damped oscillations is:

$$\omega^2 = \omega_0^2 - b^2 = 10^2 - 10^{-4} \approx 10^2 \text{ rad/s}$$

or $\qquad \omega = 10 \text{ rad/s}.$

The quality factor Q is given by

$$Q = \frac{\omega}{2b} = \frac{10}{2 \times 10^{-2}} = 500$$

(c) The fractional change in frequency is given by $\dfrac{\omega_0 - \omega}{\omega_0}$. But

$$\omega = \omega_0 \left(1 - \frac{b^2}{\omega_0^2}\right)^{1/2} = \omega_0 \left(1 - \frac{1}{4Q^2}\right)^{1/2}$$

since $\dfrac{1}{4Q^2} \ll 1$.

Hence, using binomial theorem and neglecting higher powers, we have

$$\omega = \omega_0 \left(1 - \frac{1}{8Q^2}\right)$$

then $\qquad \dfrac{\omega_0 - \omega}{\omega_0} = \dfrac{1}{8Q^2} = (8Q^2)^{-1}.$

Hence, percentage change in frequency is

$$\frac{\omega_0 - \omega}{\omega_0} \times 100 = \frac{100}{8Q^2} = \frac{100}{8(500)^2} = 0.5 \times 10^{-5}$$

From this we conclude, that for oscillations with high Q value, the change in frequency due to damping is negligibly small.

Example 2.5: According to classical electromagnetic theory, an electron in an atom executes harmonic oscillations in a straight line, and hence, undergoes acceleration. An accelerated electron radiates energy, and therefore, behaves as a damped harmonic oscillator; the damping being due to the radiation it emits. According to the theory, the electron radiates energy at the rate of $\dfrac{kq^2\omega_0^4 A^2}{3c^3}$ W, where $k = 9 \times 10^9$ N·m²·C⁻², $q = 1.6 \times 10^{-19}$ C is the electronic charge, $c = 3 \times 10^8$ ms⁻¹, the velocity of light; ω_0 is the angular frequency and A is the amplitude of oscillations. If the emitted radiation has wavelength 6000 Å. Calculate (a) the Q-value of oscillator, and (b) the radiation life time (i.e. the time for the energy to fall to e^{-1} of the original energy).

Solution: We know that the average power radiated $<p(t)>$ is related to the average energy radiated $<E(t)>$ as

$$<P(t)> = 2b <E(t)> = \frac{\omega_0}{Q} < E(t) >$$

$$<E(t)> = \frac{1}{2}mA^2\omega_0^2$$

It is given that

$$<P(t)> = \frac{kq^2\omega_0^4 A^2}{3c^3}$$

$$\therefore \quad \frac{<p(t)>}{<E(t)>} = \frac{kq^2\omega_0^4 A^2}{3c^3\, mA^2\,\omega_0^2} \times 2$$

$$= \frac{2}{3}\frac{kq^2\omega_0^2}{mc^3}$$

But

$$\frac{<p(t)>}{<E(t)>} = \frac{\omega_0}{Q}$$

$$\therefore \quad \frac{\omega_0}{Q} = \frac{2}{3}\frac{kq^2\,\omega_0^2}{mc^3}$$

or

$$Q = \frac{3}{2}\frac{mc^3}{kq^2\,\omega_0}.$$

Since $\lambda_0 = 6000\text{Å} \ \therefore \ v_0 = \frac{c}{\lambda_0}$ and $\omega_0 = 2\pi v_0 = \frac{2\pi c}{\lambda_0}$

$$\therefore \quad Q = \frac{3}{2}\frac{mc^3\lambda_0}{kq^2\,2\pi c}$$

$$= \frac{3}{2}\frac{mc^2\lambda_0}{kq^2\,2\pi} = \frac{3mc^2\lambda_0}{4\pi kq^2}$$

Substituting the values, we have

$$Q = \frac{3\times 9.1\times 10^{-31}\times(3\times 10^8)^2\,(6\times 10^{-7})}{4\pi\times 9\times 10^9\times\left(1.6\times 10^{-19}\right)^2} = 5.09\times 10^7$$

(b) We have

$$E(t) = E_0\,e^{-2bt} = E_0\,e^{-\frac{\omega_0}{Q}t}.$$

Thus, the energy decays to e^{-1} of its initial value E_0 in time τ_0

where

$$\tau_0 = \frac{Q}{\omega_0} = \frac{Q\cdot\lambda}{2\pi c}$$

$$= \frac{5.09\times 10^7\times 6\times 10^{-7}}{2\pi\times 3\times 10^8} = 1.69\times 10^{-8}\ \text{sec}$$

Hence the radiation life-time of the atom, according to the classical electromagnetic theory is of the order of 10^{-8} sec.

Example 2.6: The system shown in Fig. 2.9 is subjected to a resistive force $F = -\beta v$, where β is a constant and v is the velocity. The system is at rest initially when a velocity of 6.8 cm/s is given to it. If $k = 10\ \text{N·m}^{-1}$, $m = 10$ kg and $\beta = 8$ Ns·m^{-1}, determine the subsequent displacement and velocity of the mass.

Solution: The equation of motion is

$$m\frac{d^2x}{dt^2} + \beta\cdot\frac{dx}{dt} + 2ky = 0.$$

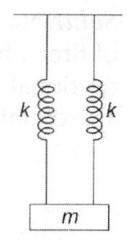

Fig. 2.9

Since the springs are connected in parallel, the effective spring constant of the system is $k + k = 2k$. The solution of the above equation is

$$x = e^{-bt} (A \cos \omega t + B \sin \omega t) \qquad\qquad (i)$$

where $\qquad 2b = \dfrac{\beta}{m} \qquad \omega^2 = \omega_0^2 - b^2$ with $\omega_0^2 = \dfrac{2k}{m}$

Substituting the values, we have

$$\omega_0 = \sqrt{\dfrac{2k}{m}} = \sqrt{\dfrac{2 \times 10}{10}} = \sqrt{2} = 1.41 \, \text{rad/s}$$

$$2b = \dfrac{\beta}{m} = \dfrac{8}{10} = 0.8 \, \text{s}^{-1}$$

$$\omega = \sqrt{\omega_0^2 - b^2} = \sqrt{2 - \dfrac{0.64}{4}} = \sqrt{1.84} = 1.36 \, \text{rad/s}$$

Differentiating equation (i), we have

$$\dfrac{dx}{dt} = -be^{-bt} (A \cos \omega t + B \sin \omega t) + e^{-bt} (-A\omega \sin \omega t + B\omega \cos \omega t)$$

using initial conditions, i.e., at $t = 0$, $x = 0$, $\dfrac{dx}{dt} = v_0 = 0.068 \, \text{m} \cdot \text{s}^{-1} \qquad\qquad (ii)$

(i) gives, $\qquad A = 0$

(ii) gives $\qquad v_0 = B\omega$ or $B = \dfrac{v_0}{\omega} = \dfrac{0.068}{1.36} = 0.05 \, \text{m}$

substituting the values in (i), we have

$$x = 0.05 \, e^{-0.4t} \sin 1.36 \, t \qquad\qquad (iii)$$

Equation (iii) determines the subsequent motion of the system. Velocity is given by

$$\dfrac{dx}{dt} = 0.05 \, e^{-0.4t} \{1.36 \cos 1.36t - 0.4 \sin 1.36t\}.$$

Here, we observe that at $t = 0$, velocity

$$\dfrac{dx}{dt} = 0.05 \times 1.36 = 0.068 \, \text{ms}^{-1}$$

and $\qquad x = 0$

which were our initial conditions and that confirms the correctness of the solution.

Example 2.7: A block is placed on a horizontal plane surface and held between two springs as shown in Fig. 2.10. The co-efficient of kinetic friction between the block and the surface is μ, which has constant value. The block is given an initial displacement x_0 from the equilibrium position and released. Investigate the subsequent motion of the block.

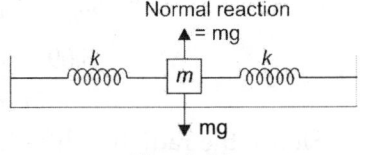

Fig. 2.10

Solution: Suppose the block is displaced to the right and its displacement at any instant of time t be x. The forces acting on the block are (i) the restoring force $= -2kx$, and (ii) the frictional force $\mu \times$ Normal reactions $= \mu \cdot mg$, this frictional force always opposes motion. The equation of motion is

$$m\dfrac{d^2x}{dt^2} = -2kx + \mu \cdot mg$$

$$= -2k\left(x - \dfrac{\mu \cdot mg}{2k}\right)$$

If we transform to a new variable

$$x' = x - \frac{\mu \cdot mg}{2k}$$

then $\quad m\dfrac{d^2x'}{dt^2} = -2kx'\quad$ or $\quad \dfrac{d^2x'}{dt^2} = -\dfrac{2k}{m}x'$

this gives $\omega = \sqrt{\dfrac{2k}{m}}$ i.e. the frequency of damped oscillations is the same as that of the undamped natural oscillations. In other words, a constant frictional force has no effect on the frequency of vibration of a system.

The solution is given by

$$x' = A\cos\omega t + B\sin\omega t$$

the motion of the particle is describedc by variable x and

$$x = x' + \frac{\mu \cdot mg}{2k}$$

$$= A\cos\omega t + B\sin\omega t + \frac{\mu \cdot mg}{2k}$$

and the velocity $\dfrac{dx}{dt} = \omega(B\cos\omega t - A\sin\omega t)$

Now, initial conditions are

At $t = 0$, $x = x_0$ and $\dfrac{dx}{dt} = 0$,

giving $\qquad A = x_0 - \dfrac{\mu \cdot mg}{2k}$

and $\qquad B = 0$

Hence the motion of the system is given by

$$x = \left(x_0 - \frac{\mu \cdot mg}{2k}\right)\cos\omega t + \frac{\mu \cdot mg}{2k}$$

$$= \left(x_0 - \frac{\mu \cdot mg}{2k}\right)\cos\frac{2\pi t}{T} + \frac{\mu \cdot mg}{2k}$$

At, $t = 0$, $x = x_0$, the maximum value of x. At the end of half a cycle, i.e. at $t = \dfrac{T}{2}$ the value of x is given by

$$x = -\left(x_0 - \frac{\mu \cdot mg}{2k}\right) + \frac{\mu \cdot mg}{2k} = -\left(x_0 - \frac{\mu \cdot mg}{k}\right)$$

At time $t = \dfrac{T}{2}$, the mass is at the extreme left position. Thus, we find that the amplitude decreases from x_0 to $(x_0 - \mu \cdot mg/k)$ during the first half cycle. Let us say that the maximum displacement in the left is x_1 i.e.

$$x_1 = x_0 - \frac{\mu \cdot mg}{k}$$

This is displacement at $t = T/2$. Applying the same reasoning the displacement at $t = T$ must decrease to $\left(x_1 - \dfrac{2\mu \cdot mg}{k}\right) = \left(x_0 - \dfrac{\mu \cdot mg}{k}\right)$. Hence, due to damping the amplitude of oscillation decreases by $\dfrac{\mu \cdot mg}{k}$ during each half cycle. Thus, we conclude that the motion is not SH (because its amplitude is not constant) but the motion is periodic with period.

$$T = 2\pi\sqrt{\dfrac{m}{2k}}$$

Example 2.8: The amplitude of vertical oscillations of the system shown in Fig. 2.11 decreases to 20% of the initial value after 5 consecutive cycles of oscillations. Determine the damping co-efficient β (frictional force $= -\beta n$) if $k = 80$ N/m and $m = 2.5$ kg.

Solution: The oscillations of the damped system are described by the equation

$$x = A_0\, e^{-bt} \sin(\omega t + \phi)$$

where $2b = \dfrac{\beta}{m}$, $\omega = \omega_0\left(1 - \dfrac{b^2}{\omega_0^2}\right)^{1/2}$, ω_0 being the angular frequency, if damping were absent. Since the springs are connected in series, the effective spring constant is given by $k/2$. Hence,

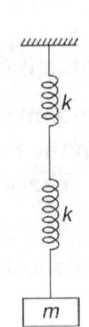

Fig. 2.11

$$\omega_0 = \sqrt{\dfrac{k}{2m}} = \sqrt{\dfrac{80}{2 \times 2.5}} = 4\ \text{rad/s}$$

The maximum amplitude occur at value of t satisfying $\sin(\omega t + \phi) = \pm 1$. The maximum amplitude are

$$A_1 = A_0\, e^{-bt_1}$$
$$A_2 = A_0\, e^{-bt_2}$$

But it is given, that $\dfrac{A_1}{A_6} = \dfrac{1}{0.2} = 5$

Logarithmic decrement is given by

$$\lambda = \dfrac{1}{n-1}\log\left(\dfrac{A_1}{A_n}\right) = \dfrac{1}{5}\log_e 5 = 0.32$$

But

$$\lambda = \dfrac{b \cdot T}{2} = \dfrac{b \cdot 2\pi}{2\omega} = \dfrac{\pi b}{\omega} = \dfrac{\pi b}{\omega_0\left(1 - \dfrac{b^2}{\omega_0^2}\right)^{1/2}}$$

or $\quad \left(1 - \dfrac{b^2}{\omega_0^2}\right)\lambda^2\,\omega_0^2 = \pi^2 b^2$

or $\quad \lambda^2\,\omega_0^2 - b^2\lambda^2 = \pi^2 b^2$

or $\quad b^2(\lambda^2 + \pi^2) = \lambda^2\,\omega_0^2$

and

$$b = \dfrac{\lambda \omega_0}{\sqrt{\lambda^2 + \pi^2}}$$

$$= \dfrac{0.32 \times 4}{\sqrt{(.32)^2 + \pi^2}} = 0.405$$

or $\qquad 2b = 0.81 \ \text{s}^{-1}$

and $\qquad 2b = \dfrac{\beta}{m}$

or $\qquad \beta = 2bm = 0.81 \times 2.5 = 2\text{Ns·m}^{-1}$

2.9 FORCED (DRIVEN) DAMPED HARMONIC OSCILLATIONS

We have so far studied the free undamped oscillations of various simple systems having one degree of freedom and investigated the effect of damping on the free oscillations of these systems and found that due to friction present in the system, the amplitude of oscillations decreases exponentially with time and the frequency of natural oscillations is slightly diminished. Usually, the change in frequency is too small to be of any significance. Now, we shall investigate the behaviour of a weakly damped harmonic oscillator when an external time dependent force is applied to the system so as to maintain the amplitude of the oscillations. We shall apply a harmonically varying driving force of frequency, not necessarily the same as that of the oscillator and investigate how the system responds to the deriving force as its frequency is gradually changed.

When the natural frequency of the driven oscillator is not the same as the frequency of the impressed force, the driven oscillator picks up energy from the driving system and oscillates. The natural frequency of oscillations dies out soon and it begins to oscillate with the frequency of impressed periodic force. The vibrations which are so maintained are called *forced vibrations*.

In the analysis of forced oscillations, here we will consider the cases in which the driven system extracts energy form the driving system, any appreciable feedback of energy from the former to the latter, i.e. the transfer of energy is essentially a one way process. This is due to (i) the coupling between the two is very weak, (ii) the driving system has such a large reservoir of energy that the energy feedback into it is negligible. Under these circumstances, the driving system remains practically unaffected by the forced oscillations of the driven system. The driving system only serves as the supplier of a periodic force. Here, we will analyse the behaviour of only those systems which satisfy the above mentioned conditions.

2.10 FORCED OSCILLATIONS OF ONE DIMENSIONAL DAMPED HARMONIC OSCILLATOR

Let us consider a system oscillating about an equilibrium position under an external periodic force. Let x be its displacement from the equilibrium position at any instant (t) during the oscillations. Its instantaneous velocity is $\dfrac{dx}{dt}$. The forces acting upon the system at this instant are:

(i) A restoring force proportional to the displacement but acting in the opposite direction. This is given by ($-kx$).

(ii) A frictional force proportional to the velocity but acting in the opposite direction given by $\left(-\beta\dfrac{dx}{dt}\right)$, where β is a positive constant called *damping constant*.

(iii) An external periodic force ($= F_0 \sin pt$) where F_0 is the amplitude or maximum value of this force and p is its angular frequency. Thus, the total force acting on the system is

$$F = -kx - \beta\frac{dx}{dt} + F_0 \sin pt$$

Using Newton's law, this force must be equal to the product of the mass m of the system and its instantaneous acceleration $\dfrac{d^2x}{dt^2}$. Hence

$$m\frac{d^2x}{dt^2} = -kx - \beta\frac{dx}{dt} + F_0 \sin pt \tag{2.32}$$

or

$$\frac{d^2x}{dt^2} + 2b\frac{dx}{dt} + \omega_0^2 x = f_0 \sin pt \tag{2.32a}$$

where

$$2b = \frac{\beta}{m},\ \omega_0^2 = \frac{k}{m}\ \text{and}\ f_0 = \frac{F_0}{m}.$$

Equation (2.32a) is the differential equation of motion of the forced harmonic oscillator. The complete solution of this equation will consist of the sum of the complementary function and particular integral.

The homogeneous part of the equation is

$$(D^2 + 2bD + \omega_0^2)x = 0$$

and the roots of the characteristic equation are

$$D = -b \pm \sqrt{b^2 - \omega_0^2}$$

and the complementary function is

$$C_F = e^{-bt}\left[Ae^{i\sqrt{(\omega_0^2 - b^2)}\,t} + Be^{i\sqrt{(\omega_0^2 - b^2)}\,t}\right]$$

$$= C_0\, e^{-bt}\, \sin\left(\sqrt{(\omega_0^2 - b^2)}\,t + \phi\right) \tag{2.33}$$

where C_0 and ϕ are new constants.

The particular integral is

$$PI = \frac{1}{(D^2 + 2bD + \omega_0^2)}\, f_0 \sin pt$$

$$= C \sin (pt - \delta) \tag{2.34}$$

where C is the amplitude and δ the phase of oscillations. The quantity δ gives the phase difference of the oscillations (displacement) and the driving force. Contrary to the case of free harmonic oscillator, where δ gives the relation between the initial position of the particle and the instantaneous displacement of the particles. In the case of driven oscillator, this initial condition is immaterial. Hence the general solution is

$$x = C_0\, e^{-bt}\, \sin\left\{\sqrt{(\omega_0^2 - b^2)}\,t + \phi\right\} + C \sin (Pt - \delta).$$

The first term is the transient term and dies away with time as e^{-bt}. During the transient state, the oscillator oscillates neither with its natural frequency nor the frequency of the impressed force. The second term is called the *steady state term* and governs the motion of the oscillator after the transient has ceased to be effective. During the steady state, the oscillator performs forced oscillations with the impressed force frequency. We are interested in analysing the steady state behaviour of a forced oscillator, i.e. analysing the motion of a damped oscillator after the force has been acting for sufficiently long time.

Steady State Behaviour of a Forced Oscillator

The steady state solution of the oscillator is given by

$$x = C \sin (pt - \delta)$$

where C and δ are yet to be determined.

Differentiating w.r.t. time, we have

$$\frac{dx}{dt} = Cp \cos (pt - \delta)$$

and

$$\frac{d^2x}{dt^2} = -Cp^2 \sin (pt - \delta)$$

Substituting the values in equation (2.32a), we have

$$-Cp^2 \sin (pt - \delta) + 2bCp\cos(pt - \delta) + \omega_0^2 C \sin (pt - \delta)$$
$$= f_0 \sin pt = f_0 \sin \{(pt - \delta) + \delta\}$$

This equation holds for all values of t. Hence, co-efficients of $\sin (pt - \delta)$ and $\cos(p - \delta)$ must separately be equal. Equating them, we have

$$C(\omega_0^2 - p^2) = f_0 \sin \delta \qquad \text{(i)}$$

and

$$2bCp = f_0 \sin \delta \qquad \text{(ii)}$$

$$\tan \delta = \frac{2bp}{\omega_0^2 - p^2} \qquad (2.35)$$

and

$$C = \frac{f_0}{\sqrt{\left(\omega_0^2 - p^2\right)^2 + 4b^2 p^2}} \qquad (2.35a)$$

Equations (2.35) and (2.35a) give us the following information about the steady state behaviour of the forced oscillator:

(i) That a phase difference (δ) exists between the displacement and the driving force, and we see that the value of δ depends upon the angular frequency of the driving force and the constants b and ω_0 of the oscillator.

(ii) The amplitude C of the forced oscillations depends on the constants, $F_0 (= f_0 \, m)$ and p, the angular frequency of the driving force and the constants b and ω_0 of the oscillator.

(iii) The motion of the oscillator is completely independent of the initial conditions, i.e. the way in which the oscillator is set into motion, the amplitude, phase and frequency of the oscillator depends only on the constant F_0 and p of the driving force and on oscillator constants m, ω_0 and b. No matter, how we start the oscillator, its motion will eventually settle down into that represented by Eq. (2.34). Thus, *steady state motion is the motion of a system that has forgotten how it started.*

Mechanical Impedance of the Oscillator

Steady state solution is

$$X = C \sin (pt - \delta) \quad \text{where} \quad C = \frac{f_0}{\sqrt{(\omega_0^2 - p^2)^2 + 4b^2 p^2}}$$

where $f_0 = \dfrac{F_0}{m}$. F_0 is amplitude of driving force.

$\therefore$
$$C = \frac{F_0}{\sqrt{(m\omega_0^2 - p^2 m)^2 + 4m^2 b^2 p^2}}$$

$$= \frac{F_0}{p\sqrt{\left(\dfrac{m\omega_0^2}{p} - pm\right)^2 + (2mb)^2}} \quad \text{and} \quad m\omega_0^2 = k$$

$$= \frac{F_0}{p\sqrt{\left(pm - \dfrac{k}{p}\right)^2 + (2mb)^2}}$$

Let $2mb = R_m$, then

$$C = \frac{F_0}{p\sqrt{R_m^2 + \left(p_m - \dfrac{k}{p}\right)^2}}$$

Here,
$$Z_m = p\sqrt{R_m^2 + \left(p_m - \dfrac{k}{p}\right)^2} \qquad (2.36)$$

Here Z_m is called *mechanical impedance of the oscillator* from the analogy of electric impedance.

Mechanical impedance is defined as the force required to produce unit velocity in the oscillator, i.e. $Zm = f/v$.

From electrical analogy

$$X_m = \left(mp - \frac{k}{m}\right)$$

is defined as mechanical reactance and $R_m = 2mp$ is defined as the mechanical resistance. Hence the expression for displacement then becomes

$$x = \frac{F_0}{p z_m} \sin(pt - \delta) \qquad (2.37)$$

As we have seen the phase in steady state is defined completely with respect to the driving force. It depends on the relative magnitudes of the driving and the natural frequencies, p and ω_0 respectively. As such three cases arise.

Case (i): Low Driving Frequency ($P \ll \omega_0$)

Under this condition the amplitude

$$C = \frac{f_0}{\sqrt{(\omega_0^2 - p^2)^2 + 4b^2 p^2}} \quad \text{as } p \to 0$$

$$C = \frac{f_0}{\omega_0^2} = \frac{F_0}{m\omega_0^2} = \frac{F_0}{k} \qquad (2.38)$$

and phase angle

$$\tan \delta = \frac{2bp}{\omega_0^2 - p^2} \to 0 \quad \text{and} \quad \delta = 0 \tag{2.39}$$

i.e., the driving force and displacement are in same phase and the amplitude at low driving frequencies thus depends on the driving force (F) and the restoring force constant K such a system is said to be restoring force controlled system.

Case (ii): p = ω₀ (Resonance)

The amplitude of the motion at this frequency will be maximum and is given by

$$C_{max} = \frac{f_0}{2bp} \tag{2.40}$$

and the phase angle

$$\delta = \tan^{-1} \frac{2bp}{\omega_0^2 - p^2} = \tan^{-1} \infty = \frac{\pi}{2} \tag{2.41}$$

The resonance response depends upon damping, the amplitude C_{max} at resonance is inversely proportional to b.

If there is no damping, i.e. $b = 0$ then C_{max} becomes ∞.

It may be remarked that in the presence of damping the maximum amplitude is attained at frequency which is slightly less than ω_0. Maximum value of amplitude C is obtained by differentiating C w.r.t. P and equating it to zero.

$$\therefore \quad \frac{dC}{dp} = \frac{d}{dp} \left\{ \frac{f_0}{\sqrt{(\omega_0^2 - p^2)^2 + 4b^2 p^2}} \right\}$$

$$= \frac{f_0}{2} \frac{2(\omega_0^2 - p^2)(-2p) + 8b^2 p}{\{(\omega_0^2 - p^2)^2 + 4b^2 p^2\}^{3/2}} = 0$$

which gives p the resonance frequency as $\omega_0^2 - p_0^2 - 2b^2 = 0$

or

$$p_0 = \omega_0 \sqrt{1 - \frac{2b^2}{\omega_0^2}}$$

$$= \omega_0 \sqrt{1 - \frac{\beta^2}{2mk}} \quad \because \frac{\beta}{m} = 2b \quad \text{and} \quad m\omega_0^2 = k$$

$$= \omega_0 \sqrt{1 - \frac{R_m^2}{z_m^2 \omega_0^2}} \quad \text{as} \quad R_m = 2mb$$

Obviously, the frequency at which the resonance occurs is slightly less than ω_0. However, lesser the damping, more near it is to the natural frequency, in that case this decrease can be neglected.

Case (iii): High Driving Frequency p >> ω₀

Under this condition, the amplitude of the resulting vibration:

$$C = \frac{f_0}{\sqrt{(\omega_0^2 - p^2)^2 + 4b^2 p^2}}$$

becomes

$$C = \frac{f_0}{\sqrt{p^4 + 4b^2 p^2}} \approx \frac{f_0}{p^2} = \frac{F_0}{mp^2} \tag{2.42}$$

as b is small quantity. Since the maximum amplitude depends on mass of oscillator, the system is said to be mass controlled.

The phase is given by

$$\tan \delta = \frac{2bp}{\omega_0^2 - p^2} \approx -\frac{2b}{p} \to -0$$

or $\qquad\qquad \delta = \pi \qquad\qquad\qquad\qquad\qquad\qquad\qquad\qquad\qquad (2.43)$

Thus, as the frequency p of the impressed force is increased, the amplitude goes on decreasing and the phase tends to π.

The dependence of the amplitude and phase angle upon the frequency p of the driving force is shown in Figs 2.12(a) and (b). The sharpness of rise of the two curves depends on the magnitude of β the damping factor. The characteristics of the forced motion are summarised as follows:

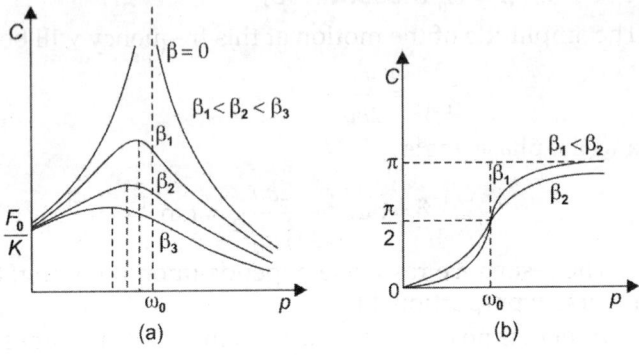

Fig. 2.12

(i) The maximum amplitude of the displacement is $\dfrac{F_0}{R_m\,\omega_0}$.

(ii) The displacement lags behind the force $f_0 \sin pt$ by an angle δ which increases continuously with p. At $p = 0$, $\delta = 0$, acquiring the value $\pi/2$ at precisely the frequency $p = \omega_0$ and to $\delta = \pi$ at $p \to \infty$.

2.11 VELOCITY RESONANCE

The steady state solution of a harmonically driven damped forced oscillator is given by Eq. (2.34).

$$x = C \sin (pt - \delta)$$

when $\qquad\qquad C = \dfrac{\dfrac{F_0}{m}}{\sqrt{(\omega_0^2 - p^2)^2 + 4b^2 p^2}} \qquad\qquad\qquad\qquad\qquad (i)$

and $\qquad\qquad \tan \delta = \dfrac{2bp}{\omega_0^2 - p^2} \qquad\qquad\qquad\qquad\qquad\qquad (ii)$

Here velocity of the oscillation is obtained by differentiating x w.r.t. (t), i.e.

$$v = cp \cos(pt - \delta)$$

$$= cp \sin\left\{(pt - \delta) + \frac{\pi}{2}\right\}$$

$$= v_0 \sin\left\{(pt - \delta) + \frac{\pi}{2}\right\} \qquad\qquad\qquad\qquad\qquad (2.44)$$

Here v_0 is the amplitude of velocity

$$v_0 = cp = \frac{\dfrac{F_0}{m}}{\left\{\dfrac{\left(\omega_0^2 - p^2\right)^2}{p^2} + 4b^2\right\}^{1/2}} \tag{2.45}$$

and phase $\phi = \left(\delta - \dfrac{\pi}{2}\right)$ $\tag{2.46}$

The comparison reveals that the velocity leads the displacement in phase by $\dfrac{\pi}{2}$ and that velocity amplitude v_0 varies with p the angular frequency of the driving force. It is apparent that $v_0 = 0$ when $p = 0$ and as p increases v_0 also increases reaching a maximum value v_{max} for $p = \omega_0$ and then decreases further. At resonance, the maximum value of velocity amplitude becomes

$$v_{max} = \frac{F_0}{2mb} = \frac{F_0}{\beta} \tag{2.47}$$

indicating that v_{max} decreases as β (damping) increases. As p is increased to a high value $(P \gg \omega_0)$, $v_0 \approx \dfrac{F_0}{mp^2}$ provided damping is not too heavy. In limit $p \to \infty$, $v_0 \to 0$.

Figure 2.13(a) shows the dependence of the velocity of the force oscillator (in steady state) on the frequency of the driving force.

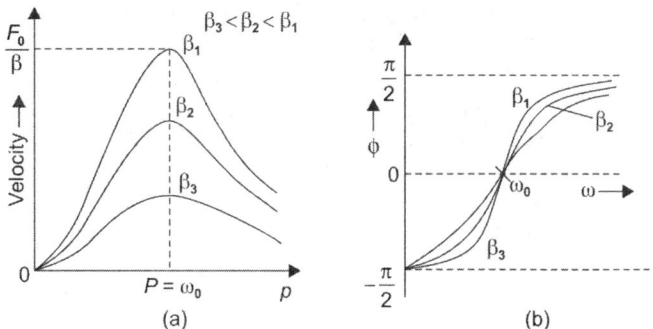

Fig. 2.13

The phase of the velocity relative to that of driving force is $\phi = \delta - \dfrac{\pi}{2}$, where δ is defined in (ii). For $p \ll \omega_0$, $\delta = 0$ and $\phi = -\dfrac{\pi}{2}$. Since ϕ is the angle by which the velocity lags behind the force. It is obvious that in range $p \ll \omega_0$, the velocity leads the force by an angle of $\pi/2$ on the other hand for $p \gg \omega_0$, $\delta = \pi$, so that $\phi = \pi/2$, implying that in the case of very high frequencies of driving force, the velocity of the oscillator lags behind the force by an angle of $\dfrac{\pi}{2}$. However, at resonance $p = \omega_0$, $\delta = \dfrac{\pi}{2}$, so that $\phi = 0$. Thus at resonance, the velocity of the driven oscillator is in phase with driving force. This is, therefore, the most favourable condition for the transfer of energy from the driving force to the oscillator. Figure 2.13(b) shows the dependence of the phase of the velocity of the oscillator on the frequency of the driving force.

2.12 POWER SUPPLIED BY THE DRIVING FORCE TO THE FORCED OSCILLATOR

We have seen that the energy of the damped oscillator decays exponentially, i.e. Eq. (2.15)

$$E(t) = E_0\, e^{-2bt}$$

In order to maintain the steady oscillations of such a system, the energy lost by the oscillator in each oscillation due to presence of resistance must be compensated by the driving force. Here, we shall show that in the steady state the amplitude and phase of a driven oscillator will adjust themselves in such a manner that the power supplied by the driving force is just compensating the losses due to the presence of friction.

The instantaneous power P_m supplied is

$$P_m = \text{Force} \times \text{Velocity}$$

Since instantaneous force

$$F = \bar{r}_0 \sin pt$$

and velocity $v = v_0 \cos(pt - \delta)$

with

$$v_0 = cp = \dfrac{\dfrac{F_0}{m}}{\left\{ \dfrac{\left(\omega_0^2 - p^2 \right)^2}{p^2} + 4b^2 \right\}^{1/2}}$$

$$= \dfrac{F_0}{\sqrt{(2mb)^2 + \left(\dfrac{m\omega_0^2}{p} - pm \right)^2}} = \dfrac{F_0}{\sqrt{(2mb)^2 + \left(mp - \dfrac{k}{p} \right)^2}}$$

$$= \dfrac{F_0}{z_m}, \text{ where } z_m \text{ is mechanical impedance.}$$

$$\therefore \qquad v_0 = \dfrac{F_0}{z_m}$$

and

$$v = \dfrac{F_0}{z_m} \cos(pt - \delta)$$

Thus,

$$P_{in} = \dfrac{F_0^2}{z_m} \sin pt \cos(pt - \delta)$$

Therefore, average power supplied per cycle is

$$P_{av} = \dfrac{\text{Power supplied in 1 cycle}}{\text{Period of cycle}}$$

$$= \dfrac{1}{T} \int_0^T P_{in}\, dt$$

$$= \dfrac{1}{T} \int_0^T \dfrac{F_0^2}{z_m} \sin pt \cos(pt - \delta)\, dt$$

$$= \dfrac{F_0^2}{T z_m} \int_0^T [\sin pt \cos pt \cos \delta + \sin^2 pt \sin \delta]\, dt$$

But we know that

$$\int_0^T \sin pt \cos pt \, dt = \int_0^T \frac{1}{2} \sin 2pt \, dt = 0$$

and

$$\int_0^T \sin^2 pt \, dt = \int_0^T \frac{(1 - \cos 2pt)}{2} \, dt = \frac{T}{2}$$

Hence average power supplied

$$P_{av} = \frac{F_0^2}{T z_m} \frac{T}{2} \sin \delta = \frac{F_0^2}{2 z_m} \sin \delta \qquad (2.47)$$

But we know that

$$\sin \delta = \frac{R_m}{z_m} \qquad \text{From (ii)}$$

Hence

$$P_{av} = \frac{F_0^2 R_m}{2 z_m^2} \qquad (2.47a)$$

Since $R_m = 2mb$ is the resistance (mechanical), i.e. resistive force per unit velocity, the total resistive force is $(R_m v)$ and the rate of doing work by this force is

$$W = (R_m v) \, v = R_m v^2$$

or

$$W = R_m \frac{F_0^2}{z_m^2} \cos^2 (pt - \delta)$$

Since average value of $\cos^2 (pt - \delta)$ for a period is $\frac{1}{2}$. Hence, average work done

$$W = \frac{R_m F_0^2}{2 z_m^2} \qquad (2.47a)$$

and this agrees with Eq. (2.47a), proving that the power supplied is equal to the power dissipated against the frictional forces.

The average power absorbed

$$P_{av} = \frac{F_0^2}{2 z_m} \sin \delta$$

will have its maximum value when $\sin \delta = 1$ that is when $\delta = \frac{\pi}{2}$ and $\omega_0^2 - p^2 = 0$ or $\omega = p$. The force and the velocity are then in phase and $z_m = R_m$. Thus,

$$P_{av} \text{(maximum)} = \frac{F_0^2}{2 R_m} \qquad (2.48)$$

The variation of P_{av} with p, the frequency of the driving force is plotted in Fig. 2.14. This curve determines the response of the oscillator to the frequency of the driving force. The height of peak is determined by R_m, the mechanical resistance since this is the only term effective at $p = \omega_0$ and the location of the peak is also at $p = \omega_0$. The frequency ω_0 is the frequency of velocity resonance since maximum power absorption takes place at it.

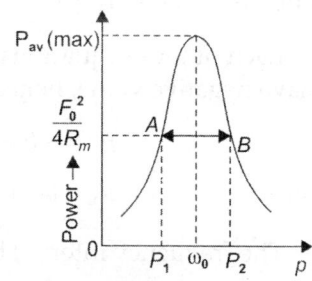

Fig. 2.14

2.13 SHARPNESS OF RESONANCE AND QUALITY FACTOR

Here, we shall study the response of the oscillator to the driving force when the driving frequency is slowly varied. It is obvious that the response of the oscillator depends on the magnitude of power, it extracts from driving force during each cycle of its oscillation, we have seen that the time average of input power is

$$P_{av} = \frac{R_m F_0^2}{2z_m^2}$$

But $R_m = 2mb$ and $z_m^2 = \left(mp - \frac{k}{p}\right)^2 + (2mb)^2$ and $k = m\omega_0^2$

$\therefore$

$$P_{av} = \frac{F_0^2 b}{m} \left\{\frac{p^2}{(\omega_0^2 - p^2)^2 + 2b^2p^2}\right\}$$

The maximum value of P_{av} will occur for the frequency $p = \omega_0$, i.e. the maximum value of power occurs at resonance

$\therefore$

$$P_{max} = \frac{F_0^2}{4mb} = \frac{F_0^2}{2R_m}$$

Hence in terms of P_{max}, P_{av} can be written as

$$P_{av} = P_{max} \left\{\frac{4b^2p^2}{(\omega_0^2 - p^2)^2 + 4b^2p^2}\right\} \tag{2.49}$$

Figure 2.14 gives the variation of average power P_{av} with p- the driving force frequency, for given values of $(2b)$ and ω_0. The values of p at which P_{av} is half its maximum value $\left(P_{av} = P_{max/2} = \dfrac{F_0^2}{4R_m}\right)$ are called *half power points*. The half power points are

$$\frac{1}{2} = \frac{4b^2p^2}{(\omega_0^2 - p^2)^2 + 4b^2p^2}$$

or

$$p^2 = \omega_0^2 \pm 2bp$$

These are two quadratic equations in p namely

$$p^2 + 2bp - \omega_0^2 = 0$$

and $\quad p^2 - 2bp - \omega_0^2 = 0$

Each of these equations have one positive root and one negative root. Since p cannot have negative value, hence retaining only the positive roots, we have

$$p_1 = -b + (\omega_0^2 + b^2)^{1/2}$$

and

$$p_2 = b + (\omega_0^2 + b^2)^{1/2}$$

The frequency interval between these two half power points is

$$\Delta p = p_2 - p_1 = 2b.$$

This frequency interval is called *full frequency width at half-maximum power* or simply *the bandwidth*.

We have already defined Γ the energy decay time, i.e. the mean decay time of forced damped oscillations is

$$\Gamma = \frac{1}{2b}$$

Hence, $\Delta p \cdot \Gamma = 1.$ (2.50)

This can be stated as, "*the frequency width of the resonance curve for driven oscillation is equal to the inverse of the mean decay lifetime for damped oscillations.*"

This equation help us to estimate decay time Γ experimentally as ΔP can be easily estimated.

Sharpness of Resonance or Q Factor

The Q factor is defined as

$$Q = \frac{\text{Resonance frequency}}{\text{Bandwidth}} = \frac{\omega_0}{p_2 - p_1} = \frac{\omega_0}{\Delta p}$$

Since, $\Delta p = 2b = \dfrac{1}{\Gamma}$, we have

$$Q = \frac{\omega_0}{2b} = \Gamma \omega_0 \tag{2.51}$$

This result can also be obtained from definition of Q-factor already given as

$$Q = 2\pi \frac{\text{Average energy stored in one cycle}}{\text{Average energy lost in one period}}$$

$$= 2\pi \frac{E}{P_{dis} T}$$

where P_{dis} is the average power dissipated is given by Eq. (2.47a) and $T = \dfrac{2\pi}{p}$ is the time period of oscillations in steady state. Thus, $P_{dis} \times T$ is the average energy lost in one period.

Total energy E at any time t is (KE + PE)

$$E(t) = \frac{1}{2} m \left(\frac{dx}{dt}\right)^2 + \frac{1}{2} kx^2$$

$$= \frac{1}{2} m \left(\frac{F_0}{z_m}\right)^2 \cos^2(pt - \delta) + \frac{1}{2} m\omega_0^2 \frac{F_0^2}{p^2 z_m^2} [\sin^2(pt - \delta)]$$

where $C = \dfrac{F_0}{pz_m}$.

We know that average values of $\cos^2(pt - \delta)$ and $\sin^2(pt - \delta)$ for one cycle is $1/2$. Hence average energy

$$E_a = \frac{1}{4} m \left(\frac{F_0}{z_m}\right)^2 + \frac{1}{4} m\omega_0^2 \frac{F_0^2}{p^2 z_m^2}$$

$$= \frac{1}{4} m \left(\frac{F_0}{z_m}\right)^2 \left\{1 + \frac{\omega_0^2}{p^2}\right\}$$

and quality factor

$$Q = 2\pi \frac{\dfrac{1}{4}m\left(\dfrac{F_0}{z_m}\right)^2 = \left\{1 + \dfrac{\omega_0^2}{p^2}\right\}}{\dfrac{2\pi}{p} \cdot \dfrac{F_0^2\, R_m}{2z_m^2}}$$

$$= \frac{p}{2}\frac{m}{R_m}\left(1 + \frac{\omega_0^2}{p^2}\right) \quad \text{but} \quad R_m = 2mb$$

$$= \frac{p}{2}\frac{m}{2mb}\left(1 + \frac{\omega_0^2}{p^2}\right)$$

$$= \frac{p}{4b}\left(1 + \frac{\omega_0^2}{p^2}\right) \tag{2.52}$$

and at resonance $\omega_0 = p$

$$Q = \frac{p}{2b} = \frac{\omega_0}{2b}.$$

For low damping b is very small compared to ω_0, and Q will be very large making resonance very sharp as quality factor Q is the measure of sharpness of resonance as already discussed.

Q as Amplification Factor

There is yet another equivalent way of defining Q as the amplification factor of the displacement amplitude. At low frequencies ($p \to 0$), the displacement amplitude C_0 when damping is small becomes

$$C_0 = \frac{F_0}{k} = \frac{f_0}{\omega_0^2} \quad \because f_0 = mF_0 \quad \text{and} \quad \omega_0^2 m = K$$

and displacement at resonance

$$X_{max} = \frac{f_0}{2b\omega_0}$$

$$\therefore \qquad \frac{X_{max}}{C_0} = \frac{\omega_0}{2b} = Q$$

or $\qquad X_{max} = C_0 Q \tag{2.53}$

i.e. the resonance displacement is Q times the displacement amplitude at low frequencies.

2.14 EXAMPLES OF RESONANCE DUE TO FORCED VIBRATIONS

The phenomenon of resonance is quite general and widespread in different branches of Physics. Whenever a system is acted upon by an external action which varies periodically with time, the response of the system as measured by its amplitude and phase or the power absorbed, undergoes rapid changes, as the frequency of the external field of force changes through a certain range of values. The response is characterised by two parameters—frequency ω_0 and the natural width of the driven system—and the resonance condition is said to be reached when the interaction between the driven and the driving system has been maximized. The maximum amplitude occurs at or near ω_0

and the most marked changes occur over a range $\pm\Gamma$ w.r.t. the maximum. It is proposed to extend the concept of resonance to other processes such as nuclear reactions and nuclear magnetic resonance, etc. in which there are favourable conditions for the transfer from one system to the other. As such the concept of energy resonance plays an important role in the description of many physical phenomenon. We will treat some of them here.

LCR Series Resonance Circuit

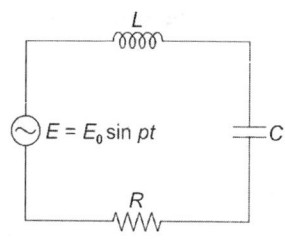

Fig. 2.15

A series LCR circuit is shown in Fig. 2.15. The circuit is connected to an external source of voltage $E = E_0 \sin pt$. The circuit will work as a driven oscillator and the frictional loss which is the resistive loss in this case, will be compensated by supply of energy. Let I be the current in the circuit at a given instant of time and q be the charge on the capacitor at that instant. The potential difference across the capacitor plates is given by

$$V_c = \frac{q}{c}.$$

Since the current is changing with time, there is a potential difference across inductance (L) and is given by

$$V_L = L\frac{dI}{dt}.$$

Finally, the potential difference V_R across the resistance R is given by

$$V_R = IR$$

since the applied voltage is $E = E_0 \sin pt$. Hence

$$E_0 \sin pt = \frac{q}{c} + L\frac{dI}{dt} + RI$$

Since, $I = \frac{dq}{dt}$, we have

$$L\frac{d^2q}{dt^2} + R\frac{dq}{dt} + \frac{q}{c} = E_0 \sin pt$$

or $\qquad \dfrac{d^2q}{dt^2} + \dfrac{R}{L}\dfrac{dq}{dt} + \dfrac{q}{LC} = \dfrac{E_0}{L}\sin pt$ $\hfill$ (2.54)

The equation is exactly the same as the mechanical equation of driven harmonic oscillator Eq. (2.32). The only difference is that here displacement x is replaced by charge q on the capacitor, $1/c$ is analogous to spring (elastic) constant K, R is analogous to damping β and L is analogous to mass. The current $I = dq/dt$ replaces velocity. Here if we put

$$2b = \frac{R}{L}, \quad \omega_0^2 = \frac{1}{LC} \quad \text{and} \quad \frac{E_0}{L} = e_0$$

then Eq. (2.54) becomes

$$\frac{d^2q}{dt^2} + 2b\frac{dq}{dt^2} + \omega_0^2 q = e_0 \sin pt$$ $\hfill$ (2.54a)

and the steady state solution of this equation would be

$$q = q_0 \sin(pt - \phi) \tag{2.55}$$

where

$$q_0 = \frac{e_0}{\{(\omega_0^2 - p^2)^2 + 4b^2 p^2\}^{1/2}}$$

Putting values, we have

$$q_0 = \frac{E_0}{p\left\{\left(Lp - \dfrac{1}{cp}\right)^2 + R^2\right\}^{1/2}} \tag{2.56}$$

and

$$\tan\phi = \frac{2bp}{\omega_0^2 - p^2} = \frac{R}{LP - \dfrac{1}{cp}} \tag{2.56a}$$

and the current is

$$I = \frac{dq}{dt} = q_0 p \cos(pt - \phi)$$

$$= q_0 p \sin\left\{pt - \left(\phi - \frac{\pi}{2}\right)\right\}$$

or

$$I = \frac{E_0}{\left\{\left(Lp - \dfrac{1}{Cp}\right)^2 + R^2\right\}} \sin\left\{pt - \left(\phi - \frac{\pi}{2}\right)\right\} \tag{2.57}$$

The phase shift of the current relative to the applied voltage is $\left(\phi - \dfrac{\pi}{2}\right)$ and the quantity

$$z = \left\{R^2 + \left(Lp - \frac{1}{cp}\right)^2\right\}^{1/2} \tag{2.58}$$

is called the *electrical impedance of the circuit* and plays a similar part in the consideration of alternating currents as that played by resistance in d.c. circuits.

Sharpness of resonance: The current amplitude

$$I_0 = \frac{E_0}{\left\{\left(Lp - \dfrac{1}{cp}\right)^2 + R^2\right\}^{1/2}} \tag{2.59}$$

will be maximum when the denominator has minimum value. The term LP and $\dfrac{1}{cp}$ are called the *inductive reactance* and *capacitive reactance* respectively. The total reactance of the circuit $\left\{Lp - \dfrac{1}{cp}\right\}$ is large, when p is high on account of inductance and also when p is low on account of capacitive reactance, so impedence is large in both when p is high or low. But when

$$\frac{1}{Cp} = Lp \quad \text{or} \quad p = \frac{1}{\sqrt{Lc}} \tag{2.60}$$

the resultant reactance will be zero and impedance reduces to resistance R so in this case for a given circuit, we will get maximum current amplitude $I_0 = \dfrac{E_0}{R}$. The value of p as required by the Eq. (2.60) for maximum current amplitude is same as ω_0, the natural frequency of the circuit. Hence, for a given lightly damped LCR circuit, if we vary the frequency p of the applied emf till it is equal to the natural frequency ω_0 of the circuit, the current amplitude (I_0) will take up the maximum value. This phenomenon is called *resonance*.

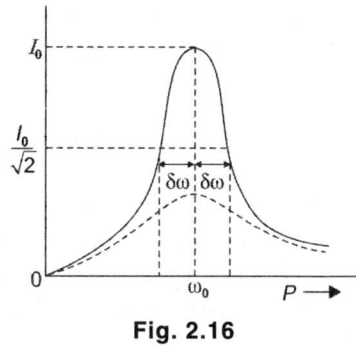

Fig. 2.16

In Fig. 2.16, the current amplitude I_0 is plotted with frequency p of the applied voltage. For a sharp resonance, peak should be narrow. The width of the resonance peak is related to the quality factor Q of the circuit.

Very near the resonance let $p = (\omega_0 + \delta\omega) = \omega_0\left(1 + \dfrac{\delta\omega}{\omega_0}\right)$ and $\delta\omega$ is very small, so that

$$\left[Lp - \frac{1}{cp}\right] = L\omega_0\left(1 + \frac{\delta\omega}{\omega_0}\right) - \frac{1}{c\omega_0\left(1 + \dfrac{\delta\omega}{\omega_0}\right)}$$

$$= L\omega_0\left(1 + \frac{\delta\omega}{\omega_0}\right) - \frac{1}{c\omega_0}\left(1 + \frac{\delta\omega}{\omega_0}\right)^{-1} \qquad \because L\omega_0 = \frac{1}{\omega_0 c}$$

$$= L\omega_0\left(1 + \frac{\delta\omega}{\omega_0}\right) - L\omega_0\left(1 - \frac{\delta\omega}{\omega_0}\right) \qquad \text{(using binomial theorem)}$$

$$= 2L\delta\omega$$

Exactly at resonance, the impedance is equal to R, the resistance of the circuit. As p is shifted away from the resonance and when $\left(Lp - \dfrac{1}{cp}\right) = R$, then we have

$$R = 2\delta\omega L$$

or $$\frac{2\delta\omega}{\omega_0} = \frac{R}{\omega_0 L}$$

and we have proved in case of LCR series damped circuit that Q factor

$$Q = \frac{\omega_0 L}{R}$$

Hence, we have

$$\frac{2\delta\omega}{\omega_0} = \frac{R}{\omega_0 L} = \frac{1}{Q} \qquad (2.61)$$

Under these conditions the amplitude of current falls to $\dfrac{I_0}{\sqrt{2}}$. At these points, the energy or power which is proportional to the square of amplitude will become $\dfrac{1}{2}$ the value at the peak. These are called *half power points,* the width of the resonance peak is given by the width between these two half power points, i.e., full width

$$\Delta\omega = 2\delta\omega = \frac{\omega_0}{Q}$$

Hence for sharpness of resonance Q should be large. In Fig. 2.16, the continuous line shows the resonance peak for a small value of R, the dotted curve is for a large value of R, i.e. the resonance is flat for circuit with higher damping (resistance).

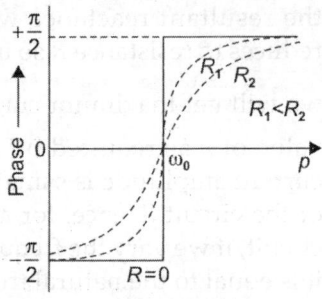

Fig. 2.17

In Fig. 2.17, the relative phase of the current is shown against the variation in p. The continuous curve is for $R = 0$, i.e. a very lightly damped circuit and the dotted curve is for higher damping.

LCR Parallel Resonance Circuit

An AC source is connected across an inductance L in parallel with a capacitor C. The resistance of inductance is R, whereas the effective resistance in series with the capacitance can be neglected because the power factor of good quality capacitors are very small. Let the instantaneous supply of emf be $E = E_0 \sin pt$ and the corresponding current be I and the current through the inductance be I_L and capacitance I_c respectively (Fig. 2.18). These currents will be almost in anti-phase if R is small, because the capacitive current leads the emf by $\pi/2$

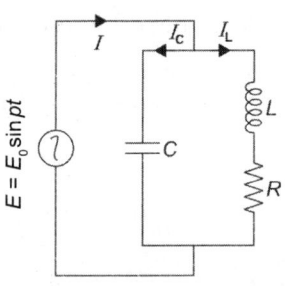

Fig. 2.18

whereas the inductive current lags behind the supply emf by $\dfrac{\pi}{2}$. In present case[1]

$$I = I_L + I_C$$

∴
$$L\frac{dI}{dt} = E_0 \sin pt$$

∴
$$I_L = -\frac{E_0}{Lp}\cos pt = \frac{E_0}{Lp}\sin\left(pt - \frac{\pi}{2}\right).$$

Here constant of integration is zero as I_L oscillates symmetrically about the zero value. The charge on the capacitor is

$$q = cE_0 \sin pt$$

and current
$$I_c = \frac{dq}{dt} = cE_0\, p\cos pt = c\, E_0 p \sin\left(pt + \frac{\pi}{2}\right)$$

and total current

$$I = E_0\left(cp - \frac{1}{Lp}\right)\cos pt. \qquad (2.62)$$

At resonance frequency $I_L = I_C$, the closed circuit containing inductance and capacitance oscillates with the natural frequency and it requires no further supply of

---

1. Here the resistance is neglected as it is very small.

energy from external source. The main current I in the circuit becomes zero and frequency is given by

$$p = \sqrt{\frac{1}{LC} - \frac{R^2}{L^2}} \qquad (2.63)$$

and in the absence of resistance

$$p = \frac{1}{\sqrt{LC}} = \omega_0 \qquad (2.63a)$$

Thus at resonance, the current and voltage are in phase. The current in the circuit is very low (unless R is very small), and variation of Z and I with p are shown in Fig. 2.19(a) and (b) respectively. In the particular case, where the circuit consists of only an inductance and a capacitance connected in parallel, we have $I = 0$. Since in the absence of resistance $Z = \infty$. In this particular case, there is no current in the circuit. Such a circuit is, therefore, used in wireless transmission line to cut-off current at a particular frequency; allowing the current at other frequencies to flow through the line. Such a parallel circuit is sometimes called a *filter* or *rejector circuit*. Notice that, though the current is very small, yet we use the word resonance because the circuit absorbs maximum power at resonance frequency, since at this frequency the current and voltage are in same phase. We can calculate the power absorbed at resonance.

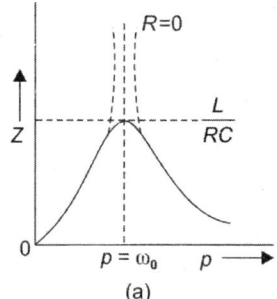
(a)

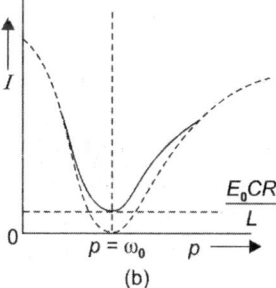
(b)

Fig. 2.19

At resonance the current in circuit is

$$I = \frac{E_0}{R} \cos pt$$

Hence, the power absorbed is

$$P_{in} = EI = I^2 R = \frac{E_0^2}{R} \cos^2 pt$$

Hence, the average power absorbed per cycle is

$$P_{av} = \langle P_{in} \rangle = \frac{E_0^2}{R} \langle \cos^2 pt \rangle$$

$$= \frac{1}{2} \frac{E_0^2}{R}$$

The time-averaged power absorbed is high if R is sufficiently low. In the absence of $R(z = \infty)$, the circuit cannot absorb any power from the applied voltage as a result of which no current flows in the circuit which, thus acts as a filter circuit. Notice that the time averaged power absorption is zero at very low and very high frequencies of driving voltage, because in both cases, the current and the voltage differ in phase by $\dfrac{\pi}{2}$.

In the case $\omega = \omega_0$, the power absorption is maximum. Thus, the condition $p = \omega_0$ is the condition of resonance in parallel LCR circuit.

Example 2.9: A damped oscillating system has an effective mass m and a natural undamped frequency ω_0 and has a damping co-efficient proportional to the velocity of magnitude $\dfrac{m\omega_0}{\sqrt{2}}$. If there is a driving force $F \cos\left(\dfrac{\omega_0 t}{\sqrt{2}}\right)$, show that the energy supplied to the system by driving force during the first quarter period of the force, i.e. during the interval $t = 0$ or $t = \dfrac{\pi}{\sqrt{2}\,\omega_0}$ is $\dfrac{F^2\,(\pi - 2)}{4m\omega_0^2}$.

Solution: The differential equation of motion is

$$\frac{d^2x}{dt^2} + \frac{\omega_0}{\sqrt{2}} \frac{dx}{dt} + \omega_0^2 x = \frac{F}{m} \cos\left(\frac{\omega_0 t}{\sqrt{2}}\right) = f \cos\left(\frac{\omega_0 t}{\sqrt{2}}\right)$$

comparing with standard equation, we have

$$2b = \frac{\omega_0}{\sqrt{2}} \quad \text{and} \quad p = \frac{\omega_0}{\sqrt{2}}$$

Hence, amplitude of motion is

$$C = \frac{f}{\sqrt{(p^2 - \omega_0^2)^2 + 4b^2 p^2}} = \frac{F/m}{\left[\left(\omega_0^2 - \dfrac{\omega_0^2}{2}\right)^2 + \dfrac{\omega_0^2}{2} \cdot \dfrac{\omega_0^2}{2}\right]^{1/2}}$$

$$= \frac{\sqrt{2}\,F}{m\omega_0^2}$$

Phase angle

$$\phi = \tan^{-1} \frac{2bp}{\omega_0^2 - p^2} = \tan^{-1} \frac{\dfrac{\omega_0^2}{2}}{\omega_0^2 - \left(\dfrac{\omega_0}{\sqrt{2}}\right)^2} = \tan^{-1}(1) = \frac{\pi}{4}$$

Hence, the solution of the differential equation is

$$x = \frac{\sqrt{2}\,F}{m\omega_0^2} \cos\left(\frac{\omega_0 t}{\sqrt{2}} + \frac{\pi}{4}\right)$$

and velocity of the system is given by

$$\frac{dx}{dt} = -\frac{F}{m\omega_0} \sin\left(\frac{\omega_0}{\sqrt{2}} t + \frac{\pi}{4}\right)$$

If the system gets displacement dx, the work done by the force is

$$d\omega = F \cdot dx$$

$\therefore$ Work done in the first quarter period is

$$W = \int F \cdot dx = \int_0^{T/4} F \cdot \frac{dx}{dt} \, dt$$

$$= \int_0^{\pi/\sqrt{2}\,\omega_0} F \cos \frac{\omega_0 t}{\sqrt{2}} \left(-\frac{F}{m\omega_0} \right) \sin \left(\frac{\omega_0 t}{\sqrt{2}} + \frac{\pi}{4} \right) dt$$

$$= -\frac{F^2}{4m\omega_0^2} (\pi - 2)$$

This work must be equal to the energy supplied by the force. Hence, the energy supplied is $\dfrac{F^2 (\pi - 2)}{4m\omega_0^2}$.

Example 2.10: Find the average values of the potential and kinetic energies of the forced oscillations of a damped oscillator and also obtain the ratio of the sum of these energies to work done by the applied force in one period, prove that for small damping, it is equal to $\dfrac{Q}{2\pi}$.

Solution: The potential energy of the oscillator is

$$u = \frac{1}{2} kx^2 = \frac{1}{2} m\omega_0^2 \frac{F_0^2}{p^2 z_m^2} \sin^2 (pt - \phi)$$

Amplitude of oscillation is $\dfrac{1}{p}$ times that of velocity oscillation

$$z_m = \left\{ \left(mp - \frac{k}{p} \right)^2 + \beta^2 \right\}^{1/2}$$

is the impedance of the oscillator

$\because$ $\qquad\qquad 2mb = \beta$

$\therefore$ $\qquad\qquad u_{av} = \frac{1}{4} m\omega_0^2 \frac{F_0^2}{p^2 z_m^2}$ $\qquad \because \langle \sin^2 (pt - \phi) \rangle = \frac{1}{2}$

Similarly, the KE has the average value

$$T_{av} = \frac{1}{2} m \langle \dot{x}^2 \rangle = \frac{1}{2} m \left\langle \frac{F_0^2}{z_m^2} \sin^2 (pt - \phi) \right\rangle$$

$$= \frac{1}{4} m \frac{F_0^2}{z_m^2}$$

The energy supplied by the applied force during one complete period is

$$W = \text{Time period} \times P_{av}$$

$$= \frac{2\pi}{p} \times \frac{1}{2} \frac{F_0^2}{z_m^2} \beta$$

Hence, the required ratio is

$$\frac{E_{av}}{W} = \frac{u_{av} + T_{av}}{W} = \frac{1}{4\pi} \frac{\left(1 + \omega_0^2/p^2\right)pm}{\beta}$$

$$= \frac{1}{4\pi}\left[1 + \frac{\omega_0^2}{p^2}\right]p\tau \quad \left(\because \tau = \frac{m}{\beta}\right)$$

or

$$\frac{E_{av}}{W} = \frac{1}{4\pi}\left[1 + \frac{\omega_0^2}{p^2}\right]Q \quad (\because Q = p\tau)$$

For a weakly damped oscillator, since it responds only to a narrow band of frequencies ω_0, i.e. $p \approx \omega_0$, hence

$$\frac{E_a}{W} = \frac{Q}{2\pi}$$

Example 2.11: An oscillator with small damping has mass 5 g and a force constant of 2.0×10^5 dynes/cm. If Q for the oscillator is 200 and the system oscillates in energy resonance with an applied periodic force, then

(a) What is the damping constant β of the oscillator?

(b) How much is the frequency of the applied force?

(c) What frequency of the applied force will produce an amplitude resonance?

(d) What is the amplitude of displacement and velocity oscillations of the oscillator if the frequency of the applied force is 90% of the natural frequency of free oscillations? Express the answer as fraction of the amplitude at energy resonance.

(e) What is the power delivered to the oscillator by the impressed force, expressed as the fraction of the power at resonance?

(f) What is the bandwidth of the given oscillator?

Solution: The natural frequency of free oscillators is

$$\omega_0 = \sqrt{\frac{k}{m}} = \sqrt{\frac{2 \times 10^5}{5}} = 200 \text{ rad/sec}$$

(a) $$\tau = \frac{Q}{\omega_0} = \frac{200}{200} = 1 \sec \quad \text{and} \quad \beta = \frac{m}{\tau} = \frac{5}{1} g/s$$

∴ $$\beta = 5 \text{ g/s} \quad \text{or} \quad 5 \times 10^{-3} \text{ kg/s}$$

(b) Since the oscillator is having energy resonance with the applied force, the frequency of the applied force equals $\dfrac{\omega_0}{2\pi} = 31.8$ Hz

(c) The frequency for amplitude resonance is

$$\frac{p_r}{2\pi} = \frac{1}{2\pi}\sqrt{\omega_0^2 - 2b^2} = \frac{1}{2\pi}\sqrt{(2 \times 100)^2 - 2(1/2)^2}$$

$$= \frac{1}{2\pi}\sqrt{(2 \times 10^2)^2} \approx 31.8 \text{ Hz}$$

(d) The amplitude of the force oscillations is

$$C_0 = \frac{\dfrac{F_0}{m}}{\sqrt{(\omega_0^2 - p^2)^2 + 4b^2 p^2}} = \frac{F_0}{m\omega_0^2 \sqrt{\left(1 - \dfrac{p^2}{\omega_0^2}\right)^2 + \dfrac{1}{\tau^2}\dfrac{p^2}{\omega_0^4}}}$$

or

$$C_0 = \frac{F_0}{k\sqrt{\left(1 - \dfrac{p^2}{\omega_0^2}\right)^2 + \dfrac{p^2}{Q^2\omega_0^2}}} = \frac{F_0}{k} \quad \text{(for resonance } p = \omega_0 \text{)}$$

$$\therefore \quad \frac{C}{C_0} = \frac{1}{Q\sqrt{\left(1 - \dfrac{p^2}{\omega_0^2}\right)^2 + \dfrac{p^2}{Q^2\omega_0^2}}}$$

$$= \frac{1}{200\sqrt{\{1 - (0.9)^2\}^2 + \dfrac{(0.9)^2}{(200)^2}}} \quad \text{for } p = 90\% \text{ of } = \omega_0$$

$$\frac{C}{C_0} = 0.0263$$

and the amplitude of the velocity oscillations is

$$v_0 = c_0\, p \quad \text{and} \quad (v_0)_{res} = (C_0)_{res}\, \omega_0$$

$$\therefore \quad \frac{v_0}{(v_0)_{res}} = \frac{C_0}{(C_0)_{res}} \times \frac{p}{\omega_0} = 0.0263 \times 0.9 = 0.0237$$

(e) The average power delivered by the impressed force is

$$P_{av} = \frac{1}{2}\beta v_0^2 \quad \text{and} \quad (P_{av})_{res} = \frac{1}{2}\beta(v_0)_{vs}^2$$

$$\therefore \quad \frac{P_{av}}{(P_{av})_{res}} = \left(\frac{v_0}{(v_0)_{res}}\right)^2 = (0.0237)^2 = 0.000561$$

In fact, the resonance is very sharp and thus the power falls off very rapidly when p charges from the resonance value.

(f) The bandwidth of the oscillator is:

$$2\Delta\omega = \frac{1}{\tau} = \frac{1}{1} = 1\,\text{rad/s}$$

Example 2.12: A vertical spring of force constant k and natural length l is fixed at one end of a horizontal table (Fig. 2.20). At the other end of the spring is attached a massless board on which a 100 g mass is kept in static equilibrium. A vertical force $F = F_0 \sin(2\omega_0 t)$ is applied to the board where

$\omega_0 = \sqrt{\dfrac{k}{m}}$. What is the maximum value of F_0 for which the mass remains in contact with the board?

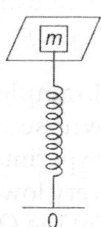

Fig. 2.20

Solution: The forces acting on the mass m are (i) weight mg, (ii) normal reaction R of the board. Taking origin at the bottom O of the spring

$$m\frac{d^2x}{dt^2} = R - mg$$

The reaction R is determined from the equation of motion of the board on which the spring force, the reaction of mass and the impressed force $F_0 \sin 2\omega_0 t$ acts. For board

$$k(x - l) + R = F_0 \sin(2\omega_0 t)$$

$$\therefore \qquad m\frac{d^2x}{dt^2} = -k(x - l) - mg + F_0 \sin(2\omega_0 t)$$

If the distance of the board as measured from the static equilibrium point be x', then $x' = x - l + mg/k$, we have

$$\frac{d^2x'}{dt^2} + \omega_0^2 x' = \frac{F_0}{m}\sin(2\omega_0 t)$$

and the solution is

$$x' = A\cos\omega_0 t + B\sin\omega_0 t - \frac{F_0}{3m\omega_0^2}\sin(2\omega_0 t)$$

at $t = 0, x' = 0$ and $v = 0$. We have

$$A = 0 \quad \text{and} \quad B = \frac{2F_0}{3m\omega_0^2}$$

$$\therefore \qquad x' = \frac{F_0}{3m\omega_0^2}\{2\sin\omega_0 t - \sin(2\omega_0 t)\}$$

and reaction

$$R = F_0 \sin(2\omega_0 t) + mg - kx'$$

$$= mg + \frac{F_0}{3}[4\sin 2\omega_0 t - 2\sin\omega_0 t]$$

for the mass to be in contact $R > 0$. Hence

$$F_0 (2\sin\omega_0 t - 2\sin 2\omega_0 t) < 3\,mg$$

The minimum value of $F_0[2\sin\omega_0 t - 4\sin 2\omega_0 t]$ may be obtained by differentiating it w.r.t. time and equating to zero. We have

$$\cos\omega_0 t - 4\cos 2\omega_0 t = 0$$

or $\qquad 8\cos^2\omega_0 t - \cos\omega_0 t - 4 = 0$

which gives $\qquad \cos\omega_0 t = 0.7723$ or $-.6473$

Substituting the correspondiong values of $\omega_0 t$, we have

$$F_0 < 0.548\,mg \quad \text{or} \quad F_0 < 0.537\,\text{Nt} \quad (m = 100\,\text{gm})$$

Example 2.13: A weakly damped harmonic oscillator is driven by a force $F = F_0 \cos pt$, whose amplitude F_0 is kept constant but its angular frequency is varied. It is experimentally observed that, the amplitude of the steady state oscillation is 0.1 mm at very low values of p and attains a maximum value of 10 cm when $\omega = 100$ rad/s. Calculate (a) the Q value of the system, (b) the time during which the energy of the oscillator falls to $1/e$ of its initial value, and (c) half-width of the power resonance.

Solution: (a) We know that if damping is weak, Q becomes very large, so that

$$\frac{C_{max}}{C_0} \cong Q$$

where C_{max} is the maximum amplitude and C_0 is the amplitude at very low frequency of the applied force

$\therefore$

$$Q = \frac{10}{0.01} = 1000$$

(b) We know that

$$Q = \omega_0 \Gamma$$

where ω_0 is the resonance frequency and Γ is the energy decay time, i.e. the time during which energy falls to $\dfrac{1}{e}$ of its initial value. Thus,

$$\Gamma = \frac{Q}{\omega_0} = \frac{1000}{100} = 10 \ s$$

(c) The full width of the power resonance curve is

$$\Delta\omega = 2b = \frac{1}{\Gamma}$$

and the half width of power resonance curve is

$$\delta\omega = \frac{\Delta\omega}{2} = \frac{1}{2\Gamma} = \frac{1}{20} = 0.05 \ rad/s$$

Example 2.14: A series LCR circuit with $L = 0.05 \ H$, $C = 50 \ \mu F$ and $R = 10\Omega$ is connected to an alternating supply at 200 V and 50 Hz. Find (a) the peak value of the current in the circuit, (b) the power factor of the circuit, (c) the average power delivered to the circuit in each cycle, and (d) the rate of production of heat in the circuit.
Solution: Impedance z of the circuit is

$$z = \left\{ R^2 + \left(\omega L - \frac{1}{\omega c} \right)^2 \right\}^2$$

$$= \left\{ (10)^2 + \left(0.05 \times 2\pi \times 50 - \frac{1}{50 \times 10^{-6} \times 100\pi} \right) \right\}$$

$$= 48.97 \ \Omega$$

(a) The peak value of current

$$I_0 = \frac{E_0}{z} = \frac{200}{48.92} = 4.08 \ A$$

(b) Power factor

$$\cos\delta = \frac{R}{z} = \frac{10}{48.97} = 0.204$$

(c) Average power delivered in each cycle is

$$P = \frac{1}{2} E_0 I_0 \cos\delta = \frac{1}{2} \times 200 \times 4.08 \times 0.204$$

$$= 83.23 \ W$$

(d) The heat is produced only in resistance R. Since the current varies as $I = I_0 \cos(\omega t - \delta)$ the average heat produced in resistance R in time t is given by

$$H = I^2 Rt$$

The rate of production is, therefore, equal to

$$I^2 R = I_0^2 R \cos^2(\omega t - \delta)$$

and the average value for cycle is

$$\frac{1}{2} I_0^2 R = 83.33 \text{ W}$$

Notice that

Average power consumed = Rate of dissipation of heat

2.15 ANHARMONIC OSCILLATOR

We have already discussed that a simple harmonic motion is characterised by the force law given by

$$F = -kx$$

where k is the force constant and x is the displacement from the equilibrium position, which in this case is taken as origin. The potential energy $u = \frac{1}{2}kx^2$.

However, when the equilibrium position is taken at x_0 instead at origin, the potential energy is

$$u = \frac{1}{2}k(x - x_0)^2$$

If this function is Taylor expanded about the stable equilibrium x_0, we get

$$u(x) = u(x_0) + \left(\frac{du}{dx}\right)_{x_0}(x - x_0) + \frac{1}{2}\left(\frac{d^2u}{dx^2}\right)_{x_0}(x - x_0)^2$$

$$+ \frac{1}{6}\left(\frac{d^3u}{dx^3}\right)_{x_0}(x - x_0)^3 + \frac{1}{24}\left(\frac{d^4u}{dx^4}\right)_{x_0}(x - x_0)^4 + \ldots$$

Since $x = x_0$ is the equilibrium position, i.e. the position of the minimum potential energy, hence

$$\left(\frac{du}{dx}\right)_{x_0} = 0 \quad \text{and} \quad \left(\frac{d^2u}{dx^2}\right)_{x_0} = +\text{ve}$$

Now, if we choose $u(x_0) = 0$, we have

$$u(x) = \frac{1}{2}kx^2 + \frac{1}{6}k_1x^3 + \frac{1}{24}k_2x^4 + \ldots \tag{2.64}$$

where

$$k = \left(\frac{d^2u}{dx^2}\right)_{x_0}, \quad k_1 = \left(\frac{d^3u}{dx^3}\right)_{x_0}, \quad k_2 = \left(\frac{d^4u}{dx^4}\right)_{x_0}$$

In case when $k_1 = k_2 = k_3 = ... = 0$, the potential well is parabolic about the point of the minimum potential energy and the force on the particle $F = -\dfrac{du}{dx} = -kx$ is a linear restoring force, and therefore, the motion is simple harmonic. This motion, we have discussed in details in Chapter 1.

Now, consider the case where potential energy function is not a parabolic function but given by equation (2.64). Here, the terms involving $k_1, k_2, k_3, ...$ etc. in the expression for potential energy $u(x)$, which give rise to non-linear contribution to the force on the particles are said to constitute the anharmonic potential energy terms and a particle whose potential energy about the point of stable equilibrium contains such terms, is called *anharmonic oscillator*. It may be noted that for small amplitude oscillations, the anharmonic terms in $u(x)$ became relatively negligible and motion, thus can be treated as

simple harmonic with force constant $k = \left(\dfrac{d^2u}{dx^2}\right)_{x_0}$ and frequency $x = \dfrac{1}{2\pi}\sqrt{\left(\dfrac{d^2u}{dx^2}\right)_{x_0}\dfrac{1}{m}}$.

But for large amplitudes, this approximation may not be valid at all and in such cases correct solution for displacement, frequency etc. should be found taking into account the effect of anharmonic terms also.

For simplicity, let us consider the potential function to be of the form

$$u(x) = \frac{1}{2}kx^2 - \frac{1}{3}Skx^3 \tag{2.65}$$

where s is called the *anharmonicity of the oscillator* and k is force constant and $\dfrac{sk}{2} = \left(\dfrac{d^3u}{dx^3}\right)_{x_0}$.

Now, the force acting on the particle is

$$F = -\frac{du}{dx} = -kx + Sk\,x^2$$

Hence, the equation of motion of the oscillator becomes

$$m\frac{d^2x}{dt^2} + kx - Sk\,x^2 = 0$$

or

$$\frac{d^2x}{dt^2} + \omega_0^2 x - S\omega_0^2 x^2 = 0 \tag{2.66}$$

where $\omega_0 = \sqrt{k/m}$ and the term containing x^2 is called the *perturbation*. Let the solution of this equation be given by

$$x = A[\cos pt + q\cos 2pt] + x_1$$

where q and x_1 are constants to be determined.

$$\frac{d^2x}{dt^2} = -Ap^2[\cos pt + 4q\cos 2pt]$$

and

$$x^2 = A^2[\cos^2 pt + q^2\cos^2 pt + ...] + x_1^2$$
$$\approx A^2\cos^2 pt$$

Here, we have neglected the terms involving q, q^2 and x_1^2 we have assumed x_1, to be very small.

Substituting the values in Eq. (2.66), we have

$$-Ap^2[\cos pt + 4q\cos 2pt] + \omega_0^2[A(\cos pt + q\cos 2pt) + x_1]$$

$$-\frac{1}{2}S\omega_0^2 A^2(1 + \cos 2pt) = 0.$$

Since this is an identity and independent of time, hence the co-efficients of $\cos pt$ and constant terms should vanish separately. As such, we have

$$-Ap^2 + \omega_0^2 A = 0 \text{ or } p = \omega_0$$

and

$$-Ap^2 \cdot 4q + \omega_0^2 Aq - \frac{1}{2}S\omega_0^2 A^2 = 0$$

or

$$q = -\frac{1}{6}SA$$

Also

$$\omega_0^2 x_1 - \frac{1}{2}S\omega_0^2 A^2 = 0 \text{ or } x_1 = \frac{1}{2}SA^2$$

Thus, the solution of anharmonic oscillator is given by

$$x = A\left[\cos\omega_0 t - \frac{1}{6}S\cos 2\omega_0 t\right] + \frac{1}{2}SA^2 \tag{2.67}$$

For this expression, the time average of position can easily be determined

$$\langle x \rangle = \frac{A\int_0^T \cos\omega_0 t\, dt}{T} - \frac{1}{6}SA\frac{\int_0^T \cos 2\omega_0 t\, dt}{T} + \frac{1}{2}S\langle A^2 \rangle$$

$$\langle x \rangle = \frac{1}{2}S\langle A^2 \rangle \tag{2.68}$$

From this equation, it is clear that the time average of position of the oscillator is not the equilibrium position of the oscillator. Thus, the effect of perturbation is to displace the average position of the oscillator from its mean position and the shift of the position in the positive direction. Further, this shift is proportional to the anharmonicity s and the square of the amplitude.

2.16 SIMPLE PENDULUM AS AN EXAMPLE OF ANHARMONIC OSCILLATOR

If the length of the pendulum is l and mass m, the equation of motions is

$$ml^2\frac{d^2\theta}{dt^2} = -l\, mg\sin\theta$$

or

$$\frac{d^2\theta}{dt^2} = -\left(\frac{g}{l}\right)\left(\theta - \frac{\theta^3}{6}\right) \quad \because \sin\theta = \theta - \frac{\theta^3}{3!} + \frac{\theta^5}{5!} - \dots$$

or

$$\frac{d^2\theta}{dt^2} = -\omega_0^2\left(\theta - \frac{\theta^3}{6}\right) \quad \text{where} \quad \omega_0^2 = \frac{g}{l}$$

or

$$\frac{d^2\theta}{dt^2} + \omega_0^2\theta = \frac{\omega_0^2\theta^3}{6} = S\theta^3 \tag{2.69}$$

Here S is a measure of the small non-linearity of the force and $\dfrac{k}{m} = \omega_0^2$ is the square of angular frequency of small harmonic oscillations about the origin. Since the motion of the particle is to be periodic, its displacement can be assumed to be a superposition of a number of simple harmonic displacement of suitable amplitude having frequencies which are integral multiples of a certain fundamental frequency $\left(\dfrac{p}{2\pi}\right)$. Hence the solution can be written as*

$$\theta = A(\cos pt + q\cos 2pt + r\cos 3pt + ...)$$

$$\frac{d^2\theta}{dt^2} = -p^2 A(\cos pt + 4q\cos 2pt + 9r\cos 3pt + ...)$$

and
$$\theta^3 = A^3 \cos^3 pt = \frac{1}{2}A^3 \cos pt(1+ \cos 2pt)$$

$$= \frac{1}{4}A^3\{3\cos pt + \cos 3pt]$$

Here we have neglected the higher terms and terms involving q, r, etc. as θ^3 itself has a small terms as the multiplier and a product of two small quantities is obviously negligible to first approximation. Hence, the solution of Eq. (2.69) can be written as

$$A\cos pt\,(\omega_0^2 - p^2)+ qA\cos 2pt\,(\omega_0^2 - 4p^2) + rA\cos 3pt\,(\omega_0^2 - 9p^2) = \frac{SA^3}{4}[3\cos pt + \cos 3pt]$$

Equating co-efficient of $\cos pt$ on both sides, we have

$$\omega_0^2 - p^2 = \frac{3SA^2}{4} \quad \text{or} \quad p^2 = \omega_0^2\left[1 - \frac{3SA^2}{4\omega_0^2}\right]$$

or
$$p = \omega_0\left[1 - \frac{3SA^2}{4\omega_0^2}\right]^{1/2}$$

$$= \omega_0\left[1 - \frac{3SA^2}{8\omega_0^2}\right] \quad \text{(using binomial theorem)}$$

Equating co-efficients of $\cos 2pt$ and $\cos 3pt$, we have

$$qA\left[\omega_0^2 - 4p^2\right] = 0 \quad \text{or} \quad q = 0 \text{ as } \omega_0 \neq 2p$$

and
$$rA\left[\omega_0^2 - 9p^2\right] = \frac{SA^3}{4} \quad \text{or} \quad r = -\frac{SA^2}{32\omega_0^2}$$

we have taken $\omega_0 = p$.

The value of p gives the change in frequency of a simple pendulum arising on account of little anharmonicity $(\alpha\theta^4)$ in the potential function. The angular displacement may now be written as:

$$\theta = A\left[\cos pt - \frac{SA^2}{32\omega_0^2}\cos 3pt\right]$$

$$= A\left[\cos pt - \frac{A^2}{192}\cos 3pt\right] \quad \because S = \frac{\omega_0^2}{6}$$

* If the oscillation is supposed to have maximum displacement at $t = 0$, the displacement, x will not include sine terms.

and
$$p = \omega_0 \left[1 - \frac{3SA^2}{8\omega_0^2} \right] = \omega_0 \left[1 - \frac{3A^2}{8\omega_0^2} \frac{\omega_0^2}{6} \right]$$

$$= \omega_0 \left[1 - \frac{A^2}{16} \right]$$

But $\omega_0 = \sqrt{\dfrac{g}{l}}$

Hence, $p = \sqrt{\dfrac{g}{l}} \left(1 - \dfrac{A^2}{16} \right)$

The period of the large amplitude simple pendulum may thus be written as

$$T = \frac{2\pi}{p} = \frac{2\pi}{\omega_0 \left[1 - \dfrac{A^2}{16} \right]} = \frac{2\pi}{\omega_0} \left[1 - \frac{A^2}{16} \right]^{-1}$$

$$= \frac{2\pi}{\omega_0} \left(1 + \frac{A^2}{16} \right) = 2\pi \sqrt{\frac{l}{g}} \left(1 + \frac{A^2}{16} \right)$$

Hence the fractional increase in the period of oscillation is

$$\frac{T - T_0}{T_0} = \frac{A^2}{16} \qquad \left(\text{here } T_0 = 2\pi \sqrt{\frac{l}{g}} \right)$$

To have an idea of the extent of modification in period of a pendulum, let us take the angular amplitude of the oscillations as $30° = \dfrac{\pi}{6}$; we have

$$\frac{T - T_0}{T_0} = \frac{1}{16} \times \frac{\pi^2}{36} = 0.017 \text{ or } 1.790\%.$$

One finds that even for such a large amplitude oscillations, the period increased by 1.7%. In fact, the increase in the period is less than 1% for amplitude of 23°.

Example 2.15: A particle of mass m placed on a horizontal frictionless table (Fig. 2.21) is held between two identical springs of force constant k and normal length l_0 whose other ends are fixed at points P_1 and P_2. If the particle is displaced sideways, an amount x_0, which is small compared with normal length l_0 of the springs and then released determine its subsequent motion. Find its frequency of oscillations and write the equation of motion.

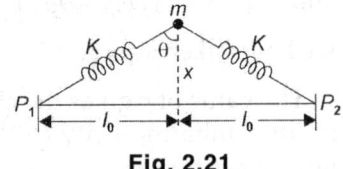

Fig. 2.21

Solution: When m has a transverse displacement x, the increase in the length of each spring is $\left(\sqrt{l_0^2 + x^2} - l_0 \right)$. The particle, therefore, experiences two equal force each of magnitude $F = k \left(\sqrt{l_0^2 + x^2} - l_0 \right)$ tending to pull it towards either of the springs. The net restoring force on m which is opposite to the displacement has magnitude.

$$2F \cos \theta = 2k \left(\sqrt{l_0^2 + x^2} - l_0 \right) \frac{x}{\sqrt{l_0^2 + x^2}}$$

$$= 2kx \left\{ 1 - \frac{l_0}{\sqrt{l_0^2 + x^2}} \right\}$$

$$= 2kx \left\{ 1 - \left(1 + \frac{x^2}{l_0^2} \right)^{-\frac{1}{2}} \right\}$$

$$= 2kx \left\{ 1 - \left(1 - \frac{x^2}{2l_0^2} + \frac{3}{8} \frac{x^4}{l_0^4} \cdots \right) \right\}$$

$$= \frac{kx^3}{l_0^2} - \frac{3}{4} \frac{kx^5}{l_0^4} + \cdots$$

The equation of motion of the particle, therefore, becomes

$$m \frac{d^2 x}{dt^2} = -\frac{kx^3}{l_0^2} + \frac{3}{4} \left(\frac{kx^5}{l_0^4} \right) + \cdots$$

or

$$\frac{d^2 x}{dt^2} = - \left(\frac{k}{m} \right) \frac{x^3}{l_0^2} + \frac{3}{4} \left(\frac{k}{m} \right) \left(\frac{1}{l_0^4} \right) x^5 + \cdots$$

Since $x \ll l_0$, we can write

$$\frac{d^2 x}{dt^2} = -\frac{k}{ml_0^2} x^3 = Sx^3 \qquad \left(\text{where } S = -\frac{k}{ml_0^2} \right)$$

The above equation is the equation of anharmonic oscillator in which the harmonic term $\omega_0^2 x$ is missing, ω_0 is thus to be taken as zero. Using the solution

$$\omega_0^2 - p^2 = \frac{3SA^2}{4}$$

or

$$p^2 = \omega_0^2 - \frac{3SA^2}{4} = 0 - \frac{3}{4} \left(-\frac{k}{ml_0^2} \right) x_0^2 \qquad (\because \text{amplitude} = x_0)$$

$$= \frac{3k}{4ml_0^2} x_0^2$$

Hence, the frequency of oscillation is

$$n = \frac{p}{2\pi} = \frac{x_0}{2\pi} \sqrt{\frac{3k}{m}} \cdot \frac{1}{2l_0} = \frac{x_0}{4\pi l_0} \sqrt{\frac{3k}{m}}$$

The displacement x may be approximately expressed as

$$x = x_0 \left\{ \cos pt - \frac{SA^2}{36p^2} \cos 3pt \right\}$$

$$= x_0 \left\{ \cos pt + \frac{kx_0^2}{ml_0^2 36} \cdot \frac{4ml_0^2}{3kx_0^2} \cos 3pt \right\}$$

$$= x_0 \left\{ \cos pt + \frac{1}{27} \cos 3pt \right\}$$

REVIEW QUESTIONS AND PROBLEMS

1. (a) What are damped oscillations? Discuss overdamping, critical damping and underdamping.
   (b) Calculate the power loss in damped oscillator.
2. (a) What is an anharmonic oscillator? Discuss.
   (b) Discuss the case of a simple pendulum as an example of an anharmonic oscillator.
3. (a) Write the differential equation for a driven harmonic oscillator and solve it.
   (b) Show that the power absorbed by a driven oscillator is maximum at resonance.
4. (a) A particle is oscillating under a damping force, show that the average power loss P is given by

   $$P = \frac{E}{\tau}$$

   where E is the average energy and τ is the relaxation time. What happens to the dissipated energy?
   (b) In a simple pendulum, mass of the bob is 1 kg and length of the string is 100 cm, then (i) Find the time period of pendulum for oscillations of small amplitude. (ii) If amplitude of oscillation is 60°; find the time period of oscillation.
   [*Ans.* (i) 2s, (ii) 2.84 s]
5. Define relaxation time of a damped oscillator, if ω_0 is the natural frequency and τ is the relaxation time, prove that the quality factor $Q = \omega_0 \tau$.
6. A damped oscillator starting from rest has amplitude 4 cm after 100 oscillations. Its first amplitude was 40 cm and its time period is 2.30 s. Calculate its relaxation time.

   $$\left[\textbf{Hint}: \lambda = \frac{2.303}{101} \log\left(\frac{40}{4}\right) \text{ and } \lambda = \frac{T}{2\tau}, \tau = 50.5 \text{ s} \right]$$

7. Obtain an expression for the period of simple pendulum as function of the amplitude of motion. Compared to the period for small oscillations, show that the period will increase by 6.25% when amplitude becomes $\frac{\pi}{3}$.
8. Mention the effects of anhormonicity on an oscillator. How can it explain thermal expansion of solids?
9. What do you mean by damped oscillations? Obtain differential equation of damped harmonic oscillator and discuss overdamped, critically damped and underdamped cases.
10. The period of a pendulum is 2 s and amplitude 2°. After 20 s its amplitude remains 1.5°. Calculate damping co-efficient.

    $$\left[\textbf{Hint.} \ A = A_0 e^{-bt}, \frac{1.5}{2} = e^{-b \times 20} \text{ or } b = 0.014 \text{ or } 2b = 0.28 \text{ s} \right]$$

11. Write differential equation of driven harmonic oscillator and solve it.
12. Show that at a resonance displacement lags behind the driving force by $\frac{\pi}{2}$, whereas velocity is in phase with the driving force.
13. The motion of a damped oscillator is given by

    $$x = x_0 \, e^{-t/2\tau} \sin \omega t$$

    where τ is relaxation time. How does the damping affect, (i) frequency of oscillations? (ii) average total energy? (iii) Q of the oscillator?

14. An inductance coil having an inductance L and resistance R is connected in parallel to a capacitor of capacity C. Determine the limiting value of R so that the circuit remains oscillatory. Why is this circuit called a rejector circuit?

15. A particle is oscillating under a damping force. Show that the average power loss P is given by $P = \dfrac{E}{\tau}$, where E is the average energy and τ relaxation time. What happens to the dissipated energy?

16. The amplitude of a driven harmonic oscillator is given by

$$A = \frac{F_0}{\sqrt{(\omega_0^2 - p^2) + 4k^2 p^2}}.$$

Show that in the limit $p \ll \omega_0$, the response is independent of mass while in the limit $p \gg \omega_0$, response is independent of the spring constant.

17. In LCR circuit, $L = 1$ mH, $C = 0.1$ μF and $R = 10\ \Omega$. Find the resonance frequency of the circuit and the bandwidth corresponding to half power points.
    [*Ans.* 1.6×10^4 Hz, $\Delta\omega = 16$ Hz]

18. Write down the equation of motion of damped harmonic oscillator and find out a general solution of this equation. Also obtain expression for mean kinetic and potential energies of damped oscillator.

19. A body of mass 100 gm is loaded on a spring which stretches the spring by 2 cm. If the relaxation time of spring is 1s, calculate the time periof of damped oscillator.
    [*Ans.* 28s]

20. What do you mean by an anharmonic oscillator? Derive an expression for the displacement of an anharmonic oscillation and prove that the displacement of oscillator in anharmonic oscillations is very small from its mean position.

21. Potential energy of an oscillator of mass 10 g is given by $1000x^2 + 500x^3 + 200x^4$ ergs. Find out the frequency of oscillations of small amplitude. If the amplitude of the oscillations of original frequency is 4 cm, also calculate the frequency and amplitude of overtones.
    [*Ans.* $A_2 = 2$ cm, $p = 32$ Hz]

22. Obtain the expression for average power dissipated in a damped harmonic oscillator and show that this average power dissipated is equal to the average work done against the damping force.

23. What is a forced harmonic oscillator? Get the differential equation for the motion of this oscillator and solve it.

24. In a series L–C–R circuit, it is given that $L = 2$ mH, $C = 5$ μF and $R = 0.2\ \Omega$. For its damped oscillation, calculate the frequency of the circuit and quality factor.

$$\left[Ans.\ \omega_0 = 1600\ \text{Hz},\ Q = \frac{\omega_0 L}{R} = \frac{1600 \times 2 \times 10^{-3}}{0.2} = 32 \right]$$

25. A mass is subjected to a resistive force $-bv$ but not spring like restoring force.
    (a) Show that its displacement as a function of time is of the form

$$x = c - \frac{V_0}{r}e^{-rt} \qquad \text{where} \qquad r = \frac{b}{m}$$

    (b) At $t = 0$, the mass is at rest at $x = 0$. At this instant a driving force $F = F_0 \cos pt$ is switched on. Find the values of A and δ in steady state solution $x = A \cos(pt - \delta)$.
    (c) Write down the general solution [sum of parts (a) and (b)] and find the value of

c and k from the condition that $x = 0$ and $\dfrac{dx}{dt} = 0$ at $t = 0$. Sketch x as a function of t.

$$\left[Ans.\ (b)\ A = \frac{F_0}{mp(p^2 + r^2)^{1/2}},\ \tan \delta = -\frac{r}{p} \right]$$

26. The power input to maintain forced vibrations is the mean rate of doing work against the resistive force $-bv$.
    (a) Show that the instantaneous rate of doing work against this resistive force is $-bv^2$.
    (b) Using $x = A \cos (\omega t - \delta)$, show that the mean rate of doing work is $\dfrac{b\omega^2 A^2}{2}$.

27. Consider a damped oscillator with $m = 0.2$ kg, $b = 4$ N·m^{-1}s and $k = 80$ N·m^{-1}. Suppose this oscillator is driven by a force $F = F_0 \cos (pt - \delta)$, where $F_0 = 2$N and $p = 30$ rad/s.
    (a) What are the values of A and δ of the steady-state response described by $x = A \cos (\omega t - \delta)$?
    (b) How much energy is dissipated against the resistive force in one cycle?
    (c) What is mean power input?   [*Ans.* (a) 1.3 cm 130°, (b) 0.063 J, (c) 0.3 W]

28. An object of mass 2 kg hangs from a spring of negligible mass. The spring is extended by 2.5 cm when the object is attached. The top end of the spring is oscillating up and down in SHM with an amplitude of 1 mm. The Q of the system is 15.
    (a) What is ω_0 for this system?
    (b) What is the amplitude of the forced oscillations at $\omega = \omega_0$?
    (c) What is the mean power input to maintain the forced oscillations at a frequency 2% greater than ω_0?   [*Ans.* (a) 19.8 rad/s,￼(b) 1.5 cm, (c) 0.086 W]

29. A rod of length L oscillates about a horizontal axis passing through an end. A body having the same mass as the rod can be clamped to the rod at a distance h from the axis.
    (a) Obtain the period of system as a function of h and L.
    (b) Is there a value of h for which the period is same as if there were no mass?

$$\left[Ans.\ (a)\ 2\pi = \sqrt{\frac{h^2 + \dfrac{L^2}{3}}{g\left(h + \dfrac{L}{2}\right)}},\ (b)\ \text{Yes},\ h = \frac{2L}{3} \right]$$

30. A particle of mass m is at the point of stable equilibrium corresponding to the potential function $u(x) = -\dfrac{U_0}{16}(x^4 - 8x^2)$ where U_0 is positive. Discuss the motion of the particle if it is left after giving an initial unit displacement. Write down the equation of motion and calculate the period of oscillation correct to within 0.5%.

[*Ans.* Motion is periodic about origin, $\ddot{x} + \dfrac{U_0}{m}x = \dfrac{U_0}{4m}x^3$. $T = 2\pi\sqrt{\dfrac{m}{\mu_0}}$ or $1.108T_0$ where T_0 is the period for small oscillations.]

SHORT ANSWER QUESTIONS

1. Amplitude of a pendulum of time period 2 sec is 2°. After 20 s, this amplitude becomes 1.5° due to the damping. Find the damping co-efficient.

   Ans. $A = A_0 e^{-bt}$ or $\dfrac{1.5}{2} = e^{-b \times 20}$ or $b = 0.014$ and $2b = 0.028$ s^{-1}.

2. A simple pendulum of 1 m effective length uses a bob of radius 1.0 cm made of iron (density 7.8 g·cm^{-3}). Calculate percentage reduction in amplitude due to air viscosity in 5 minutes if the co-efficient of viscosity of air at room temperature is 1.8×10^{-4} g·cm^{-1}s^{-1}.

   Ans. $2b = \dfrac{6\pi\eta a}{m} = 1 \times 10^{-4}$ s^{-1}, $A = A_0 e^{-bt}$,

   $\dfrac{A}{A_0} = e^{-0.15} = 0.985$, $\dfrac{A_0 - A}{A_0} \times 100 = \dfrac{0.015}{0.985} \times 100 = 1.57$.

3. In above problem, find Q factor of the pendulum.

   Ans. $Q = \dfrac{\omega_0}{2b} = \dfrac{1}{2b}\sqrt{\dfrac{g}{l}} = \dfrac{\sqrt{10}}{10^{-4}} = 3.16 \times 10^4$.

4. Find the difference of the frequency from resonance fre-quency for which the amplitude of velocity oscillations (lightly damped) have one-half the value at resonance.

   Ans. $V_0 = \dfrac{\dfrac{F_0}{m}}{\left\{\dfrac{(\omega_0^2 - \omega^2)^2}{\omega^2} + 4b^2\right\}^{1/2}}$ for resonance

   $\omega = \omega_0$ or $(v_0)_{max} = \dfrac{F_0}{2mb}$ for $v_0 = (v_0)_{max}/2$,

   $\dfrac{(\omega_0^2 - \omega^2)^2}{\omega^2} + 4b^2 = 16b^2$

   or $(\omega_0^2 - \omega^2)^2 = 12b^2\omega^2$ $(\omega - \omega_0)^2 (\omega_0 + \omega)^2 = 12b^2\omega^2$ for small damping $\omega \approx \omega_0$

   $\therefore (\omega_0 - \omega)^2 = 3b^2$ or $\omega_0 - \omega = \sqrt{3}b$.

5. A capacitor of 0.1 μF an inductance of 0.1 H and a resistance of 800 Ω are connected in series. This combination is connected with a battery. Is the current in the circuit oscillatory?

   Ans. For current to be oscillator $\dfrac{R^2}{4L^2} < \dfrac{1}{LC}$ or $R < 2\sqrt{\dfrac{L}{C}}, \sqrt{\dfrac{L}{C}} = \sqrt{\dfrac{0.1}{10^{-7}}} = 1000$. Hence, the current is oscillatory.

6. A simple pendulum has a period of 1 s and an amplitude of 10°. After 10 complete oscillations, its amplitude reduces to 5°. What is its relaxation time?

   Ans. $A = A_0 e^{-bt}$ $\tau = \dfrac{1}{2b}$ $\therefore A = A_0 e^{\frac{-t}{2\tau}}, (2)^{-1} = e^{\frac{-t}{2\tau}}, \tau = 7.25$ s.

7. Find the fractional change in the resonant frequency ω_0 of a damped SH mechanical oscillator in terms of Q factor.

Ans. It will oscillate when it is under damped and its frequency is

$$\omega = \sqrt{\omega_0^2 - b^2}, \omega^2 = \omega_0^2 - \frac{1}{4\tau^2} \text{ or } \frac{\omega^2}{\omega_0^2} = 1 - \frac{1}{4\omega_0^2\tau^2}$$

$$\text{or } \frac{\omega}{\omega_0} = \left(1 - \frac{1}{4\omega_0^2\tau^2}\right)^{1/2} = 1 - \frac{1}{8Q^2} \text{ or } \frac{\omega_0 - \omega}{\omega_0} = \left(\frac{1}{8Q^2}\right).$$

8. After how long time will the charge oscillations decay to half amplitude, if $L = 10 \text{ mH}, C = 1.0 \,\mu\text{F}$ and $R = 0.1 \,\Omega$?

   Ans. $q = q_0 e^{-\frac{Rt}{2L}}$ or $(2)^{-1} = e^{-\frac{RT}{2L}}$ or $t = 0.14\,\text{s}$.

9. The quality factor of an undamped tuning fork of frequency 256 is 10^3. After how many oscillations, its energy is reduced to $\dfrac{1}{e}$ of its energy in absence of damping.

   Ans. $E = E_0 e^{-\frac{t}{\tau}}$ $\therefore t = \tau = \dfrac{Q}{\omega_0} = \dfrac{10^3}{2\pi \times 256} = 0.62 \text{ s}$

   $\therefore$ No. of oscillations $= \left(\dfrac{\omega_0}{2\pi}\right)\tau = \dfrac{Q}{2\pi} = \dfrac{10^3}{2\pi} = 159$

10. An excited atom (as a damped oscillator) radiates energy at wavelength 7000 Å and has the Q-value 10^7 at resonance. Estimate the full width of the spectral line.

    Ans. $v = \dfrac{c}{\lambda}$ or $dv = -\dfrac{c}{\lambda^2} \, d\lambda, Q = \dfrac{v}{dv} = \left|\dfrac{\lambda}{d\lambda}\right|$

    or full width $= d\lambda = \lambda_1 - \lambda_2 = \dfrac{\lambda}{Q} = \dfrac{7 \times 10^{-7}}{10^7} = 7 \times 10^{-14}$ m.

11. In ballistic galvanometer, first throw is 15 cms and 10th throw is 3 cms. Find the correct throw.

    Ans. Correct throw is $\theta = \theta_1 \left(1 + \dfrac{\lambda}{2}\right)$

    and $\lambda = \dfrac{1}{n-1} \log_e \dfrac{\theta_1}{\theta_n}, \lambda = 0.18, \theta = 15\,(1 + 0.09) = 16.35$ cm.

12. Define bandwidth and write an expression for bandwidth in terms of quality factor.

    Ans. The frequency difference between half power points is called *bandwidth*. $\Delta\omega = \omega_2 - \omega_1$, ω_2 and ω_1 are respectively the freqnencies for which power dissipated is half the power dissipated at resonance $Q = \dfrac{\omega_0}{\Delta\omega}$ or $\Delta\omega = \dfrac{\omega_0}{Q}$.

13. What is the difference between mechanical and electrical impedance?

    Ans. Mechanical impedance of a physical system subjected to a driving force is the ratio of driving force to the associated velocity of the system, i.e. $Z_m = \dfrac{f}{v}$. In presence of inductance L or capacitance c, or both along with the resistance R in the circuit introduces a phase difference between the applied energy and the current. Ohm's law then takes the vector form $Z_e = \dfrac{E}{i}$, where Z_e is the electrical impedance of the circuit. In fact, it is the vector sum of resistance R, inductive reactance (X_L) and capacitive reactance (X_C).

14. A massless spring suspended from rigid support carries a mass of 200 gm at its lower end. It is observed that the system oscillates with time period of 0.2 sec and amplitude of oscillation reduces to half its value in 30 sec. Calculate its quality factor.

Ans. $A = A_0 e^{-bt}, (2)^{-1} = e^{-bt}$ or $b = \dfrac{0.693}{30}$ and $\tau = \dfrac{1}{2b}$

$\therefore \ Q = \omega_0 \tau = \dfrac{2\pi}{0.2} \times \dfrac{30}{2 \times 0.693} = 680.$

15. The Q value of an under damped harmonic oscillator of frequency 480 Hz is 80000. Calculate the time in which its amplitude reduces to $\dfrac{1}{e}$ of its initial values. How many oscillations does it make in this time?

Ans. $A = A_0 e^{-bt} = A_0 e^{-\frac{t}{2\tau}}$

$\therefore \ e^{-1} = e^{-\frac{t}{2\tau}}, t = 2\tau = \dfrac{2\tau\omega_0}{\omega_0} = \dfrac{2Q}{\omega_0}$ or $t = 53$ s

$n = 53 \times \dfrac{1}{T} = 53 \times f = 53 \times 480 = 2.5 \times 10^{+4}.$

MULTIPLE CHOICE QUESTIONS

1. A circuit contains $L = 10$ mH, $C = 1.0\ \mu$F and $R = 0.1\ \Omega$. The frequency of the damped oscillations is (in rads/s):
   (a) 10^2 (b) 10^3 (c) 10^4 (d) 10^5

2. In an oscillatory circuit $L = 0.2$ H, $C = 0.0012\ \mu$F. The maximum value of the resistance for the circuit to be oscillatory is (in ohms):
   (a) $2.58 \times 10^2\ \Omega$ (b) $2.58 \times 10^3\ \Omega$ (c) $2.58 \times 10^4\ \Omega$ (d) $2.58 \times 10^5\ \Omega$

3. A simple pendulum has period of 2 s and an amplitude of 10°. After 5 complete oscillations its amplitude reduces to 5°. The quality factor of the pendulum is:
   (a) 12   (b) 15   (c) 23   (d) 45

4. In case of a weakly damped oscillator, average $<E(t)>$ varies with time. Correct relation is:
   (a) $<E(t)> = E_0 e^{-bt}$ (b) $<E(t)> = E_0 e^{-2bt}$

   (c) $<E(t)> = E_0 \sin pt$ (d) $<E(t)> = E_0 e^{-\frac{bt}{2}}$

5. A circuit has an inductance of $\dfrac{1}{\pi}$ H and resistance 100 Ω and AC supply of 50 cycles is applied to it. The reactance and impedance of the circuit respectively is:
   (a) $100\ \Omega, 100\ \Omega$ (b) $100\ \Omega, 141.4\ \Omega$ (c) $141.4\ \Omega, 100\ \Omega$ (d) $141.4\ \Omega, 141.4\ \Omega$

6. A series LCR circuit has $L = 1$ mH, $C = 0.1\ \mu$F and $R = 10\ \Omega$. The separation between two half power points is (in rad/s):
   (a) 10^4 (b) 10^3 (c) 10^2 (d) 10

7. A sinusoidal driving force is applied to a body whose natural frequency is 100 Hz and relaxation time 0.5 sec. If the frequency is changed to 10 Hz, its relative amplitude becomes:
   (a) 10^{-1} (b) 10^0 (c) 10^1 (d) 10^2

8. The amplitude of forced oscillation is
   $$C = \frac{f_0}{[(\omega_0^2 - \omega^2)^2 + 4b^2\omega^2]^{\frac{1}{2}}} \text{ if } Q = \frac{\omega_0}{2b} = \omega_0\tau = 100. \text{ The ratio } \frac{C_{max}}{C} \text{ for } \frac{\omega}{\omega_0} = 1.01 \text{ is:}$$
   (a) 0.455 (b) 1 (c) 1.5 (d) 0.667

9. The energy of a piano string of frequency 256 Hz reduces to half its initial value in 2 s. Q-value for the string is:
   (a) 46.43 (b) 464.3 (c) 4643 (d) 46430

10. The quality factor of a sonometer wire of frequency 500 Hz is 5000. The time in which its energy reduces to $1/e$ of its value in the absence of damping is (approximately):
    (a) 1 sec (b) 1.5 sec (c) 2 sec (d) 3 sec

11. The amplitude of a weakly damped oscillator reduces to half its initial value in approximate time given by:
    (a) $\tau \log_e 2$ (b) $\dfrac{\tau}{\log_e 2}$ (c) $2\tau \log_e 2$ (d) $\dfrac{2\tau}{\log_e 2}$

12. An object of mass 0.1 kg is hung from spring whose spring constant is 40 N·m^{-1}. The object is subject to force $F = -\beta v$, where β is a constant and v is the velocity of the mass. It is observed that the frequency of the damped oscillation is 99/100 of the undamped value. Q value of the oscillator is:
    (a) 5×10^4 (b) 5×10^3 (c) 5×10^2 (d) 5×10^1

13. In above problem, by what factor is the amplitude of oscillations reduced after 10 complete oscillations:
    (a) 2×10^1 (b) 2×10^0 (c) 2×10^{-2} (d) 2×10^{-3}

14. A massless spring of constant 9 N·m^{-1} is suspended from a rigid support and carries a mass of 1 kg at its lower end. The system oscillates in a liquid that exerts a viscous force $F = -\beta v$ where β is constant and v velocity. It is observed that the energy decays to half its original value in 5 sec, value of β is (in Ns·m^{-1}):
    (a) 14 (b) 1.4 (c) 0.14 (d) 0.014

15. In above problem, changes in frequency due to damping in (in Hz):
    (a) 4×10^{-5} Hz (b) 4×10^{-4} Hz (c) 4×10^{-3} (d) 4×10^{-2}

ANSWERS

1. (c) 2. (c) 3. (c) 4. (b) 5. (b) 6. (a) 7. (b) 8. (a) 9. (c) 10. (b) 11. (a)
12. (c) 13. (d) 14. (d) 15. (a)

Coupled Oscillations

3.1 INTRODUCTION

So far we have discussed the simple vibrating systems such as simple harmonic oscillator, damped or forced oscillator with one degree of freedom. In last chapter, we discussed in details the response of driven oscillator to an externally applied periodic force. There, we assumed that the driving system remains practically unaffected by the forced oscillations of the driven system. Former only serving as a source of a periodic force with a negligible feedback of energy from the latter. Here in this chapter, we shall study the oscillations of coupled systems in which we cannot neglect the feedback of energy from the driven system to the driver. Some examples of coupled oscillating systems are:

(a) Two masses attached to each other by three springs [Fig. 3.1(a)]. In this figure the middle spring provides the coupling between the masses.

(b) Two simple pendulum with their bobs attached to each other by means of a spring [Fig. 3.1(b)].

(c) Two coupled LC circuits.

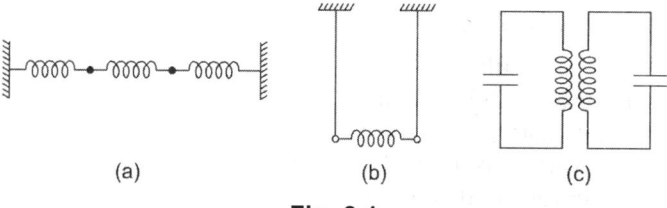

| (a) | (b) | (c) |

Fig. 3.1

Moving parts of such a system oscillate in a plane and two variables are required to specify the motion completely. Such systems are said to have two degrees of freedom. In case of two coupled masses or two coupled pendulums, the two variables are the displacements of the two masses or bobs and in case of two coupled LC circuits, the two variables are the charges on the two capacitors or currents in two circuits. In these examples, the two oscillators are coupled together so tightly that we cannot neglect the feedback of the energy from the driven oscillator to the driver. In fact, we cannot distinguish and label one as driver and the other as driven. The two oscillators have now to be treated on an equal footing.

In physics, we come across a variety of coupled systems that oscillate. In fact, oscillations rarely exist in complete isolation. For example, a solid body is composed of many atoms or molecules. Every atom may behave as an oscillator vibrating about an equilibrium position. But the motion of each atom affects its neighbours so that, in effect all the atoms of the solid are coupled together. Here we will discuss in details, as to how does the coupling affects the behaviour of individual oscillators.

We shall begin by discussing in some detail the properties of a system of just two coupled oscillators, we will then use the method to discuss the problems of an arbitrary large (but finite) number of N oscillators. In this way, starting from quite simple beginnings, we can end up with a significant insight into the dynamical properties of something as complicated as a crystal lattice.

3.2 TWO COUPLED OSCILLATORS

The simplest example of coupled oscillators is two identical pendulums connected with a spring whose *relaxed length* is exactly equal to the distance between the pendulum bobs as shown in Fig. 3.2(a). Draw a pendulum A aside while holding B fixed and then release both of them. What happens?

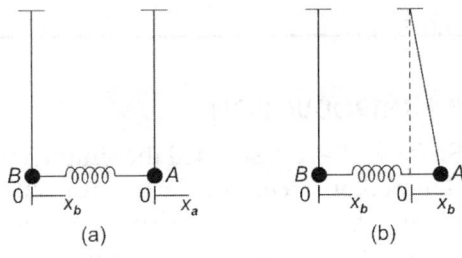

(a)   (b)

Fig. 3.2

The amplitude of oscillations of A will continuously decrease while that of B, initially zero will begin to increase, i.e. it will start oscillating. Soon A and B will have same amplitudes. The process will continue till the amplitude of A becomes zero and that of B becomes equal to that of A initially i.e., the starting condition is almost reversed. Now, the motion of B will be transferred back to A and so it continues to oscillate. The energy originally given to A (and to the spring) does not remain confined to the oscillation of A, but is transferred gradually to B and continues to oscillate back and forth between A and B. It is the spring which is responsible for the transfer of energy between the two pendulums.

(a) Normal Models of Vibrations

Suppose we draw both pendulums A and B aside by equal amounts [Fig. 3.3(a), and then release them. The distance between them equals the relaxation length of the coupling spring, and therefore, the spring exerts no force on either pendulum. A and B will oscillate in phase and with equal amplitudes always maintaining the

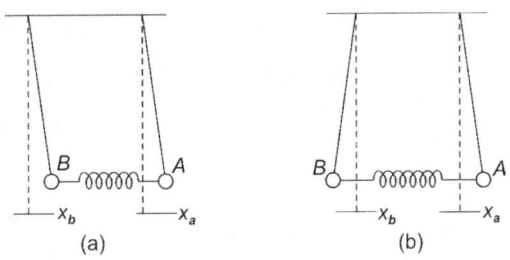

(a)   (b)

Fig. 3.3

same separation. Each pendulum might just as well be free (i.e. uncoupled). Each oscillates with its free natural frequency $\omega_0 = \left(\sqrt{\dfrac{g}{l}} \right)$. The equations of motion are

$$x_a = c \cos \omega_0 t \qquad x_b = c \cos \omega_0 t \tag{3.1}$$

where x_a and x_b are the displacements of each pendulum from its equilibrium position. This represents a normal mode of the coupled system. Both masses vibrate with the same frequency and each has a constant amplitude (same for both).

There is one more normal mode of vibration for the system. Draw A and B aside by equal amounts but in opposite direction [Fig. 3.3(b)] and then release them. Now, the coupling spring is stretched and a half cycle latter it will be compressed and it does exert forces. The symmetry consideration reveals that the motion of A and B will be mirror images of each other.

If the pendulum were free and either one was displaced a small distance x, the restoring force would be $m\omega_0^2 x$. But in present situation, the coupling spring is stretched (or compressed) a distance $2x$ and exerts a restoring force $2kx$, where k is the spring constant. The equation of motion for A is

$$m\frac{d^2 x_a}{dt^2} + m\omega_0^2 x_a + 2kx_a = 0 \qquad (3.2)$$

or $$\frac{d^2 x_a}{dt^2} + \left(\omega_0^2 + 2\omega_c^2\right)x_a = 0 \quad \text{where} \quad \omega_c^2 = \frac{k}{m}. \qquad (3.2a)$$

This equation is the equation for simple harmonic motion of frequency ω' given by:

$$\omega' = \left(\omega_0^2 + 2\omega_c^2\right)^{1/2} = \left(\frac{g}{l} + \frac{2k}{m}\right)^{1/2} \qquad (3.3)$$

Hence for given initial conditions, the solution of differential equation (3.2) is

$$x_a = A \cos \omega' t \qquad (3.4)$$

and since the motion of B is mirror image of A, and therefore:

$$x_b = -A \cos \omega' t \qquad (3.4a)$$

Thus, each pendulum oscillates with harmonic motion but the action of the coupling spring has been to increase the restoring force, and therefore, to increase the frequency over that of the uncoupled oscillations. The motions of A and B are clearly always $180°$ out of phase. In this type of oscillation which constitute the second normal mode, the importance of the normal modes of vibration is that they are entirely independent of each other. The energy associated with a normal mode is never exchanged with another mode; this is why, we can add the energies of the modes to give the total energy. If only one mode vibrates, the second mode of our system will always be at rest, acquiring no energy from the vibrating mode.

We define a mode by which a system can acquire energy. *The number of different ways by which a system may take up energy, defines its number of degrees of freedom.* Each mode may be associated with more than one degree of freedom, e.g. a harmonic oscillator has two degrees of freedom because it takes up both potential and kinetic energies.

(b) Superposition of Normal Modes

In above two cases, the motion once begun will, in the absence of damping forces continue without change. No transfer of energy occurs from some one mode of oscillation to another. An important reason for introducing these two easily solved cases is that any motion of the pendulum, in which each starts from rest, can be described as combination of these two. Let us see how that can be done.

Let at any arbitrary time (t), the displacement of pendulum A is x_a and that of B be x_b (Fig. 3.4). The stretching of the spring at this instant is $(x_a - x_b)$. Thus, the magnitude of the restoring force on A is $m\omega_0^2 x_a + k(x_a - x_b)$ and on B is $m\omega_0^2 x_b - k(x_a - x_b)$. Hence the equations of motion of A and B are

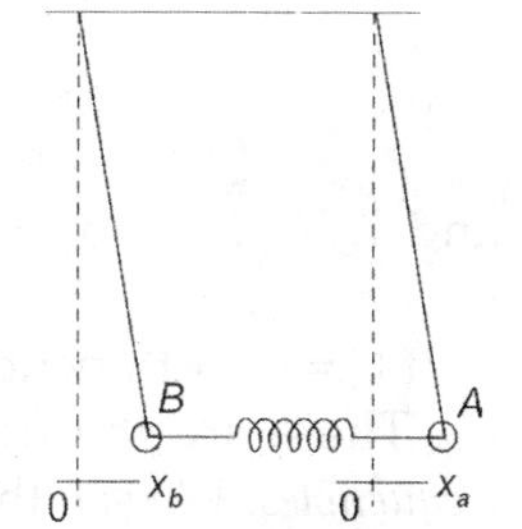

Fig. 3.4

$$m\frac{d^2 x_a}{dt^2} + m\omega_0^2 x_a + k(x_a - x_b) = 0$$

or
$$\frac{d^2 x_a}{dt^2} + \left(\omega_0^2 + \omega_c^2\right)x_a - \omega_c^2 x_b = 0 \tag{3.5}$$

and for B is
$$\frac{d^2 x_b}{dt^2} + \left(\omega_0^2 + \omega_c^2\right)x_b - \omega_c^2 x_a = 0 \tag{3.5a}$$

where $\omega_c^2 = \dfrac{k}{m}$.

The Eq. (3.5) describing the acceleration of A contains a term x_b and equation (3.5a) contains a term x_a. These two equations cannot be solved indenepdently but must be solved simultaneously. A motion given to A does not stay confined to A but effects B and vice versa.

These equations can be solved very easily. Adding the two, we have

$$\frac{d^2}{dt^2}(x_a + x_b) + \omega_0^2(x_a + x_b) = 0 \tag{3.6}$$

and subtracting, we have

$$\frac{d^2}{dt^2}(x_a - x_b) + \left(\omega_0^2 + 2\omega_c^2\right)(x_a - x_b) = 0 \tag{3.6a}$$

These are familiar equations for simple harmonic oscillations. In the first, the variable is $(x_a + x_b)$ and the frequency is ω_0 and in the second, the variable is $(x_a - x_b)$ and the frequency $\omega' = (\omega_0^2 + 2\omega_c^2)^{1/2}$. These two frequencies correspond precisely to those of the two normal modes that we identified previously. If we put $x_a + x_b = x_1$ and $x_a - x_b = x_2$, we have two independent equations as follows:

$$\frac{d^2 x_1}{dt^2} + \omega_0^2 x_1 = 0 \quad \text{and} \quad \frac{d^2 x_2}{dt^2} + \omega'^2 x_2 = 0$$

possible solutions (though not most general ones) are
$$x_1 = C\cos(\omega_0 t + \phi_1)$$

and
$$x_2 = D\cos(\omega' t + \phi_2)$$

where C and D are constants which depend upon the initial conditions.

Here we have two independent oscillations, they represent another description of normal modes, as represented by oscillations of variables x_1 and x_2 respectively and these variables are consequently called *normal co-ordinates*. Change in the value of x_1 occur independently of x_2 and vice versa.

In terms of original co-ordinates x_a and x_b, the solutions are

$$x_a = \frac{1}{2}(x_1 + x_2) = \frac{1}{2}C\cos(\omega_0 t + \phi_1) + \frac{1}{2}D\cos(\omega' t + \phi_2)$$

and
$$x_b = \frac{1}{2}(x_1 - x_2) = \frac{1}{2}C\cos(\omega_0 t + \phi_1) - \frac{1}{2}D\cos(\omega' t + \phi_2).$$

If $C = 0$, both pendulums oscillate with frequency ω' and if $D = 0$ with the frequency ω_0. These are the frequencies of the individual normal modes and are called *normal frequencies*. We see that a characteristic of a normal frequency is that both, x_a and x_b can oscillate with that frequency.

If we take initial conditions as $x_a = a_0$, $\dfrac{dx_a}{dt} = 0, x_b = 0, \dfrac{dx_b}{dt} = 0$ at $t = 0$ and $\phi_1 = \phi_2 = 0$.
Then

$$a_0 = x_a = \frac{1}{2}(C + D) \quad \text{and} \quad 0 = x_b = \frac{1}{2}(C - D)$$

$\therefore$ $\qquad\qquad C = D = a_0.$

Hence under these initial conditions, we have

$$x_a = \frac{1}{2}a_0 (\cos \omega_0 t + \cos \omega' t)$$

and

$$x_b = \frac{1}{2}a_0 (\cos \omega_0 t + \cos \omega' t)$$

or

$$x_a = a_0 \cos \frac{(\omega_0 + \omega')t}{2} \cos \frac{(\omega_0 - \omega')t}{2}$$

and

$$x_b = a_0 \sin \frac{(\omega_0 + \omega')t}{2} \sin \frac{(\omega_0 - \omega')t}{2}$$

Each of these is a sinusoidal oscillation of angular frequency $\left(\dfrac{\omega' + \omega_0}{2}\right)$ modulated in amplitude in the way discussed earlier. This amplitude associated with one of the pendulum is zero at the instant when the amplitude associated with the other is maximum, although the actual displacement of the latter at subsequent times depends on the value $\left(\dfrac{\omega' + \omega_0}{2}\right)t$.

3.3 GENERAL METHOD OF FINDING NORMAL MODES

Many times it is not easy to discover the normal modes from symmetry consideration, nor it is easy to choose the normal co-ordinates to solve the differential equations of the coupled system. As an example, let us consider the case of two coupled pendulums of equal string length l but of unequal bob masses m_a and m_b. Now, the equations of motion are:

$$\frac{d^2 x_a}{dt^2} - \omega_0^2 x_a + \frac{k}{m_a}(x_b - x_a) = 0 \tag{3.7}$$

and

$$\frac{d^2 x_b}{dt^2} - \omega_0^2 x_b + \frac{k}{m_b}(x_a - x_b) = 0 \tag{3.7a}$$

To solve such equations, we make use of the fact that in normal mode, all the moving parts of the system execute simple harmonic motions at the same frequency and the same phase constant. The only restriction is that the differential equations governing the motion of the system must be linear. In fact, this method is a general one and is applicable to any coupled system with any number of degrees of freedom.

Let us assume the normal mode to exist at an angular frequency ω and phase angle ϕ. This implies that both pendulums execute simple harmonic motion with the same angular frequency and the same phase angle. Hence, the solution for that mode is

$$x_a = A \cos(\omega t + \phi) \quad \text{and} \quad x_b = B \cos(\omega t + \phi) \tag{3.8}$$

substitution in Eqs. (3.7) and (3.7a), we have

$$\left\{ \omega_0^2 + \frac{k}{m_a} - \omega^2 \right\} x_a = \frac{k}{m_a} x_b \qquad (3.9)$$

and

$$\left\{ \omega_0^2 + \frac{k}{m_b} - \omega^2 \right\} x_b = \frac{k}{m_b} x_a \qquad (3.9a)$$

from these equations, we have

$$\frac{x_a}{x_b} = \frac{\dfrac{k}{m_a}}{\omega_0^2 + \dfrac{k}{m_a} - \omega^2}$$

and

$$\frac{x_a}{x_b} = \frac{\left(\omega_0^2 + \dfrac{k}{m_b} - \omega^2 \right)}{\dfrac{k}{m_b}}$$

Since x_a and x_b both are not zero, the right hand side of these equations must be equal. Thus,

$$\frac{\dfrac{k}{m_a}}{\omega_0^2 + \dfrac{k}{m_a} - \omega^2} = \frac{\omega_0^2 + \dfrac{k}{m_b} - \omega^2}{\dfrac{k}{m_b}}$$

giving

$$\frac{k^2}{m_a m_b} = \left(\omega_0^2 + \frac{k}{m_a} - \omega^2 \right) \left(\omega_0^2 + \frac{k}{m_b} - \omega^2 \right)$$

$$= \omega^4 - \left(2\omega_0^2 + \frac{k}{m_a} + \frac{k}{m_b} \right) \omega^2 + \left(\omega_0^2 + \frac{k}{m_a} + \frac{k}{m_b} \right) \omega_0^2 + \frac{k^2}{m_a m_b}$$

or

$$\omega^4 - \left(2\omega_0^2 + \frac{k}{m_a} + \frac{k}{m_b} \right) \omega^2 + \left(\omega_0^2 + \frac{k}{m_a} + \frac{k}{m_b} \right) \omega_0^2 = 0$$

and the roots of this equation are

$$\omega_1^2 = \omega_0^2 \quad \text{and} \quad \omega_2^2 = \omega_0^2 + k\left(\frac{1}{m_a} + \frac{1}{m_b} \right).$$

Thus, we come to the conclusion that there are just two modes. Substituting the value ω_1 for ω, we have

$$\frac{x_a}{x_b} = \frac{A}{B} = +1$$

and setting $\omega = \omega_2$, we have

$$\frac{x_a}{x_b} = \frac{A}{B} = \frac{m_b}{m_a}.$$

If m_a were equal to m_b, $\dfrac{x_a}{x_b} = \dfrac{A}{B} = -1$, the same result we have already arrived at. The displacement of the oscillators in mode 1 is given by

$$(x_a)_1 = A_1 \cos(\omega_1 t + \phi_1)$$
$$(x_b)_1 = B_1 \cos(\omega_1 t + \phi_1)$$

with
$$B_1 = A_1$$

and for mode 2 the displacements are given by

$$(x_a)_2 = A_2 \cos(\omega_2 t + \phi_2)$$
$$(x_b)_2 = B_2 \cos(\omega_2 t + \phi_2)$$

with
$$B_2 = -\frac{m_a}{m_b} A_2.$$

The most general solution of Eqs. (3.7) and (3.7a) are then given by superposition of two normal modes i.e.

$$\begin{aligned} x_a &= (x_a)_1 + (x_a)_2 \\ &= A_1 \cos(\omega_1 t + \phi_1) + A_2 \cos(\omega_2 t + \phi_2) \end{aligned} \tag{3.10}$$

and
$$\begin{aligned} x_b &= (x_b)_1 + (x_b)_2 \\ &= B_1 \cos(\omega_1 t + \phi_1) + B_2 \cos(\omega_2 t + \phi_2) \end{aligned} \tag{3.10a}$$

Notice that whereas we have complete freedom in choosing the four constant A_1, A_2, ϕ_1 and ϕ_2. We have no freedom in choosing B_1 and B_2 at all, since they are determined from A_1 and A_2. The four unknown constants A_1, A_2, ϕ_1 and ϕ_2 are to be determined from the given initial conditions.

3.4 ENERGY EXCHANGE IN COUPLED OSCILLATIONS

Let us again consider the case of two identical pendulums A and B connected with a spring, whose relaxed length is exactly equal to the distance between the pendulum bobs [Fig. 3.1(a)]. A is displaced while holding B and then both are released from rest simultaneously ($t = 0$). As the pendulum A swings, its amplitude decreases continuously. Pendulum B which initially was undisplaced begins to oscillate and its amplitude increases. After some time, the amplitude of A is momentarily zero and that of B becomes equal to the displacement initially given to A i.e. the starting conditions are reversed. The motion of B is transferred back to A and the sequence of events continues. The energy is transferred from one pendulum to the other. The energy originally given to A is not confined only to A but is gradually transferred to B and continues to shuttle back and forth between A and B. One oscillation for the energy from A to B and back to A is called a *beat*. The time for one round trip for energy is called *beat period* and its inverse as *beat frequency*.

If mass of bob A (m_a) is different from mass m_b of bob B, a similar behaviour is observed. The energy keeps bouncing back and forth between A and B except that the amplitude of A never becomes zero i.e. the energy exchange is not complete. These results are predicted from the displacement Eqs. (3.10) and (3.10a). From these, the instantaneous velocity of each pendulum is

$$\frac{dx_a}{dt} = -A_1 \omega_1 \sin(\omega_1 t + \phi_1) - A_2 \omega_2 \sin(\omega_2 t + \phi_2)$$

and
$$\frac{dx_b}{dt} = -B_1 \omega_1 \sin(\omega_1 t + \phi_1) - B_2 \omega_2 \sin(\omega_2 t + \phi_2)$$

Initial condiitions are $x_a = a$, $x_b = 0$, $\dot{x}_a = 0$, $\dot{x}_b = 0$ at $t = 0$.

Setting $B_1 = A_1$ and $B_2 = -\dfrac{m_a}{m_b} A_2$, we have

$$a = A_1 \cos\phi_1 + A_2 \cos\phi_2 \tag{i}$$

$$0 = A_1 \cos\phi_1 - \frac{m_a}{m_b} a_2 \cos\phi_2 \tag{ii}$$

$$0 = -A_1\omega_1 \sin\phi_1 - A_2\omega_2 \sin\phi_2 \tag{iii}$$

$$0 = -A_1\omega_1 \sin\phi_1 + \frac{m_a}{m_b} \omega_2 A_2 \sin\phi_2 \tag{iv}$$

These four equations determine the four constanta A_1, A_2, ϕ_1 and ϕ_2. These equations give

$$\sin\phi_1 = \sin\phi_2 = 0$$

$$A_1 \cos\phi_1 = \frac{m_a}{m_b} A_2 \cos\phi_2 = \frac{am_a}{m_a + m_b}$$

$$A_2 \cos\phi_2 = \frac{am_b}{m_a + m_b}.$$

Substituting these values, we have

$$x_a = \frac{a}{m_a + m_b}(m_a \cos\omega_1 t + m_b \cos\omega_2 t) \tag{3.11}$$

and

$$x_b = \frac{am_a}{m_a + m_b}(\cos\omega_1 t - \cos\omega_2 t) \tag{3.11a}$$

We observe that x_a and x_b do not vary harmonically with time. In fact, x_a and x_b represent superposition of two harmonic oscillations at frequencies ω_1 and ω_2 of the two modes. It is obvious that there are beats in the system. If we introduce two frequencies ω_m and ω_a such that

$$\omega_m = \frac{1}{2}(\omega_2 - \omega_1) \quad \text{and} \quad \omega_a = \frac{1}{2}(\omega_2 + \omega_1)$$

then Eq. (3.11) becomes

$$\begin{aligned}
x_a &= a\cos\omega_m t \cos\omega_a t + \frac{a(m_a - m_b)}{m_a + m_b}\sin\omega_m t \\
&= A_m \cos(\omega_a t - \theta)
\end{aligned} \tag{3.12}$$

where

$$\begin{aligned}
A_m &= \left\{ a^2 \cos^2\omega_m t + \frac{a^2(m_a - m_b)^2}{(m_a + m_b)}\sin^2\omega_m t \right\}^{1/2} \\
&= \frac{a}{m_a + m_b}\left\{ m_a^2 + m_b^2 + 2m_a m_b \cos 2\omega_m t \right\}^{1/2}
\end{aligned} \tag{3.13}$$

and

$$\tan\theta = \frac{m_a - m_b}{m_a + m_b}\tan\omega_m t \tag{3.13a}$$

Similarly, Eq. (3.11a) becomes

$$x_b = B_m \sin\omega_a t \tag{3.12a}$$

where

$$B_m = \frac{2am_a}{(m_a + m_b)}\sin\omega_m t \tag{3.13b}$$

Case of Weak Coupling

In Eqs. (3.12) and (3.12a), A_m and B_m vary with time as such these equations do not represent simple harmonic motion. But if coupling is very weak, i.e. k is very small, then ω_2 will be slightly higher than ω_1 and $\omega_m = \dfrac{1}{2}(\omega_2 - \omega_1)$ will be very small, so that A_m and B_m will take very long time to change appreciably. As such for small values of k, A_m and B_m remain appreciably constant during oscillations of frequency ω_a and the motion associated with Eqs. (3.12) and (3.12a) can be treated as almost harmonic. Hence, the energy E_a of the pendulum A can be written as

$$E_a = \frac{1}{2}kA_m^2 = \frac{1}{2}m_a\omega_a^2 A_m^2$$

substituting the value of A_m from (3.13), and the energy of the pendulum at $t = 0$ as $E = \dfrac{1}{2}m_a\omega_a^2 a^2$. We have, (using $2\omega_m = \omega_2 - \omega_1$)

$$E_a = E\left\{\frac{m_a^2 + m_b^2 + 2m_a m_b \cos(\omega_2 - \omega_1)t}{(m_a + m_b)^2}\right\}. \tag{3.14}$$

Similarly, the energy of pendulum B is given by

$$E_b = \frac{1}{2}m_b\omega_a^2 B_m^2 = \frac{2m_a m_b}{(m_a + m_b)^2}\left\{1 - \cos(\omega_2 - \omega_1)t\right\} \tag{3.14a}$$

These equations reveal that at time $t = 0$, $E_a = E$ and $E_b = 0$, which agrees with our initial conditions. As time passes on, E_a begins to decrease and E_b begins to increase until E_a becomes minimum at time such that $\cos(\omega_2 - \omega_1)t = -1$ and

$$E_a(\min) = \frac{E(m_a - m_b)^2}{(m_a + m_b)}$$

and E_b becomes maximum such that

$$E_b(\max) = \frac{E(4m_a - m_b)}{(m_a + m_b)^2}$$

After this value of t, E_a increases until it again becomes equal to E, the energy pendulum A had at $t = 0$ and E_b becomes zero and the process repeats. The value of t when E_a becomes maximum, equal to E are given by

$$\cos(\omega_2 - \omega_1)t = +1$$

i.e. $\quad (\omega_2 - \omega_1)t = 0, 2\pi, 4\pi, ..., 2n\pi$

i.e.
$$t = \frac{0}{\omega_2 - \omega_1}, \frac{2\pi}{\omega_2 - \omega_1}, \frac{4\pi}{\omega_2 - \omega_1}, ..., \frac{2n\pi}{\omega_2 - \omega_1}$$

$$= 0, \frac{2\pi}{\omega_2 - \omega_1}, \frac{4\pi}{\omega_2 - \omega_1}, ..., \frac{2n\pi}{\omega_2 - \omega_1}.$$

Thus, time period of beats is

$$t_b = \frac{2\pi}{\omega_2 - \omega_1} = \frac{1}{v_2 - v_1}$$

and heat frequency is

$$v_b = v_2 - v_1$$

which is equal to the difference between the two normal mode frequencies of the system.

We also observe that the total energy of the system $E_a + E_b$ remains constant $(= E)$. This is so because we have assumed the friction to be zero. Thus, we conclude that the energy keeps bouncing back and forth between the two pendulums at a frequency $v_b = v_2 - v_1$, since $(E_a)_{min} \neq 0$, the energy of pendulum A is not completely transferred to B. In other words, energy exchange is not complete.

Case of Identical Pendulums

In case of two identical pendulums $m_a = m_b = m$, then

$$x_a = \frac{a}{2}(\cos\omega_1 t + \cos\omega_2 t)$$

$$x_b = \frac{a}{2}(\cos\omega_1 t - \cos\omega_2 t)$$

$$E_a = \frac{E}{2}[1 + \cos(\omega_2 + \omega_1)t]$$

$$E_b = \frac{E}{2}[1 + \cos(\omega_2 - \omega_1)t]$$

Notice that $(E_a)_{min} = 0$ and $(E_b)_{max} = E$. Thus, if the bob masses are equal, the energy exchange is complete. The energy of each pendulum oscillates between a minimum value of zero and maximum value of E. Figure 3.5 shows the variation of x_a, x_b, A_m, B_m, E_a and E_b with time. Notice that x_a and x_b do not vary harmonically with time.

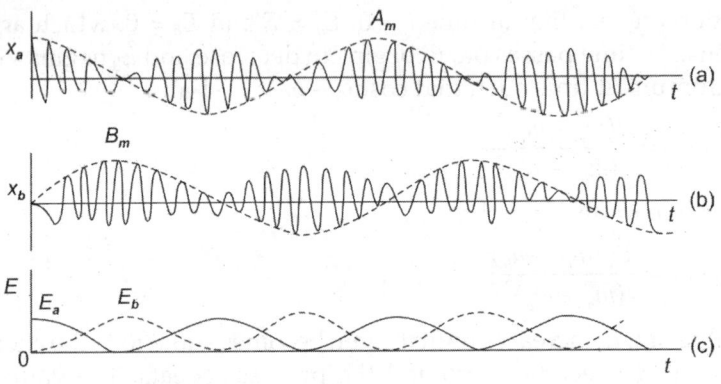

Fig. 3.5

It is worthwhile to mention that though we have taken the example of two coupled pendulums, the mathematical technique developed here is very general and is applicable to any coupled system with two degrees of freedom.

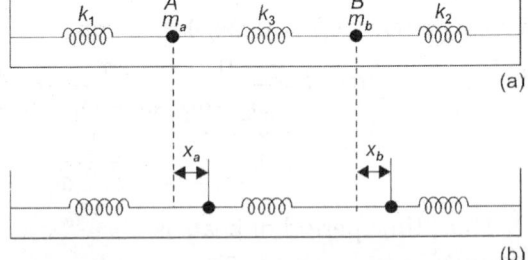

Fig. 3.6

3.5 TWO COUPLED MASSES

Here we will consider two masses connected by means of three springs (Fig. 3.6). We will first analyse the case where two masses are same ($m_a = m_b = m$) and all the three springs are identical ($k_1 = k_2 = k_3 = k$). This system can oscillate longitudinally as well as transversely.

Normal Modes of Longitudinal Oscillations

Figure 3.6(a) shows the equilibrium state of the system. Each spring is assumed to be massless and of normal length l. Figure 3.6(b) depicts the system in oscillations. Let x_a and x_b be the displacements of masses A and B at any instant of time $(x_a < x_b)$, then the equation of motion of the system is

$$m\ddot{x}_a = -kx_a + k(x_b - x_a) \tag{i}$$

$$m\ddot{x}_b = -kx_b + k(x_b - x_a) \tag{ii}$$

If we assume the existence of normal modes at angular frequency ω and phase constant ϕ, the solutions can be written as

$$x_a = a\cos(\omega t + \phi)$$

$$x_b = b\cos(\omega t + \phi).$$

Substituting these values in (i) and (ii), gives

$$\frac{x_b}{x_a} = \frac{2\dfrac{k}{m} - \omega^2}{\dfrac{k}{m}} \tag{iii}$$

$$\frac{x_b}{x_a} = \frac{\dfrac{k}{m}}{2\dfrac{k}{m} - \omega^2} \tag{iv}$$

which gives

$$\frac{k^2}{m^2} = \left(\frac{2k}{m} - \omega^2\right)^2 \quad \text{or} \quad \left(\frac{2k}{m} - \omega^2\right) = \pm\frac{k}{m}$$

or

$$\omega_1^2 = \frac{k}{m} \quad \text{and} \quad \omega_2^2 = \frac{3k}{m}.$$

These are the angular frequencies of two normal modes. Substitution of $\omega_1^2 = \dfrac{k}{m}$ in (iii) or (iv), we have

$$\frac{x_b}{x_a} = +1$$

This is mode 1, in which the two masses always move in same direction.

And substitution of ω_2 gives

$$\frac{x_b}{x_a} = -1$$

This is mode 2, in which the two masses always more in opposite direction.

Normal Modes of Transverse Oscillations

Figure 3.7(a) depicts a coupled system consisting of three identical massless springs of normal length l_0 and spring constant k. Two equal masses (m) A and B are attached to them. This is the equilibrium position of the system. Figure 3.7(b) shows two

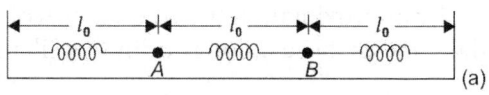
(a)

(b)

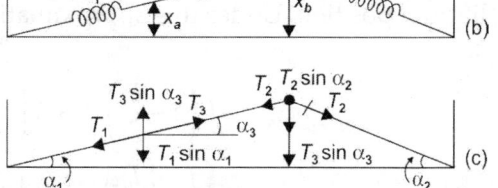
(c)

Fig. 3.7

particles in oscillating position. They are oscillating in transverse configuration. At a certain moment (t), the displacements of the two masses are x_a and x_b. Figure 3.7(c) shows the forces acting on the two masses at that moment (t). The new lengths of the springs at this moment are l_1, l_2, l_3 respectively. Hence, the tensions in three springs are respectively.

$$T_1 = k(l_1 - l_0),\ T_2 = k(l_2 - l_0)\ \text{and}\ T_3 = k(l_3 - l_0).$$

The restoring forces acting on the two masses are $(-T_1 \sin \alpha_1 + T_3 \sin \alpha_3)$ and $(-T_2 \sin \alpha_2 + T_3 \sin \alpha_3)$ where $\alpha_1, \alpha_2, \alpha_3$ are the angles subtended by the springs with X-axis. The equations of motion of the two masses are

$$m\ddot{x}_a = -T_1 \sin \alpha_1 + T_3 \sin \alpha_3 = -T_1 \frac{x_a}{l_1} + T_3 \frac{(x_b - x_a)}{l_3}$$

$$m\ddot{x}_b = -T_2 \sin \alpha_2 - T_3 \sin \alpha_3 = -T_2 \frac{x_a}{l_2} - T_3 \frac{(x_b - x_a)}{l_3}$$

On substituting the values of T_1, T_2 and T_3, we have

$$m\ddot{x}_a = -k\left(1 - \frac{l_0}{l_1}\right)x_a + k\left(1 - \frac{l_0}{l_3}\right)(x_b - x_a) \tag{i}$$

$$m\ddot{x}_b = -k\left(1 - \frac{l_0}{l_2}\right)x_b + k\left(1 - \frac{l_0}{l_3}\right)(x_b - x_a) \tag{ii}$$

To find normal modes, we have to make approximations. We shall consider the following two approximations:

(i) Slinky Approximation

Under this approximation the terms like l_0/l_1 are of second order in smallness ($\because l_0 < l_1$) and can therefore be neglected compared to unity. Under this approximation Eqs. (i) and (ii) reduce to

$$m\ddot{x}_a = -kx_a + k(x_b - x_a)$$

or

$$m\ddot{x}_b = -kx_b - k(x_b - x_a)$$

These are the same equations which we discussed in case of longitudinal oscillations. Therefore, the two normal mode frequencies are

$$\omega_1 = \sqrt{\frac{k}{m}} \quad \text{and} \quad \omega_2 = \sqrt{\frac{3k}{m}}$$

and the motion in two modes is such that in one $x_a = x_b$ and in other $x_a = -x_b$. Thus, under slinky approximation the frequencies of the normal modes of transverse oscillations are the same as those of longitudinal oscillations. We say that there is a form of degeneracy.

(ii) Small Oscillation Approximation

In this approximation, we assume that the system is oscillating very close to the equilibrium position. Under this approximation $l_1 \approx l_2 \approx l_3 \approx l$. Hence, the equations of motion are

$$m\ddot{x}_a = -k\left(1 - \frac{l_0}{l}\right)x_a + k\left(1 - \frac{l_0}{l}\right)(x_b - x_a)$$

and

$$m\ddot{x}_b = -k\left(1 - \frac{l_0}{l}\right)x_b + k\left(1 - \frac{l_0}{l}\right)(x_b - x_a)$$

These equations are again the same as obtained in case of longitudinal motion with the difference that here K is replaced by k^*, given by

$$K^* = k\left(1 - \frac{l_0}{l}\right).$$

Therefore, the two frequencies of normal modes are

$$\omega_1 = \sqrt{\frac{k^*}{m}} = \sqrt{\frac{k}{m}\left(1 - \frac{l_0}{l}\right)}$$

and

$$\omega_2 = \sqrt{\frac{3k^*}{m}} = \sqrt{\frac{3k}{m}\left(1 - \frac{l_0}{l}\right)}$$

with $x_a = x_b$ in first mode and $x_a = -x_b$ in second mode. It is apparent that the mode frequencies of transverse small oscillations are smaller than their corresponding frequencies of the longitudinal small oscillations.

3.6 ENERGY RELATION IN COUPLED OSCILLATIONS

We have established a very important property of normal modes, i.e. the general motion of the system is determined if its normal modes are known. The general displacement of each part of the system under any arbitrary initial condition is simply a linear superposition of its displacement in normal modes of vibration. Here, we will establish another important property of the normal modes, i.e. total energy of a coupled system started in any way is just equal to the sum of the energies of its normal modes of vibrations. This property will be more helpful in analysis of coupled oscillations as it is very easy to find the energy of normal modes since the motion associated with a normal mode is simple harmonic. To prove this property of normal modes, we will consider the oscillations of two masses coupled together by means of three springs (Fig. 3.6). But, we will consider the most general case in which the masses and springs are not identical. We will consider only longitudinal oscillations.

Let masses of A and B be m_a and m_b respectively and mass A is connected to the springs of constants k_1 and k_3 and that B is connected to the springs of constants k_2 and k_3. The spring k_3 provides the coupling.

Figure 3.6(a) shows the equilibrium state of the system. The masses are pulled along a frictionless horizontal surface and released. They begin to oscillate. Let x_a and x_b ($x_b > x_a$) respectively, be the displacement of A and B at a certain instant of time (t), then equations of motion of A and B are

$$m_a \ddot{x}_a = -k_1 x_a + k_3(x_b - x_a) = -(k_1 + k_3)x_a + k_3 x_b$$

and

$$m_b \ddot{x}_b = -k_2 x_b - k_3(x_b - x_a) = -(k_2 + k_3)x_b + k_3 x_a.$$

These equations are reduced to a simpler form if we substitute

$$x = \sqrt{m_a}\,.x_a \quad \text{and} \quad y = \sqrt{m_b}\,x_b$$

where x and y are called *reduced co-ordinates*.

In terms of reduced co-ordinates x and y, the equations of motion of particles A and B becomes

$$\ddot{x} = -a_{11}x - a_{12}y \tag{3.15}$$

and

$$\ddot{y} = -a_{22}y - a_{21}x \tag{3.15a}$$

where $$a_{11} = \frac{k_1 + k_3}{m_a}, \quad a_{22} = \frac{k_2 + k_3}{m_b}$$

and $$a_{12} = a_{21} = -\frac{k_3}{\sqrt{m_a m_b}}.$$

It may be remembered that the equations of motion of the moving parts of any system with two degrees of freedom can always be reduced to the above general form with appropriate values of the co-efficients a_{11}, a_{22}, a_{12} and a_{21}. The significance of different co-efficients is as under:

$\sqrt{a_{11}} = (\omega_a)$ is the natural angular frequency of harmonic oscillations of A if mass B is not allowed to move. This can be understood as if B is clamped so that $x_b = 0$ and A is moved to right by x_a. In that case, restoring force on A is $-(k_1 + k_3)x_a$ and hence, its angular frequency

$$\omega_a = \sqrt{\frac{k_1 + k_3}{m_a}}$$

Similarly, if A is clamped, the angular frequency of mass B is

$$\omega_b = \sqrt{\frac{k_2 + k_3}{m_b}} = \sqrt{a_{22}}$$

A displacement of mass A produces a force on mass B. This force is $k_3 x_a$. From symmetry, the force on A by a displacement of B is $k_3 x_b$. This is what, we call as coupling.

Frequencies of the Normal Modes

We assume that we have oscillation in single mode. This means that both degrees of freedom, namely x and y oscillate with same frequency simple harmonically and have same phase constant ϕ, i.e.

$$x = A \cos(\omega t + \phi)$$

and $$y = B \cos(\omega t + \phi)$$

Then, we have
$$\ddot{x} = -\omega^2 x \quad \text{and} \quad \ddot{y} = -\omega^2 y$$

Substituting values in Eq. (3.15) and (3.15a) yield two homogeneous linear equations in x and y.

$$(a_{11} - \omega^2)x + a_{12}y = 0 \tag{3.16}$$
$$(a_{21}x) + (a_{22} - \omega^2)y = 0. \tag{3.16a}$$

Each of these give the ratio $\dfrac{y}{x}$.

$$\frac{y}{x} = \frac{\omega^2 - a_{11}}{a_{12}}$$

and $$\frac{y}{x} = \frac{a_{21}}{\omega^2 - a_{22}}$$

For consistency, the right hands of the two equations are equal, i.e.
$$(a_{11} - \omega^2)(a_{22} - \omega^2) = a_{12} \cdot a_{21}$$

and $$\omega^4 - (a_{11} + a_{22})\omega^2 + (a_{11}a_{22} - a_{12} \cdot a_{21}) = 0$$

This is quadratic equation in ω^2 and has two solutions

$$\omega_1^2 = \frac{1}{2}(a_{11} + a_{22}) + \frac{1}{2}\left\{(a_{11} - a_{22})^2 + 4a_{12} \cdot a_{21}\right\}^{1/2}$$

$$\omega_2^2 = \frac{1}{2}(a_{11} + a_{22}) - \frac{1}{2}\left\{(a_{11} - a_{22})^2 + 4a_{12} \cdot a_{21}\right\}^{1/2}$$

The angular frequency of mode (1) is ω_1 and that of the mode (2) is ω_2. These values give

$$\left(\frac{y}{x}\right)_1 = \left(\frac{B}{A}\right) = \frac{B_1}{A_1} = \frac{\omega_1^2 - a_{11}}{a_{12}} = \frac{a_{21}}{\omega_1^2 - a_{22}} \tag{3.17}$$

and

$$\left(\frac{y}{x}\right)_2 = \left(\frac{B}{A}\right) = \frac{B_2}{A_2} = \frac{\omega_2^2 - a_{11}}{a_{12}} = \frac{a_{21}}{\omega_2^2 - a_{22}} \tag{3.17a}$$

and the displacements in two normal modes are given by

$$x_1 = A_1 \cos(\omega_1 t + \phi_1)$$
$$x_2 = B_1 \cos(\omega_1 t + \phi_1)$$

where A_1 and B_1 are related by Eq. (3.17). Similarly, for mode (2), we have

$$x_2 = A_2 \cos(\omega_2 t + \phi_2) \quad \text{and} \quad y_2 = B_2 \cos(\omega_2 t + \phi_2)$$

and A_2 and B_2 are related by Eq. (3.17a) and the most general solution is

$$x = x_1 + x_2 = A_1 \cos(\omega_1 t + \phi_1) + A_2 \cos(\omega_2 t + \phi_2)$$

and

$$y = y_1 + y_2 = B_1 \cos(\omega_1 t + \phi_1) + B_2 \cos(\omega_2 t + \phi_2)$$

Here we have four undetermined constants A_1, A_2, ϕ_1 and ϕ_2, which are determined from the initial conditions.

Normal Co-ordinates

The normal co-ordinates of the system can be obtained if the amplitude ratio of the two modes is known. Here[1]:

$$\frac{A_1}{B_1} = -\frac{B_2}{A_2} = \tan\alpha \quad \text{(say)} \tag{3.18}$$

Now, putting

$$A_1 = c_1 \sin\alpha \qquad A_2 = c_2 \cos\alpha$$
$$B_1 = c_1 \cos\alpha \qquad B_2 = -c_2 \sin\alpha$$

equation for most general solution is

$$x = c_1 \sin\alpha \cos(\omega_1 t + \phi_1) + c_2 \cos\alpha \cos(\omega_2 t + \phi_2)$$
$$y = c_1 \cos\alpha \cos(\omega_2 t + \phi_2) - c_2 \sin\alpha \cos(\omega_2 t + \phi_2)$$

Now, let

$$X = c_1 \cos(\omega_1 t + \phi_1) \quad \text{and} \quad Y = c_2 \cos(\omega_2 t + \phi_2)$$

then

$$x = X\sin\alpha + Y\cos\alpha \tag{3.19}$$
$$y = X\cos\alpha - Y\sin\alpha \tag{3.19a}$$

---

1. It is obtained by adding ω_1^2 and ω_2^2: $\omega_1^2 + \omega_2^2 = a_{11} + a_{22}$
$\therefore \omega_1^2 - a_{11} = -\left(\omega_2^2 - a_{22}\right)$ and Eqs. (3.17) and (3.17a) give the result.

These equations represent the transformation of coordinates of a point in plane with respect to axes X and Y to a new set of axes x and y inclined at an angle α to the first set as shown in Fig. 3.8

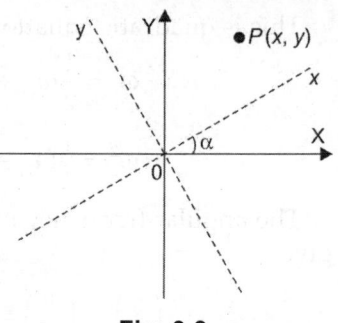

Notice that the motion associated with x or y is not even periodic but that associated with X as well as Y is harmonic. The simple harmonic motion associated with X has angular frequency ω_1 which is the frequency of first normal mode. The system will oscillate simple harmonically with frequency ω_1 only if the ratio of the amplitudes of motions of the two masses are related by

Fig. 3.8

Eq. (3.17). Similarly, the motion associated with Y has angular frequency ω_2 which is the frequency of the second normal mode. Here also, the system will oscillate simple harmonically with frequency ω_2, if the amplitude ratio of motions of two masses is given by Eq. (3.17a). If the motion is started in any other way, there will be no permanent ratio between the amplitudes of the two masses and the motion will not even be periodic.

Referring to Fig. 3.8, let us suppose that we represent the position of the system at any time t by a point P on this plane, whose co-ordinates at this instant are (x, y). Here x is the measure of the displacement of the first mass m_a. The actual displacement of this is $\left(x_a = \dfrac{x}{\sqrt{m_a}}\right)$ and y measures the displacement of second mass $m_b\left(x_b = \dfrac{x}{\sqrt{m_b}}\right)$. The motion of the system corresponds to the motion of the point in this plane with the time, the point P moves and its motion is governed by the general oscillation of the system. The projection of P on the X-axis moves back and forth in a complicated non-periodic fashion and so, does the projection of P on Y-axis. However, the projection of P or X-axis always moves back and forth simple harmonically with frequency ω_1 and amplitude c_1. Similarly, the projection of P on the Y-axis also moves simple harmonically with frequency ω_2 and amplitude c_2. The plane in which the point P moves is called the *configuration plane of the system* and the axes X and Y are called *normal co-ordinates of the system.*

The normal co-ordinates X and Y are

$$X = x\sin\alpha + y\cos\alpha \quad \text{and} \quad Y = x\cos\alpha - y\sin\alpha$$

In terms of x_a and x_b, these co-ordinates are

$$X = \sqrt{m_a}\,\sin\alpha \cdot x_a + \sqrt{m_b}\,\cos\alpha\, x_b \tag{3.20}$$

and

$$Y = \sqrt{m_a}\,\cos\alpha\, x_a - \sqrt{m_b}\,\sin\alpha\, x_b \tag{3.20a}$$

It is apparent that dimensions of X and Y are displacement (mass)$^{1/2}$. They are suitably normalised to give the dimensions of displacement if necessary. The general motion of the system is given by

$$x_a = \frac{x}{\sqrt{m_a}} = \frac{c_1}{\sqrt{m_a}}\sin\alpha\cos(\omega_1 t + \phi_1) + \frac{c_2}{\sqrt{m_a}}\cos\alpha\cos(\omega_2 t + \phi_2) \tag{3.21}$$

$$x_b = \frac{y}{\sqrt{m_b}} = \frac{c_1}{\sqrt{m_b}}\cos\alpha\cos(\omega_2 t + \phi_1) - \frac{c_2}{\sqrt{m_b}}\sin\alpha\cos(\omega_2 t + \phi_2) \tag{3.21a}$$

angle α is already defined as

$$\tan\alpha = \frac{A_1}{B_1} = \frac{\omega_1^2 - a_{22}}{a_{21}} = \frac{a_{12}}{\omega_1^2 - a_{11}}$$

The general solution overall involves 4 constants c_1, c_2, ϕ_1 and ϕ_2 which are determined from initial conditions.

Energy Relations

Here, we will see that when the motion of coupled system is described in terms of the normal co-ordinates X and Y rather than in terms of x and y, the expression for the energy of the system becomes remarkably simple.

The kinetic energy of the system is equal to the kinetic energy of the mass m_a and mass m_b. If x_a and x_b are their displacements at any time t, then kinetic energy of the system at that time is given by

$$KE = \frac{1}{2} m_a (\dot{x}_a)^2 + \frac{1}{2} m_b (\dot{x}_b)^2$$

In order to obtain the expression for potential energy of the system at time t, we have to find work, necessary to bring the system from equilibrium to a position where the displacements are x_a and x_b. We will first consider the motion of mass m_a keeping m_b fixed. In this case, work done is

$$W_1 = -\int_0^{x_a} (k_1 x + k_3 x) \, dx = \frac{1}{2} (k_1 + k_3) x_a^2$$

In next step, we will keep mass m_a fixed and consider the motion of mass m_b

$$W_2 = \int_0^{x_b} (k_2 x \, dx + \int_0^{x_b} k_3 (x - x_a) \, dx = \frac{1}{2} (k_2 + k_3) x_b^2 - k_3 x_a x_b$$

The sum W_1 and W_2 gives the instantaneous potential energy of the system. Thus, total potential energy

$$PE = W_1 + W_2 = \frac{1}{2} (k_1 + k_3) x_a^2 + \frac{1}{2} (k_2 + k_3) x_b^2 - k_3 x_a x_b$$

and the total instantaneous energy is

$$E = KE + PE = \frac{1}{2} m_a (\dot{x}_a)^2 + \frac{1}{2} m_b (\dot{x}_b)^2 + \frac{1}{2} (k_1 + k_2) x_a^2 + \frac{1}{2} (k_2 + k_3) x_b^2 - k_3 x_a x_b$$

or

$$E = \frac{1}{2} \{ \dot{x}^2 + \dot{y}^2 + a_{11} x^2 + a_{22} y^2 + 2 a_{12} xy \} \qquad (3.22)$$

In fact, this expression is complicated due to the presence of product term $x_a x_b$. We shall see that this expression becomes much simpler in terms of co-ordinates X and Y. Using the relation, we have

$$(\dot{x})^2 + (\dot{y})^2 = (\dot{X} \sin \alpha + \dot{Y} \cos \alpha)^2 + (\dot{X} \cos \alpha - \dot{Y} \sin \alpha)^2$$
$$= \dot{X}^2 + \dot{Y}^2$$

Similarly, the last three terms can be written as

$$a_{11} x^2 + a_{22} y^2 + 2 a_{12} xy = a_{11} \{ X \sin \alpha + Y \cos \alpha \}^2 + a_{22} \{ X \cos \alpha - Y \sin \alpha \}^2$$
$$+ 2 a_{12} \{ X \sin \alpha + Y \cos \alpha \} \{ X \cos \alpha - Y \sin \alpha \}$$
$$= \{ a_{11} \sin^2 \alpha + a_{22} \cos^2 \alpha + 2 a_{12} \sin \alpha \cos \alpha \} X^2$$
$$+ \{ a_{11} \cos^2 \alpha + a_{22} \sin^2 \alpha - 2 a_{12} \sin \alpha \cos \alpha \} Y^2$$
$$+ 2 \{ a_{11} \sin \alpha \cos \alpha - a_{22} \sin \alpha \cos \alpha$$
$$- a_{12} \sin^2 \alpha + a_{12} \cos^2 \alpha \} XY$$

$$= \left\{ \frac{1}{2}a_{11}(1 - \cos 2\alpha) + \frac{1}{2}a_{22}(1 + \cos 2\alpha) + a_{12}\sin 2\alpha \right\} X^2$$

$$+ \left\{ \frac{1}{2}a_{11}(1 + \cos 2\alpha) + \frac{1}{2}a_{22}(1 - \cos 2\alpha) = a_{12}\sin 2\alpha \right\} Y^2$$

$$+ \{(a_{11} - a_{12})\sin 2\alpha + 2a_{12}\cos \alpha\} XY$$

Now, using the result

$$\tan \alpha = \frac{a_{12}}{\omega_1^2 - a_{11}} = -\frac{\omega_2^2 - a_{11}}{a_{12}}$$

giving

$$\tan^2\alpha = -\frac{\omega_2^2 - a_{11}}{\omega_1^2 - a_{11}}$$

and

$$\tan 2\alpha = \frac{2\tan \alpha}{1 - \tan^2 \alpha} = \frac{2a_{11}}{a_{22} - a_{11}}$$

Here, we have used the relation $\omega_1^2 + \omega_2^2 = a_{11} + a_{22}$

We have,

$$\sin^2 2\alpha = \frac{\tan^2 2\alpha}{1 + \tan^2 2\alpha} \quad \text{and} \quad \cos^2 2\alpha = \frac{1}{1 + \tan^2 2\alpha}$$

putting the values and on simplification, we get

$$\sin 2\alpha = -\frac{2a_{12}}{\left\{ (a_{11} - a_{22})^2 + 4a_{12}^2 \right\}^{1/2}}$$

$$\cos 2\alpha = -\frac{a_{11} - a_{22}}{\left\{ (a_{11} - a_{22})^2 + 4a_{12}^2 \right\}^{1/2}}.$$

Using the result

$$a_{12} = a_{21}$$

and

$$\{(a_{11} - a_{22})^2 + 4a_{12}^2\}^{1/2} = \omega_2^2 - \omega_1^2$$

$$\sin 2\alpha = \frac{2a_{12}}{\omega_2^2 - \omega_1^2} \quad \text{and} \quad \cos 2\alpha = \frac{a_{11} - a_{22}}{\omega_2^2 - \omega_1^2}$$

Using these results, we see that the co-efficient of X^2 is

$$\frac{1}{2}a_{11}(1 - \cos 2\alpha) + \frac{1}{2}a_{22}(1 + \cos 2\alpha) + a_{12}\sin 2\alpha = \omega_1^2$$

Similarly, co-efficient of Y^2 is

$$\frac{1}{2}a_{11}(1 + \cos 2\alpha) + \frac{1}{2}a_{22}(1 - \cos 2\alpha) + a_{12}\sin 2\alpha = \omega_2^2$$

and co-efficient of XY is

$$(a_{11} - a_{22})\sin 2\alpha + 2a_{12}\cos 2\alpha = 0$$

Hence

$$a_{11}x^2 + a_{22}y^2 + 2a_{12}xy = \omega_1^2 X^2 + \omega_2^2 Y^2$$

and the expression for energy now becomes

$$E = \frac{1}{2}(\dot{X}^2 + \dot{Y}^2 + \omega_1^2 X^2 + \omega_2^2 Y^2) \qquad (3.23)$$

We notice that now there is no XY term. Now, since

$$X = c_1 \cos(\omega_1 t + \phi_1)$$

we have,

$$\dot{X}^2 + \omega_1^2 X^2 = c_1 \omega_1^2$$

Similarly,

$$\dot{Y}^2 + \omega_2^2 Y^2 = c_2 \omega_2^2$$

and Eq. (3.23) now becomes

$$E = \frac{1}{2} c_1 \omega_1^2 + \frac{1}{2} c_2 \omega_2^2 \qquad (3.24)$$

We know that dimensionally $c_1 = \sqrt{m_a} X$ amplitude of mass m_a in mode 1 and $c_2 = \sqrt{m_b} X$ amplitude of mass m_b in mode 2. The Eq. (3.24) now simply state that the total energy of the coupled system is simply equal to the sum of the energies of harmonic vibrations along the normal co-ordinates. Thus, the problem of finding the energy of the system simply reduces to that of finding the energy associated with each normal mode. Since the oscillations of the system in a normal mode is harmonic, this indeed is a significant result.

3.7 ELECTRICALLY COUPLED CIRCUITS

If there are two separate circuits, then variation of current in one of them will set up an induced e.m.f. in the other. The magnitude of this induced e.m.f. being dependent on the extent to which the flux produced by the first circuit called the *primary circuit* is linked with the second circuit called the *secondary circuit*. Here, we will consider two ways of coupling two LC circuits.

(a) Capacitive Coupling

In Fig. 3.9, two L–C circuits are shown. The resistance of the circuits is neglected. For simplicity, here we have assumed the identical values of inductance L and capacitance C. The two circuits are coupled by means of a capacitor whose value is also taken to be C. This coupling capacitor is connected in the middle. Let Q_1, Q_2 and Q_3 be the charges on the capacitors as shown in Fig. 3.9 at any time t. If I_a and I_b be the instantaneous values of the currents in two circuits, then e.m.f. across the left hand inductance is $L\dfrac{dI_a}{dt}$. A charge Q_1 on the capacitor in this circuit causes a potential difference of $-\dfrac{Q_1}{C}$ which tries to decrease I_a and a charge Q_3 on the coupling capacitor produces a potential difference $\dfrac{Q_3}{C}$ which tries to increase the value of I_a. Thus,

$$L\frac{dI_a}{dt} = -\frac{Q_1}{C} + \frac{Q_3}{C}.$$

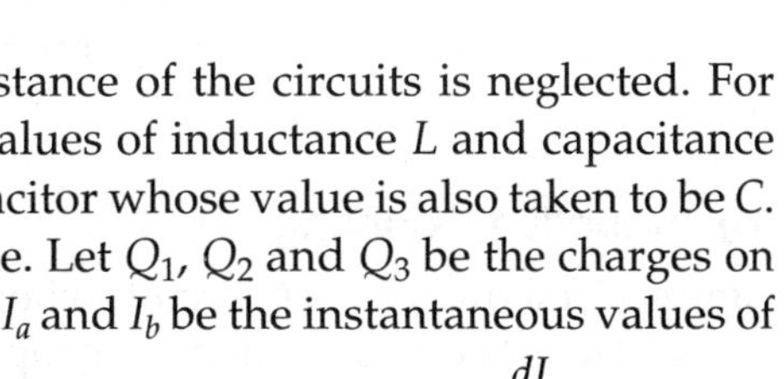

Fig. 3.9

Similarly,

$$L\frac{dI_b}{dt} = -\frac{Q_2}{C} - \frac{Q_3}{C}$$

differentiating these equations w.r.t. time, we have

$$L\frac{d^2I_a}{dt^2} = -\frac{1}{C}\frac{dQ_1}{dt} + \frac{1}{C}\frac{dQ_3}{dt}$$

and

$$L\frac{d^2I_b}{dt^2} = -\frac{1}{C}\frac{dQ_2}{dt} - \frac{1}{C}\frac{dQ_3}{dt}$$

Since charge is conserved, hence $\frac{dQ_1}{dt} = I_a$ and $\frac{dQ_2}{dt} = I_b$ and $\frac{dQ_3}{dt} = I_b - I_a$. Hence, we have the two equations in terms of currents as

$$L\frac{d^2I_a}{dt^2} = -\frac{1}{C}I_a + \frac{1}{C}(I_b - I_a) \tag{3.25}$$

and

$$L\frac{d^2I_b}{dt^2} = -\frac{1}{C}I_b - \frac{1}{C}(I_b - I_a) \tag{3.25a}$$

If we replace L by m; $\frac{I}{C}$ by k and I by x these equations become identical to equations of motions of two coupled masses (Fig. 3.6 Art. 3.5). Hence the two normal modes of the system are

$$\text{mode 1} \quad I_b = I_a \quad \text{and} \quad \omega_1 = \sqrt{\frac{1}{LC}}$$

$$\text{and} \quad \text{mode 2} \quad I_b = -I_a \quad \text{and} \quad \omega_2 = \sqrt{\frac{3}{LC}}$$

Notice that in mode 1, the middle capacitor (coupling capacitor) has no charge i.e. $Q_3 = 0$ and it could be removed without affecting the current in the circuits which are always equal. In mode 2, the current in one circuit is equal and opposite to that in the other and the charge Q_3 is twice in magnitude compared to either Q_1 or Q_2 which are equal.

(b) Inductive Coupling

Figure 3.10 shows two L-C circuits whose resistance is neglected. The two circuits are coupled through mutual inductance of two coils L_1 and L_2. When current changes, the magnetic flux linked with coils changes producing induced e.m.f. in both circuits. What happens in one circuit affects the second circuit and the two circuits are said to be coupled. This is called *inductive coupling*. Let L_1 and L_2 be the inductances of the circuits and c_1 and c_2 their capacitance. If M is the mutual inductance then the strength of the coupling is measured by the *coupling co-efficient* μ, which is defined as

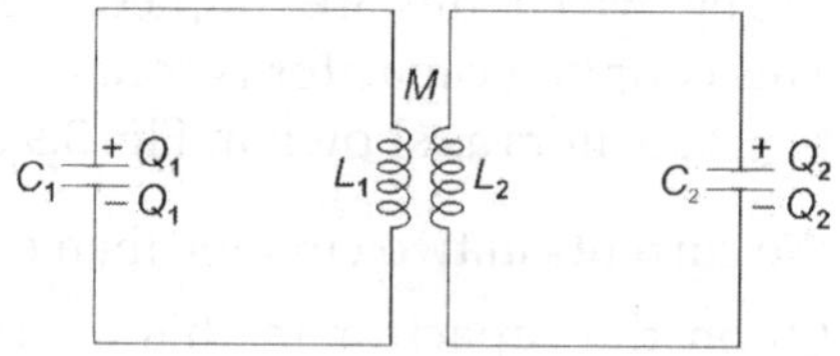

$$\mu = \frac{M}{\sqrt{L_1 L_2}}$$

Fig. 3.10

We shall now obtain the equation of motion of charges in each circuit. Let Q_1 and Q_2 be the charges on capacitors c_1 and c_2 respectively at any instant of time. Let I_a and I_b be the instantaneous values of the currents in the two circuits. The e.m.f. across inductance L_1 is $L_1 \dfrac{dI_a}{dt}$. A charge Q_1 on capacitor c_1 produces a potential difference $-\dfrac{Q_1}{c_1}$ which tends to decrease I_a. The e.m.f. produced in this circuit due to the current I_b in the other circuit is $M \dfrac{dI_b}{dt}$, which tends to increase I_a. Thus, the equation governing the balance of voltages in the circuit involving c_1 and L_1 is

$$L_1 \frac{dI_a}{dt} = -\frac{Q_1}{c_1} + M \frac{dI_b}{dt}$$

and differentiating with respect to time and setting $\dfrac{dQ_1}{dt} = I_a$, we have

$$\frac{d^2 I_a}{dt} = -\omega'^2 I_a + \frac{M}{L_2} \frac{d^2 I_b}{dt^2} \tag{3.26}$$

when $\omega' = \dfrac{1}{\sqrt{L_1 c_1}}$ is the natural angular frequency of the circuit.

Similarly, the circuit involving L_2 and c_2, we have

$$\frac{d^2 I_b}{dt^2} = -\omega''^2 I_b + \frac{M}{L_2} \frac{d^2 I_a}{dt^2} \tag{3.26a}$$

where $\omega'' = \dfrac{1}{\sqrt{L_2 c_2}}$ is the natural angular frequency of the second circuit.

Equations (3.26) and (3.26a) are the two coupled equations. In order to obtain the normal modes of current oscillations, we assume as before that, there exists a normal mode at angular frequency ω and phase constant ϕ. Therefore

$$I_a = A\cos(\omega t + \phi) \quad \text{and} \quad I_b = B\cos(\omega t + \phi)$$

giving
$$\frac{d^2 I_a}{dt^2} = -\omega^2 I_0 \quad \text{and} \quad \frac{d^2 I_b}{dt^2} = -\omega^2 I_b$$

Substituting in Eq. (3.26), we have

$$\frac{I_b}{I_a} = \frac{L_1}{M} \frac{(\omega^2 - \omega'^2)}{\omega^2}$$

and in Eq. (3.26a), we have

$$\frac{I_b}{I_a} = \frac{M}{L_2} \frac{\omega^2}{(\omega^2 - \omega''^2)}$$

Since left hand sides of the above equations are equal, hence, right hand sides must be same. Equating them, we have

$$(\omega^2 - \omega'^2)(\omega^2 - \omega''^2) = \frac{M^2}{L_1 L_2} \omega^4 = \mu^2 \omega^4 \tag{3.27}$$

when μ is the coupling co-efficient. This is a quadratic equation in ω^2, the roots of this equation give the normal mode frequencies. If the two circuits were identical i.e., $L_1 = L_2 = L$ and $c_1 = c_2 = c$, then

$$\omega' = \omega'' = \omega_0 = \frac{1}{\sqrt{LC}}$$

Under this condition, Eq. (3.27) reduces to

$$(\omega^2 - \omega_0^2)^2 = \mu^2\omega^4$$

or $\qquad \omega^2 - \omega_0^2 = \pm\mu\omega^2$

Since frequency cannot be negative, hence, the two allowed frequencies are

$$\omega_1 = \frac{\omega_0}{\sqrt{1+\mu}} \quad \text{and} \quad \omega_2 = \frac{\omega_0}{\sqrt{1-\mu}}$$

These are the angular frequencies of the normal modes of identical L-C circuits coupled inductively. If the coupling is very weak ($\mu \to 0$), $\omega_1 = \omega_2 = \omega_0$, the natural frequency of either circuit. It is clear that in mode 1 with angular frequency ω_1, $I_b = -I_a$, i.e. the currents in these two circuits are always equal and in opposite directions. However, in mode 2 with angular frequency ω_2, $I_b = I_a$, i.e. currents in the two circuits are always equal and in the same direction.

3.8 A FORCED COUPLED OSCILLATOR

Let us consider a system of two identical pendulums A and B each having a mass m suspended on a light rigid rod of length l and coupled by a light spring of stiffness constant k (Fig. 3.11). Suppose a harmonic driving force $F_0 \cos \omega t$ is applied to pendulum A while the motion of pendulum B is controlled only by its restoring force and the coupling spring. Let x_a and x_b be the displacements of pendulums A and B at any instant $x_a > x_b$. The spring is stretched by an amount $x_a - x_b$, the net force on A is

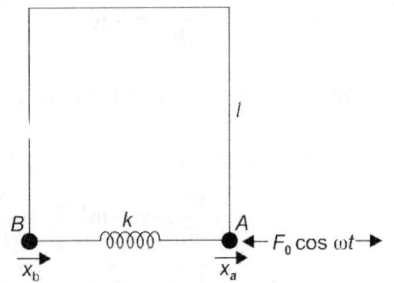

Fig. 3.11

$$-\frac{mg}{l}x_a - k(x_a - x_b) + F_0 \cos \omega t$$

and that on B is

$$-\frac{mg}{l}x_b + k(x_a - x_b).$$

The equations of motion of two are

$$m\frac{d^2x_a}{dt^2} = -\frac{mg}{l}x_a - k(x_a - x_b) + F_0 \cos \omega t$$

$$m\frac{d^2x_b}{dt^2} = -\frac{mg}{l}x_b + k(x_a - x_b)$$

writing $\dfrac{g}{l} = \omega_0^2$, $\dfrac{k}{m} = \omega_c^2$, we have

$$\frac{d^2x_a}{dt^2} + (\omega_0^2 + \omega_c^2)x_a - \omega_c^2 x_b = \frac{F_0}{m}\cos \omega t \qquad (3.28)$$

and $\qquad \dfrac{d^2x_b}{dt^2} + (\omega_0^2 + \omega_c^2)x_b - \omega_c^2 x_a = 0. \qquad (3.28a)$

Adding the two equations of motion, we have

$$\frac{d^2}{dt^2}(x_a + x_b) + \omega_0^2(x_a + x_b) = \frac{F_0}{m}\cos \omega t \qquad (3.29)$$

and subtracting Eq. (3.28a) from (3.28), we have

$$\frac{d^2}{dt^2}(x_a - x_b) + (\omega_0^2 + 2\omega_c^2)(x_a - x_b) = \frac{F_0}{m}\cos\omega t \qquad (3.29a)$$

Introducing normal co-ordinates $X = x_a + x_b$ and $Y = x_a - x_b$, we have

$$\frac{d^2X}{dt^2} + \omega_0^2 X = \frac{F_0}{m}\cos\omega t$$

and

$$\frac{d^2Y}{dt^2} + \omega'^2 Y = \frac{F_0}{m}\cos\omega t$$

where $\omega'^2 = \omega_0^2 + 2\omega_c^2$. These are the equations of undamped forced oscillations of natural frequencies ω_0 and ω'. Frequencies ω_0 and ω' are the normal mode frequencies of the system.

The steady state solutions of these equations are

$$X = C\cos\omega t \quad \text{and} \quad Y = D\cos\omega t$$

where $C = \dfrac{\frac{F_0}{m}}{\omega_0^2 - \omega^2}$ and $D = \dfrac{\frac{F_0}{m}}{\omega'^2 - \omega^2}$

The amplitudes C and D show resonances just like a single oscillator. In terms of original co-ordinates, the solutions are

$$x_a = \frac{1}{2}(X + Y) = \frac{1}{2}(C + D)\cos\omega t$$

and

$$x_b = \frac{1}{2}(X - Y) = \frac{1}{2}(C - D)\cos\omega t$$

The individual amplitudes of the two pendulums are thus given by

$$A_1 = \frac{1}{2}(C + D) = \frac{F_0}{m}\frac{\frac{1}{2}(\omega_0^2 + \omega'^2) - \omega^2}{(\omega_0^2 - \omega^2)(\omega'^2 - \omega^2)}$$

But

$$\omega'^2 = \omega_0^2 + 2\omega_c^2$$

$\therefore$

$$A_1 = \frac{F_0}{m}\frac{(\omega_0^2 + \omega_c^2) - \omega^2}{(\omega_0^2 - \omega^2)(\omega'^2 - \omega^2)}$$

$$= \frac{F_0}{m}\frac{(\omega_0^2 + \omega_c^2) - \omega^2}{(\omega_0^2 - \omega^2)(\omega_0^2 + 2\omega_c^2 - \omega^2)}$$

Similarly,

$$A_2 = \frac{1}{2}(C - D) = \frac{F_0}{m}\frac{\frac{1}{2}(\omega'^2 - \omega_0^2)}{(\omega_0^2 - \omega^2)(\omega'^2 - \omega^2)}$$

$$= \frac{F_0}{m}\frac{\omega_c^2}{(\omega_0^2 - \omega^2)(\omega_0^2 + 2\omega_c^2 - \omega^2)}$$

The variation of A_1 and A_2, the (forced) amplitudes of the two coupled pendulums with driving frequency ω are shown in Fig. 3.12(a) and (b) respectively. For frequencies up to lower resonance ($\omega_0 = \omega$), A_1 and A_2 are always of the same sign i.e. the pendulum oscillate in same phase. For frequencies beyond the higher resonance ($\omega = \omega'$), A_1 and A_2 are of opposite sign i.e. the pendulums oscillate 180° out of phase. At a certain frequency

$\omega_1 = \left[\left(\omega_0^2 + \omega'^2\right)^{1/2}\right]$ between the resonances $A_1 = 0$ and A_2 is non-zero. ω_1 is precisely the natural frequency of a single pendulum with coupling spring attached and the other pendulum held fixed.

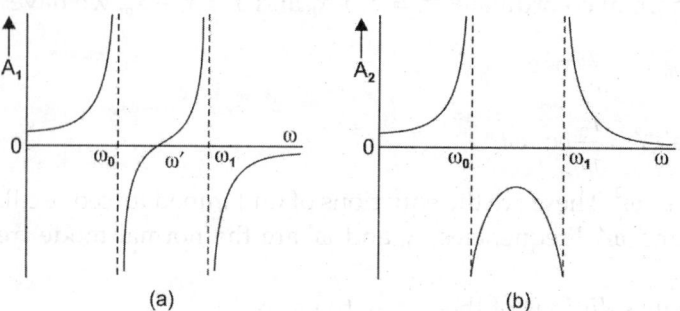

(a)   (b)

Fig. 3.12

3.9 EXAMPLES OF FORCED COUPLED OSCILLATORS

(a) Inductively Coupled Circuits

Consider the general case of an alternating supply e.m.f. represented by $E = E_0 \cos \omega t$ across the primary of a mutual inductance M, where the primary coil has inductance L_1 and resistance R_1, whereas the secondary is a closed circuit of inductance L_2 and resistance R_2 (Fig. 3.13).

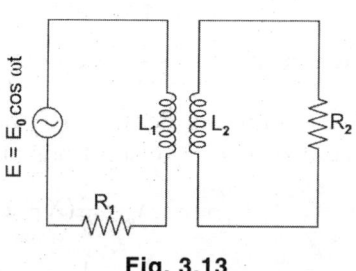

Fig. 3.13

The e.m.f. equations for the primary and secondary circuits are then respectively.

$$E_0 \cos \omega t = L_1 \frac{dI_1}{dt} + M \frac{dI_2}{dt} + I_1 R_1 \tag{3.30}$$

and

$$0 = L_2 \frac{dI_2}{dt} + M \frac{dI_1}{dt} + I_2 R_2 \tag{3.30(a)}$$

where I_1 and I_2 are the instantaneous primary and secondary currents respectively at time t.

Since $E_0 \cos \omega t$ is the real part of $E_0 e^{i\omega t}$, consequently Eq. (3.30) can be written in the form

$$\left(E_0 e^{i\omega t}\right)_{real} = L_1 \frac{dI_1}{dt} + M \frac{dI_2}{dt} + I_1 R_1$$

In order to find the solution for the instantaneous current I_1 and I_2 which are of the same frequency as the supply, let us put $I_1 = $ real part of $I_0 e^{i\omega t}$ and $I_2 = $ real part of $i_0 e^{i\omega t}$, where I_0 and i_0 are the peak primary and secondary currents respectively. Then

$$\frac{dI_1}{dt} = i\omega I_0 e^{i\omega t} \quad \text{and} \quad \frac{dI_2}{dt} = i\omega i_0 e^{i\omega t}.$$

Substituting the values in emfs equations, we have

$$E_0 e^{i\omega t} = i\omega I_0 L_1 e^{i\omega t} + i\omega M i_0 e^{i\omega t} + R_1 I_0 e^{i\omega t} \tag{3.31}$$

and

$$0 = i\omega i_0 L_2 e^{i\omega t} + i\omega M I_0 e^{i\omega t} + R_2 i_0 e^{i\omega t} \tag{3.31a}$$

$$\therefore \qquad i_0 = \frac{i\omega M I_0}{R_2 + i\omega L_2}$$

Putting the values of i_0 in Eq. (3.31), we have

$$E_0 = i\omega L_1 I_0 - \frac{i^2 \omega^2 M^2 I_0}{R_2 + i\omega L_2} + I_0 R_1$$

and rationalising, we have

$$E_0 = i\omega L_1 I_0 + \frac{M^2 \omega^2 (R_2 - i\omega L_2) I_0}{(R_2^2 + \omega^2 L_2^2)} + I_0 R_1$$

or

$$I_0 = \frac{E_0}{\left\{ R_1 + \dfrac{M^2 \omega^2 R_2}{(R_2^2 + \omega^2 L_2^2)} \right\} + i\omega \left\{ L_1 - \dfrac{M^2 \omega^2 L_2}{(R_2^2 + \omega^2 L_2^2)} \right\}}$$

$$= \frac{E_0}{R_1' + i\omega L_1'} \tag{3.33}$$

where

$$R_1' = R_1 + \frac{M^2 \omega^2 R_2}{(R_2^2 + \omega^2 L_2^2)} \quad \text{and} \quad L_1' = L_1 - \frac{M^2 \omega^2 L_2}{(R_2^2 + \omega^2 L_2^2)}$$

R_1' being the effective primary resistance, which is greater by $\dfrac{\omega^2 M^2 R_2}{R_2^2 + \omega^2 L_2^2}$ than R_1 due to so called *reflected resistance* from the secondary and L_1' being the effective primary inductance which is less than L_1 by $\dfrac{\omega^2 M^2 L_2}{R_2^2 + \omega^2 L_2^2}$ due to reflected inductance from the secondary.

Substituting the values of I_0 in equation $I_1 = i_0 e^{i\omega t}$, we have

$$I_1 = \frac{E_0}{R_1' + i\omega L_1'} e^{i\omega t} \tag{3.34}$$

But

$$R_1' + i\omega L_1' = \sqrt{R_1'^2 + \omega^2 L_1'^2} (\cos\theta + i\sin\theta) = \sqrt{R_1'^2 + \omega^2 L_1'^2} e^{i\theta}$$

where

$$\theta = \tan^{-1} \frac{\omega L_1'}{R_1'}$$

$\therefore$ equation (3.34) becomes

$$I_1 = \frac{E_0}{\sqrt{R_1'^2 + \omega^2 L_1'^2}} \cdot \frac{e^{i\omega t}}{e^{i\theta}} = \frac{E_0}{\sqrt{R_1'^2 + \omega^2 L_1'^2}} e^{i(\omega t - \theta)}$$

$$= \frac{E_0}{Z_1'} e^{i(\omega t - \theta)} \tag{3.34a}$$

where Z_1' is the magnitude of the effective primary impedence whilst the peak primary current

$$I_0 = \frac{E_0}{Z_1'} e^{-i\theta} \tag{3.35}$$

whereas the phase lag of current on voltage for primary circuit alone is $\tan^{-1} \dfrac{\omega L_1}{R_1}$, this

phase difference becomes $\tan^{-1} \dfrac{\omega L_1'}{R_1'}$ on coupling to the secondary circuit. Since

$R_1' > R_1$ and $L_1' < L_1$, so the phase difference is reduced when such coupling takes place. On the other hand, the cosine of the phase angle becomes larger as the angle becomes smaller. Therefore the power factor, and so the power taken from the AC source increases on coupling the secondary to the primary.

The peak current in the secondary is

$$i_0 = \frac{i\omega M I_0}{R_2 + i\omega L_2} = -\frac{i\omega I_0 M (R_2 - i\omega L_2)}{R_2^2 + \omega^2 L_2^2}$$

$$= -\frac{i\omega M I_0}{R_2^2 + i\omega^2 L_2^2} \sqrt{R_2^2 + \omega^2 L_2^2} \, (\cos\alpha - i\sin\alpha)$$

$$= -\frac{i\omega M I_0}{\sqrt{R_2^2 + \omega^2 L_2^2}} e^{-i\alpha} \tag{3.36}$$

where

$$\tan\alpha = \frac{\omega L_2}{R_2}$$

$\therefore$

$$i_0 = \frac{\omega M I_0}{\sqrt{R_2^2 + \omega^2 L_2^2}} e^{-i\left(\frac{\pi}{2} + \alpha\right)} = \frac{\omega M I_0}{Z_2} e^{-i\left(\frac{\pi}{2} + \alpha\right)} \tag{3.37}$$

Since $-i = \cos\dfrac{\pi}{2} - i\sin\dfrac{\pi}{2} = e^{\frac{-i\pi}{2}}$ and $Z_2 = \sqrt{R_2^2 + \omega^2 L_2^2}$ the impedance of secondary. Substituting the values of I_0 from Eq. (3.35), we have

$$i_0 = \frac{\omega M E_0}{Z_1' Z_2} e^{-i\left(\frac{\pi}{2} + \alpha + \theta\right)}$$

and the instantaneous secondary current I_2 at any instant is given by

$$I_2 = i_0 e^{i\omega t} = \frac{\omega M E_0}{Z_1' Z_2} e^{i\left\{\omega t - \theta - \left(\alpha + \frac{\pi}{2}\right)\right\}}. \tag{3.38}$$

Therefore, the secondary current lags in phase by an angle of $\left(\alpha + \dfrac{\pi}{2}\right)$ on primary current.

(b) Coupled LCR Circuit

As a preamble to the study of transmission lines, which may be visualised as a series of coupled electrical oscillators with the same values of inductance and capacitance, we consider energy transfer between two electrical circuits which are coupled through mutual inductance.

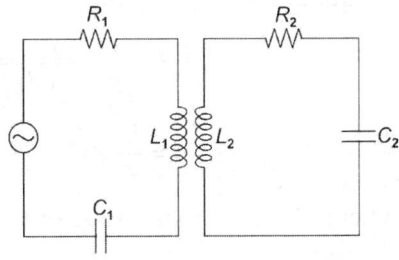

Fig. 3.14

Let us consider coupled LCR circuit (Fig. 3.14), the e.m.f. equations for the primary and secondary circuits are

$$E_0 = i\omega L_1 I_0 + i\omega M i_0 + I_0 R_1 - \frac{i I_0}{\omega C_1} \tag{3.39}$$

and

$$0 = i\omega L_2 i_0 + i\omega M I_0 + i_0 R_2 - \frac{i i_0}{\omega C_2} \tag{3.39a}$$

where E_0 is the peak value of primary e.m.f. and I_0 and i_0 are peak values of current in primary and secondary circuits respectively. Putting

$$X_1 = \omega L_1 - \frac{1}{\omega C_1} \quad \text{and} \quad X_2 = \omega L_2 - \frac{1}{\omega C_2}$$

for the total primary and secondary reactances respectively, we have

$$i_0 = -\frac{i\omega M I_0}{R_2 + i x_2} = -\frac{i\omega M I_0 (R_2 - i x_2)}{R_2^2 + x_2^2}$$

$$= \frac{i\omega M I_0 (R_2 - i x_2)}{Z_2^2} \tag{3.40}$$

where Z_2 is the impedence of of the secondary. Substituting i_0 in Eq. (3.39), we have

$$E_0 = I_0 R_1 + i I_0 x_1 + \frac{\omega^2 M^2 I_0 (R_2 - i x_2)}{Z_2^2}$$

$$\therefore \quad Z_1' = \frac{E_0}{I_0} = \left\{ R_1 + \frac{\omega^2 M^2}{Z_2^2} R_2 \right\} + i \left\{ x_1 - \frac{\omega^2 M^2}{Z_2^2} X^2 \right\}$$

Here Z_1' is called *effective primary impedance* and has magnitude

$$|Z_1'| = \sqrt{ \left\{ R_1 + \frac{\omega^2 M^2}{Z_2^2} R_2 \right\}^2 + \left\{ X_1 - \frac{\omega^2 M^2}{Z_2^2} X_2 \right\}^2 }$$

Here effective primary resistance $R_1' = R_1 + \dfrac{\omega^2 M^2}{Z_2^2} R_2$, which is greater by $\dfrac{\omega^2 M^2 R^2}{Z_2^2}$, than R_1 due to the so called *reflected resistance* from the secondary and the effective reactance $X_1' = X_1 - \dfrac{\omega^2 M^2}{Z_2^2} X_2$, which is less than X_1 by $\dfrac{\omega^2 M^2}{Z_2^2} X_2$ due to reflected reactance from secondary and $Z_r = \dfrac{\omega^2 M^2}{Z_2^2} (R_2 - i X_2)$ is called the *reflected impedance* from the secondary.

Substituting the value of I_0 in Eq. (3.40), we have

$$i_0 = -\frac{i\omega M E_0}{Z_1' Z_2^2} (R_2 - i X_2)$$

and its magnitude is

$$|i_0| = \frac{\omega M E_0}{Z' Z_2^2} \sqrt{R_2^2 + X_2^2} = \frac{\omega M E_0}{Z_1' Z_2^2} Z_2 = \frac{\omega M E_0}{Z_1' Z_2} \tag{3.41}$$

If the source of supply to the primary has its frequency $f = \dfrac{\omega}{2\pi}$ varied continuously from zero to high value, the secondary current i_0 and so the output e.m.f. across C_2 will be a maximum when the denominator of Eq. (3.41) is minimum. The circuit will be then at resonance, with supply. The numerator will increase steadily with ω and so is not related to the resonance.

Thus, the resonance occurs when $z_1'z_2$ is minimum. Substituting the magnitude of z_1' and putting $z_2 = \sqrt{R_2^2 + x_2^2}$, we have magnitude of $z_1'z_2$ is

$$|z_1'z_2| = \sqrt{\left\{\left(R_1 + \frac{\omega^2 M^2}{Z_2^2}R^2\right)^2 + \left(X_1 - \frac{\omega^2 M^2}{Z_2^2}r_2\right)^2\right\}(R_2^2 + X_2^2)}$$

$$= \sqrt{(R_1^2 + X_1^2)(R_2^2 + X_2^2) + 2\omega^2 M^2(R_1 R_2 - X_1 X_2) + \omega^4 M^4}$$

The reactive opposition in the expression for $Z_1'Z_2$ vanishes when

$$X_1^2 X_2^2 - 2\omega^2 M^2 X_1 X_2 + \omega^4 M^4 = (X_1 X_2 - \omega^2 M^2)^2$$

Here, the resistances R_1 and R_2 are neglected because in practice R_1 and R_2 are small compared with reactances X_1 and X_2.

$$\therefore \qquad X_1 X_2 = \omega^2 M^2$$

or

$$\omega^2 = \frac{X_1 X_2}{M^2} = \frac{\left(\omega L_1 - \dfrac{1}{\omega C_1}\right)\left(\omega L_2 - \dfrac{1}{\omega C_2}\right)}{M^2}$$

or

$$\omega^4(M^2 - L_1 L_2) + \omega^2\left(\frac{L_1}{C_2} + \frac{L_2}{C_1}\right) - \frac{1}{C_1 C_2} = 0$$

and

$$\omega^2 = -\frac{\left(\dfrac{L_1}{C_1} + \dfrac{L_2}{C_2}\right) \pm \sqrt{\left(\dfrac{L_1}{C_1} + \dfrac{L_2}{C_2}\right)^2 + \dfrac{4(M^2 - L_1 L_2)}{C_1 C_2}}}{2(M^2 - L_1 L_2)}$$

A commonly encountered case is when both primary and secondary circuits have the same inductance L and capacitance C which gives coupled tuned circuit, the resonance frequency of which are given by putting $L_1 = L_2 = L$ and $C_1 = C_2 = C$.

$$\therefore \qquad \omega^2 = \frac{-\dfrac{L}{C} \pm \sqrt{\dfrac{M^2}{C^2}}}{M^2 - L^2} = \frac{-\dfrac{L}{C} \pm \dfrac{M}{C}}{(M^2 - L^2)}$$

This is usually expressed in terms of the coupling co-efficient $\mu = \dfrac{M}{\sqrt{L_1 L_2}} = \dfrac{M}{L}$ and $\omega_0 = 2\pi f_0 = \dfrac{1}{\sqrt{LC}}$ where f_0 is the resonance frequency of either tuned circuit considered separately

$$\omega^2 = \frac{-\dfrac{L}{c} \pm \dfrac{\mu L}{c}}{\mu^2 L^2 - L^2} = \frac{-\dfrac{1}{LC} \pm \dfrac{\mu}{LC}}{\mu^2 - 1}$$

$$= \frac{-\omega_0^2 \pm \mu\omega_0^2}{\mu^2 - 1} = \frac{\omega_0^2(1 \pm \mu)}{(1 - \mu^2)} = \frac{\omega_0^2}{(1 \pm \mu)}$$

$$\therefore \qquad f = \sqrt{\frac{f_0}{\sqrt{1 \pm \mu}}} \qquad\qquad (3.42)$$

Therefore, there are two resonant frequencies. The graph for secondary current against supply frequency is shown in Fig. 3.15. It exhibits two peaks—a phenomenon termed "double humped" tuning. These two frequencies will be separated by an amount which increases with μ. If μ is large, tight coupling is experienced and when μ is small, loose coupling occurs. The optimum coupling is one when the secondary current is a maximum. Let us find the condition for optimum coupling. We have,

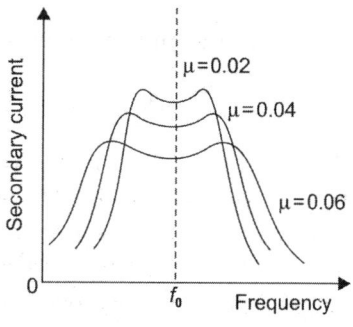

Fig. 3.15

$$|i_0| = \frac{\omega M E_0}{Z_1' Z_2} = \frac{\omega M E_0}{\left\{\left(R_1 + \frac{\omega^2 M^2}{Z_2^2} R_2\right)^2 + \left(X_1 - \frac{\omega^2 M^2}{Z_2^2} X_2^2\right)^2\right\}^{\frac{1}{2}} Z_2}$$

where $Z_2 = \sqrt{R_2^2 + X_2^2}$

If two circuits are at resonance then $X_1 = X_2 = 0$.

$\therefore$
$$i_0 = \frac{\omega M E_0}{\left(R_1 + \frac{\omega^2 M^2}{R_2}\right) R_2}$$

This peak secondary current will be maximum when the mutual inductance M is in accordance with $\frac{di_0}{dM} = 0$, i.e.

$$\frac{d}{M}\left[\frac{M}{R_1 R_2 + \omega^2 M^2}\right] = 0$$

or
$$\frac{R_1 R_2 + \omega^2 M^2 - M(2\omega^2 M)}{(R_1 R_2 + \omega^2 M^2)^2} = 0$$

$\therefore$
$$R_1 R_2 = \omega^2 M^2 \quad \text{or} \quad \omega M = \sqrt{R_1 R_2}.$$

This gives the value of M for optimum coupling between two tuned circuits. The frequency separation is given by

$$\frac{f_0}{\sqrt{1-\mu}} - \frac{f_0}{\sqrt{1+\mu}} = f_0\left\{1 + \frac{1}{2}\mu - \left(1 - \frac{1}{2}\mu\right)\right\} = \mu f_0 \quad \text{for } \mu \ll 1.$$

This decides the width of the band of frequencies between the two resonant frequencies. Since the fall of secondary current between these two frequencies is not great. If μ is small, *band pass circuit* is of great value in radio communication which is obtained for small μ.

Example 3.1: Two equal masses m are connected with two identical massless springs of spring constant K as shown in Fig. 3.16. Show that the angular frequencies of the two normal modes of vertical oscillations are given by $\omega^2 = (3 \pm \sqrt{5})\dfrac{k}{2m}$. Also show that in the slower mode, the

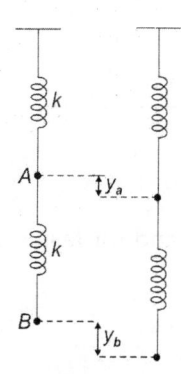

Fig. 3.16

ratio of the amplitude of mass A to that of mass B is $\frac{1}{2}(\sqrt{5}-1)$, while in faster mode, this ratio is $\frac{1}{2}(\sqrt{5}+1)$.

Solution: From Fig. 3.16, it is clear that the restoring force on mass A is $-ky_a + k(y_b - y_a)$. Here, y_a and y_b are the displacement of masses at a certain instant of time. Here, the gravitational forces acting on the two masses at a certain instant are not taken into account as they are independent of the displacements and hence, do not contribute to the restoring forces responsible for the oscillations. The equations of motion are

$$m\ddot{y}_a = -ky_a + k(y_b - y_a)$$

or

$$\ddot{y}_a = -\frac{2k}{m}y_a + \frac{k}{m}y_b \qquad (i)$$

and

$$\ddot{y}_b = -\frac{k}{m}y_b + \frac{k}{m}y_a. \qquad (ii)$$

Normal modes with frequency ω and phase constant ϕ are

$$y_a = A\cos(\omega t + \phi) \quad \text{and} \quad y_b = B\cos(\omega t + \phi)$$

where A and B are the amplitudes of the two masses. Using these values in (i) and (ii), we have

$$\frac{y_a}{y_b} = \frac{A}{B} = \frac{\dfrac{k}{m}}{2\dfrac{k}{m} - \omega^2} = \frac{\dfrac{k}{m} - \omega^2}{\dfrac{k}{m}}$$

or

$$\left(\frac{2k}{m} - \omega^2\right)\left(\frac{k}{m} - \omega^2\right) = \frac{k^2}{m}$$

or

$$\omega^4 - \frac{3k}{m}\omega^2 + \frac{k^2}{m^2} = 0$$

The angular frequencies of the normal modes are the two roots of this equation which are given by

$$\omega^2 = (3 \pm \sqrt{5})\frac{k}{2m}$$

Now, to find the ratio of the amplitudes of the two modes, we substitute $\omega^2 = (3 - \sqrt{5})\dfrac{k}{2m}$ for slower mode. The ratio for this mode is

$$\frac{A}{B} = \frac{\dfrac{k}{m} - (3 - \sqrt{5})\dfrac{k}{2m}}{\dfrac{k}{m}} = \frac{1}{2}(\sqrt{5} - 1)$$

and for faster mode, we substitute $\omega^2 = (3 + \sqrt{5})\dfrac{k}{2m}$, and the ratio is

$$\frac{A}{B} = \frac{\dfrac{k}{m} - (3 + \sqrt{5})\dfrac{k}{2m}}{\dfrac{k}{m}} = -\frac{1}{2}(\sqrt{5} + 1).$$

The minus sign indicates that this is out of phase mode. The displacements y_a and y_b are oppositely directed. The amplitude ratio of these displacement is just $\frac{1}{2}(\sqrt{5}+1)$.

Example 3.2: Two pendulums are suspended one below the other to form a double pendulum as shown in Fig. 3.17. If $m_1 = m_2 = m$ and $l_1 = l_2 = l$ show that the frequencies of the two normal modes for small oscillations are given by $\omega^2 = (2 \pm \sqrt{2})\frac{g}{l}$.

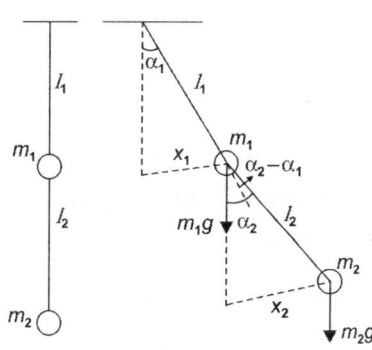

Fig. 3.17

Solution: Referring to Fig. 3.17, the equations of motion of the two pendulums are

$$m_1 \ddot{x}_1 = -m_1 g \sin\alpha_1 + m_2 g \sin(\alpha_2 - \alpha_1)$$

$$m_2 \ddot{x}_2 = -m_2 g \sin\alpha_2 + m_1 \ddot{x}_1 \cos(\alpha_2 - \alpha_1)$$

For small displacements

$$\sin\alpha_1 = \alpha_1 = \frac{x_1}{l_1}, \sin\alpha_2 = \alpha_2 = \frac{x_2}{l_2}$$

and

$$\sin(\alpha_2 - \alpha_1) = \alpha_2 - \alpha_1 = \frac{x_2}{l_2} - \frac{x_1}{l_1}$$

and

$$\cos(\alpha_2 - \alpha_1) = 1$$

For small displacements and $m_1 = m_2 = m$, $l_1 = l_2 = l$, the equations of motion become

$$\ddot{x}_1 = -\frac{g}{l}x_1 + \frac{g}{l}(x_2 - x_1)$$

and

$$\ddot{x}_2 = -\frac{g}{l}x_2 - \ddot{x}_1$$

The coupling between these equations is not clear as the equations are not symmetric. Symmetry can be achieved if we replace the first equation by sum of the two equations. In the symmetric form, the equations can be written as

$$\ddot{x}_1 = -\frac{g}{l}x_1 - \frac{1}{2}\ddot{x}_2 \tag{i}$$

and

$$\ddot{x}_2 = -\frac{g}{l}x_2 - \ddot{x}_1 \tag{ii}$$

The coupling between the two equations is now obvious. The acceleration of one pendulum affects the motion of the other pendulum. Such a coupling is called *inertial coupling*.

In normal mode of frequency ω, we have

$$x_1 = A\cos(\omega t + \phi) \quad \text{and} \quad x_2 = B\cos(\omega t + \phi)$$

Substituting in (i) and (ii), we have

$$\left(\frac{g}{l} - \omega^2\right)A - \frac{1}{2}\omega^2 B = 0$$

and

$$\left(\frac{g}{l} - \omega^2\right)B - \omega^2 A = 0$$

These equations yield a quadratic equation in ω^2 which reads

$$\omega^4 - \left(\frac{4g}{l}\right)\omega^2 + \frac{2g^2}{l^2} = 0$$

and the two roots are

$$\omega^2 = (2 \pm \sqrt{2})\frac{g}{l}$$

These are the two frequencies of the normal modes.

Example 3.3: Determine the normal coordinates of the system shown in Fig. 3.16 (Example 3.1), if $k = 10$ N/m and $m = 1$ kg.

Solution: Referring to example 3.1, the normal mode frequencies are

$$\omega_1 = \left\{(3 - \sqrt{5})\frac{k}{2m}\right\}^{\frac{1}{2}} = 1.95 \text{ rad/s}$$

$$\omega_2 = \left\{(3 + \sqrt{5})\frac{k}{2m}\right\}^{\frac{1}{2}} = 5.12 \text{ rad/s}$$

The general solution of the system is given by

$$y_a = A_1 \cos(\omega_1 t + \phi_1) + A_2 \cos(\omega_2 t + \phi_2)$$
$$y_b = B_1 \cos(\omega_1 t + \phi_1) + B_2 \cos(\omega_2 t + \phi_2)$$

where,

$$\frac{B_1}{A_1} = \frac{\dfrac{k}{m}}{\dfrac{k}{m} - \omega_1^2} = \frac{10}{10 - (1.95)^2} = 1.62$$

and

$$\frac{B_2}{A_2} = \frac{\dfrac{k}{m}}{\dfrac{k}{m} - \omega_2^2} = \frac{10}{10 - (5.12)^2} = -0.62$$

Hence, we have

$$y_a = A_1 \cos(1.95t + \phi_1) + A_2 \cos(5.12t + \phi_2)$$
$$y_b = 1.62A_1 \cos(1.95t + \phi_1) - 0.62A_2 \cos(5.12t + \phi_2)$$

Defining new co-ordinates X and Y such that

$$X = A_1 \cos(1.95t + \phi_1) \quad \text{and} \quad Y = A_2 \cos(5.12t + \phi_2)$$

The motion associated with coordinates X and Y are simple harmonic at frequencies $\omega_1 = 1.95$ rad/s and $\omega_2 = 5.12$ rad/s respectively. Hence, X and Y are two normal co-ordinates of the system. Now, $y_a = X + Y$ and $y_b = 1.62X - 0.62Y$ which gives

$$X = 0.28y_a + 0.45y_b \quad \text{and} \quad Y = 0.72y_a - 0.45y_b$$

Example 3.4: The mass B of the system shown in Fig. (3.16) (Example 3.1) is held at rest at a position 10 cm from its mean position and mass A is at its equilibrium position. The system is then released. Determine the subsequent motion of each mass, if $k = 10$ N/m and $m = 1$ kg.

Solution: The general displacement is given by

$$y_a = A_1 \cos(\omega_1 t + \phi_1) + A_2 \cos(\omega_2 t + \phi_2) \tag{i}$$

$$y_b = B_1 \cos(\omega_1 t + \phi_1) + B_2 \cos(\omega_2 t + \phi_2). \tag{ii}$$

The velocities of the two masses is given by

$$\dot{y}_a = -A_1 \omega_1 \sin(\omega_1 t + \phi_1) - A_2 \omega_2 \sin(\omega_2 t + \phi_2) \tag{iii}$$

$$\dot{y}_b = -B_1 \omega_1 \sin(\omega_1 t + \phi_1) - B_2 \omega_2 \sin(\omega_2 t + \phi_2) \tag{iv}$$

where $B_1 = 1.62 A_1$, $B_2 = -0.62 A_2$, $\omega_1 = 1.95$ rad/s, and $\omega_2 = 5.12$ rad/s. The initial conditions are $y_a = 0$, $y_b = 0.1$ m, $\ddot{y}_a = 0$, $\ddot{y}_b = 0$ at $t = 0$. Using these conditions, we have

$$0 = A_1 \cos\phi_1 + A_2 \cos\phi_2$$

$$0.1 = B_1 \cos\phi_1 + B_2 \cos\phi_2 = 1.62 A_1 \cos\phi_1 - 0.624 A_2 \cos\phi_2$$

$$0 = A_1\omega_1 \sin\phi_1 + A_2\omega_2 \sin\phi_2$$

$$0 = 1.62 A_1 \omega_1 \sin\phi_1 - 0.62 A_2\omega_2 \sin\phi_2$$

Hence

$$\sin\phi_1 = \sin\phi_2 = 0$$

$$\therefore \quad A_1 \cos\phi_1 = 0.446 \approx 0.45$$

$$A_2 \cos\phi_2 = -0.446 \approx 0.45$$

Substituting these values in (i) and (ii), we have

$$y_a = 0.045(\cos 1.95t - \cos 5.12t)$$

and

$$y_b = 0.072(\cos 1.95t + 0.028 \cos 5.12t)$$

Example 3.5: Two resonant circuits are identical, each consisting of a coil having inductance 100 μH and Q-factor of 100, tuned by 200 pF capacitor (the losses of which are negligible).

An e.m.f. of 1 volt is injected into one circuit and the current measured in the other. If the co-efficient of coupling between the circuits is 0.005, find the frequency for which the secondary current is maximum and the value of the primary and secondary currents at this frequency.

Solution: Referring to Fig. 3.14 and Eqs (3.39 and 3.39a) in which $L_1 = L_2 = L$, $C_1 = C_2 = C$ and $R_1 = R_2 = R$, the Eq. (3.39) becomes (for primary circuit)

$$E_0 = i\omega I_0 + i\omega M i_0 + I_0 R - \frac{iI_0}{\omega C}$$

where E_0 is the applied peak e.m.f. The Eq. (3.39a) for the secondary circuit becomes

$$0 = i\omega L i_0 + i\omega M I_0 + I_0 R - \frac{ii_0}{\omega C}$$

where $M = \mu\sqrt{L_1 L_2} = \mu L$. Here μ is the co-efficient of coupling. Since both circuits are at resonance, it follows that

$$\omega L - \frac{1}{\omega C} = 0$$

Then these equations become

$$E_0 = i\omega M i_0 + I_0 R \tag{i}$$

and

$$0 = i\omega M I_0 + i_0 R \tag{ii}$$

$$\therefore \qquad i_0 = -\frac{i\omega M}{R}I_0$$

$$\text{and} \qquad E_0 = \frac{\omega^2 M^2}{R}I_0 + I_0 R \quad \text{or} \quad I_0 = \frac{E_0}{R + \dfrac{\omega^2 M^2}{R}}$$

$$\text{and} \qquad i_0 = -\frac{i\omega M E_0}{R^2 + \omega^2 M^2}$$

This secondary current will be maximum at frequency $= \dfrac{\omega}{2\pi}$ given by $\dfrac{di_0}{d\omega} = 0$

$$\therefore \qquad \frac{d}{d\omega}\left[-\frac{i\omega M E_0}{R^2 + \omega^2 M^2}\right] = 0$$

$$\text{or} \qquad M E_0 \frac{(\omega^2 M^2 + R^2) - \omega \cdot 2\omega M^2}{(\omega^2 M^2 + R^2)^2} = 0$$

$$\therefore \qquad \omega^2 M^2 = R^2 \quad \text{or} \quad \omega^2 = \frac{R^2}{M^2}$$

Putting $M = \mu L = 0.005 \times 100 \times 10^{-6} = 5 \times 10^{-7}$, where Q is given by

$$Q = \frac{\omega L}{R} = \frac{\dfrac{L}{\sqrt{LC}}}{R} = \frac{\sqrt{\dfrac{L}{C}}}{R}$$

$$\text{or} \qquad R = \frac{1}{Q}\sqrt{\frac{L}{C}} = \frac{1}{100}\sqrt{\frac{100 \times 10^{-6}}{200 \times 10^{-12}}} = \frac{10}{\sqrt{2}}\,\Omega$$

$$\text{Here,} \qquad \omega^2 = \frac{R^2}{M^2} = \frac{R^2}{(\mu L)^2} = \frac{\left(\dfrac{10}{\sqrt{2}}\right)^2}{(5 \times 10^{-7})^2} = 2 \times 10^{14} \quad \left[\because \mu = \frac{M}{\sqrt{L_1 L_2}}\right]$$

$$\text{or} \qquad \omega = 1.414 \times 10^7 \text{ rad/sec}$$

$$\text{and} \qquad f = \frac{\omega}{2\pi} = \frac{1.414}{2\pi} \times 10^7 = 2.25 \text{ MHz}$$

i.e. the secondary current is maximum when the supply frequency is 2.25 MHz. The r.m.s. primary current $(I_0)_{\text{rms}}$ at this frequency is given by

$$(I_0)_{\text{rms}} = \frac{E_0}{\dfrac{\omega^2 M^2}{R} + R} = \frac{E_0 R}{\omega^2 M^2 + R^2} = \frac{E_0 R}{2R^2} = \frac{E_0}{2R}$$

$$= \frac{1}{2 \times \dfrac{10}{\sqrt{2}}} = \frac{1}{\sqrt{2} \times 10} = 0.0707 \text{ amp} = 70.7\,\text{mA}$$

$$\text{whilst} \qquad (i_0)_{\text{rms}} = -\frac{i\omega M E_0}{\omega^2 M^2 + R^2}$$

$\therefore$ Magnitude $(i_0)_{\text{rms}}$ is

$$|(i_0)_{\text{rms}}| = \frac{R E_0}{2R^2} = \frac{E_0}{2R} = 70.7\,\text{mA}.$$

Thus, the frequency at which the secondary current is maximum is 2.25 MHz and the primary and secondary currents at this frequency are both 70.7 mA.

Example 3.6: The CO_2 molecule may be represented by a system consisting of a central mass m_2 connected by identical springs of constant k to two masses m_1 and m_3 (with $m_1 = m_3 = m$) as shown in Fig. 3.18. Write down the equations of motion of each mass and solve them for the two normal modes in which the masses oscillate along the

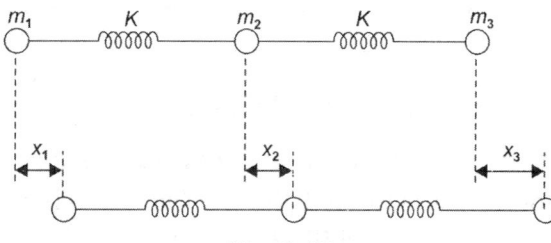

Fig. 3.18

line joining their centres. Take $m_1 = m_3 = 16$ and $m_2 = 12$ units and assuming that this classical model is applicable, find the ratio of the frequencies of the two modes.

Solution: In Fig. 3.18, lower part shows the general configuration of the system where x_1, x_2 and x_3 are the displacements of three masses from their respective equilibrium position shown in upper part. Assuming that $x_3 > x_2 > x_1$, the equations of motion are

$$m_1 \ddot{x}_1 = k(x_2 - x_1) \tag{i}$$

$$m_2 \ddot{x}_2 = k(x_3 - x_2) - k(x_2 - x_1) \tag{ii}$$

$$m_3 \ddot{x}_3 = -k(x_3 - x_2) \tag{iii}$$

Setting $m_1 = m_3 = m$, we have from Eq. (i) and (iii)

$$m(\ddot{x}_3 - \ddot{x}_1) = -k(x_3 - x_1) \quad \text{or} \quad m\ddot{x} = -kx$$

where $x = (x_3 - x_1)$ is the relative displacement of mass m_1 and m_3. This equation gives one of the normal modes of the system with angular frequency

$$\omega_1^2 = \frac{k}{m} \tag{iv}$$

To find other mode, let us assume that

$$x_1 = A\cos(\omega t + \phi), \, x_2 = B\cos(\omega t + \phi)$$

and $$x_3 = c\cos(\omega t + \phi)$$

Substituting the value in (i), (ii) and (iii), we have

$$m_1 \omega^2 A = k(A - B)$$

$$m_2 \omega^2 B = k(2B - A - C)$$

$$m_3 \omega^2 C = k(B - C)$$

Eliminating A, B, C from these equations, we obtain the angular frequency of the second mode which is

$$\omega_2^2 = k\frac{(m_2 + 2m_1)}{m_1 m_2}$$

and the frequency ratio of the two modes is

$$\frac{\omega_2}{\omega_1} = \frac{(m_2 + 2m_1)^{\frac{1}{2}}}{(m_2)^{\frac{1}{2}}} = \left(\frac{12 + 2 \times 16}{12}\right)^{\frac{1}{2}} = 1.91$$

3.10 DYNAMICS OF NUMBER OF OSCILLATORS

Any real physical system, such as a piece of string or a volume of fluid, or a continuous solid contains many particles bound (or coupled) to each other by forces of cohesion. We are, therefore, interested in tackling the problem of an arbitrary number of similar oscillators coupled together. We shall investigate this problem by using the ideas developed in the preceding sections relating to two coupled oscillators.

(a) Mono Atomic One-dimensional Lattice

Consider an infinite *lattice* of identical particles of mass M, the particles in equilibrium are separated by a distance a along the X-axis and we shall take the oscillations of the particles to be longitudinal.

Number the particles by calling the particle at origin as O, the next particle to the right 1, etc. (Fig. 3.19). The displacement of the nth particle is denoted by x_n. Whenever the vibrational motion is excited, they will execute periodic motion about their equilibrium position.

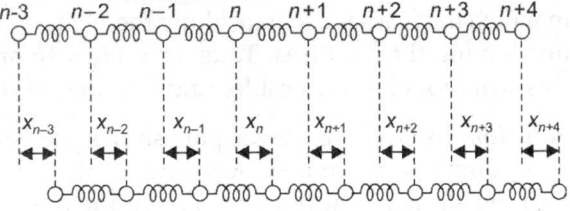

Fig. 3.19

To set-up force equation, assume that the force between neighbouring particles are the linear forces, that is the force required to produce the displacement is proportional to the displacement itself. This is reasonable assumption, since the particles (atoms) are being considered as being bound by ideal elastic springs. Further, assume that the only significant forces between atoms are due to direct *nearest-neighbour interaction*. With these assumptions, we can write the net force acting on the pth particle (atom) in terms of the extension of the two springs which bind it to the $(p + 1)^{th}$ and $(p - 1)^{th}$ particle, as

$$F_p = f(x_{p+1} - x_p) - f(x_p - x_{p-1})$$
$$= f(x_{p+1} + x_{p-1} - 2x_p)$$

where f is taken for spring constant. It may be mentioned that the influence of the atoms other than nearest neighbour is regarded as negligible. The equation of motion then is

$$M\frac{d^2 x_p}{dt^2} = f(x_{p+1} + x_{p-1} - 2x_p) \tag{3.43}$$

where $\dfrac{d^2 x_p}{dt^2}$ is the acceleration of the pth particle.

Let us seek the periodic solution of this equation, of the form

$$x_p = x_0 c^{i(\omega t - kpa)} \tag{3.44}$$

where K is the wave vector and ω the angular frequency. Then, we should expect

$$x_{p+1} = x_0 e^{i\{\omega t - (p+1)ka\}}$$

$$x_{p-1} = x_0 e^{i\{\omega t - (p-1)ka\}}$$

Substituting the values in Eq. (3.43), we have

$$-M\omega^2 = f(e^{ika} + e^{-ika} - 2) = 2f(\cos ka - 1)$$

$$= -4f \sin^2 \frac{ka}{2}$$

or
$$\omega = \pm \sqrt{\frac{4f}{M}} \sin \frac{ka}{2} \tag{3.45}$$

We arrange the sign of the square root so that the frequency ω is always positive for suitable lattice. Figure 3.20 gives the plot of $\dfrac{\omega^2}{4f}$ versus ka and Fig. 3.21, ω versus k. Both curves are periodic functions of k with period $\dfrac{2\pi}{a}$.

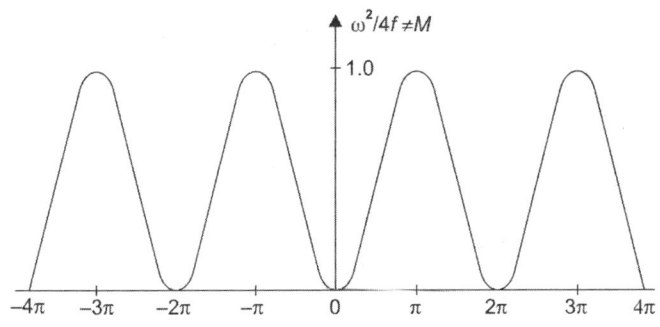

Fig. 3.20

Another important result may be obtained by comparing the solution (3.44) with another in which k is replaced by

$$k_a = k + 2\pi \cdot \frac{n}{a} \quad \text{with} \quad n = \pm 1, \pm 2, \pm 3, ..., \text{etc.}$$

We have, for the displacement of the pth particle

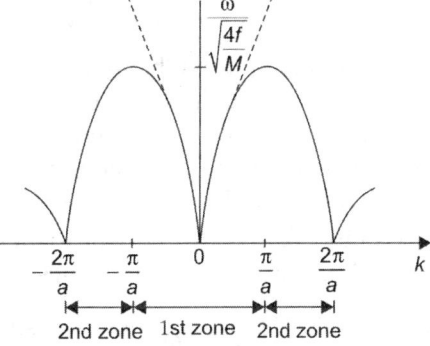

Fig. 3.21

$$x'_p = x_0 e^{i\{\omega t - k_n pa\}} = x_0 e^{i\left\{\omega t - \left(k + \frac{2\pi n}{a}\right)pa\right\}}$$

$$= x_0 e^{i\{\omega t - kpa\}} = x_p \quad (\because e^{i2\pi np} = 1)$$

and

$$\omega' = \pm\sqrt{\frac{4f}{M}}\,\sin\frac{a}{2}(kn) = \pm\sqrt{\frac{4f}{M}}\,\sin\frac{a}{2}\left(k + \frac{2\pi n}{a}\right)$$

$$= \pm\sqrt{\frac{4f}{M}}\,\sin\frac{ka}{2} = \omega$$

It follows that the solution (3.43) and frequency ω corresponding to modes K and k_n are identical i.e. the state of vibration of the array of mass points corresponds to a wave vector. k is the same as that for any of the wave vector $k + \dfrac{2\pi n}{a}$. In order to obtain unique relationship between the state of vibration of the lattice and the wave vector k, the latter must be confined to a range of values of $\dfrac{2\pi}{a}$. Since we want both positive and negative values of k because wave can propagate to the right as well as to the left, the range of independent values of k can be specified by $-\pi \le ka \le \pi$ or $-\dfrac{\pi}{a} \le k \le \dfrac{\pi}{a}$.

This range of values of k is referred to as the *first Brillouin zone* of the linear lattice. The second zone consists of two intervals of half a period each, one on each side of the first zone as indicated in Fig. 3.21. Higher order zones are defined in similar manner.

If we now restrict ourselves to values of k which are much less than π/a and thus, to the value of λ which are much greater than twice the inter atomic distance (lattice constant) i.e., $\lambda \gg a$.

ω is approximately linear with k since then

$$\sin \frac{ka}{2} \approx \frac{ka}{2}$$

and

$$\omega = ka\sqrt{\frac{f}{M}}$$

In this long wavelength approximation, the phase velocity will be essentially constant because

$$\text{phase velocity} \quad v_p = \frac{\omega}{k} = a\sqrt{\frac{f}{M}} = v_0 \quad \text{(say)} \tag{3.46}$$

and group velocity v_g is seen to be

$$v_g = \frac{d\omega}{dk} = a\sqrt{\frac{f}{M}} = v_0 \tag{3.47}$$

constant and equal to the phase velocity. For very long wavelengths then the dispersion effects are negligible and the medium acts like a continuous and homogeneous elastic medium. This is of course most reasonable from physical point of view, since for such long wavelength the atomic nature of the chain is of little importance in so far as the dynamical behaviour of the system is concerned.

As k increases, however, the dispersion effects become more important and ω no longer varies linearly with k. Under these conditions, we find from Eq. (3.45) that

$$v_p = \frac{\omega}{k} = 2a\sqrt{\frac{f}{M}} \cdot \frac{\sin \dfrac{ka}{2}}{ka} = v_0 \frac{\sin \dfrac{ka}{2}}{\dfrac{ka}{2}} \tag{3.48}$$

and

$$v_g = \frac{d\omega}{dk} = 2\sqrt{\frac{f}{M}} \cos \frac{ka}{2} \cdot \frac{a}{2} = v_0 \cos \frac{ka}{2} \tag{3.49}$$

where $v_0 = a\sqrt{\dfrac{f}{M}}$ the long wavelength limit of both v_g and v_p. Figures 3.22 and 3.23 show plots of v_p and v_g as function of k. It is observed from Eq. (3.49) that $v_g \to 0$ as

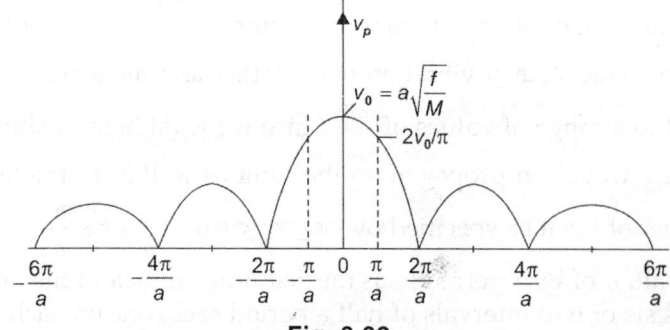

Fig. 3.22

$k = \pi/a$ and thus $\lambda \to 2a$. In this case, the phase of the vibration of neighbouring atom differs by π radians and the character of the motion is simply a standing wave. (Note that a standing wave is an example of wave motion with finite phase velocity and zero group velocity). The actual physical character of motion is illustrated in Fig. 3.24 for long wavelength case and Fig. 3.25 illustrates the condition $\lambda = 2a$ in which the stationary waves are set-up.

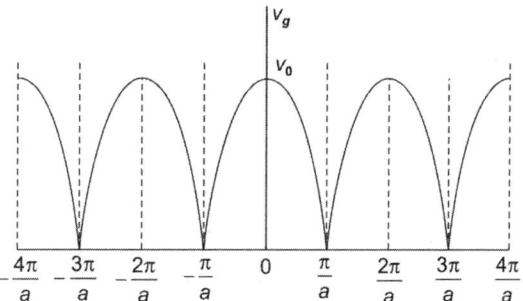

Fig. 3.23

It is thus interesting to note that only the frequencies $\omega \le \sqrt{\dfrac{f}{M}} = \dfrac{2v_0}{a}$ will propagate through the chain. Hence, a mono-atomic linear lattice can be considered a low-pass filter which transmits only the frequency range between zero and $\dfrac{2v_0}{a}$. In contrast with this, the continuous string has no frequency limit. The maximum frequency of the chain of atoms occurs when $k = \dfrac{\pi}{a}$ i.e. $\lambda_m = 2a$. The

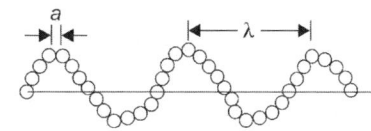

Fig. 3.24

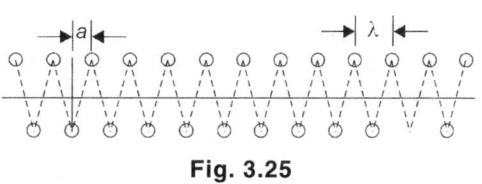

Fig. 3.25

knowledge of velocity of sound in solids ($10^5 \sim 10^6$ cm·s^{-1} can be used to find the order of maximum frequency using $a \approx 10^{-8}$ cm.

$$v_{max} = \frac{10^5}{10^{-8}} \approx 10^{+13} \text{ Hz}$$

Hence, the cut-off frequency lies in infrared region.

(b) Vibrational Modes of a Finite One-dimensional Lattice of Identical Atoms

In the preceding section, the discussion referred to an infinite lattice. In the present section, we shall see how many independent solutions (Modes of Vibration) of the Eq. (3.43), which satisfy a given set of boundary conditions can be associated with a linear finite chain of atoms.

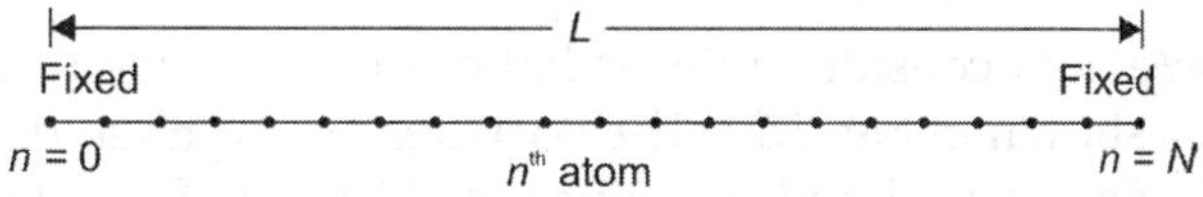

Let there be $N + 1$ atoms in the array numbered from 0 to N. We further suppose that the two end atoms are rigidly fixed so that the

$$\xleftarrow{\hspace{3cm}} L \xrightarrow{\hspace{3cm}}$$

Fixed .... Fixed

$n = 0 \qquad\qquad n^{th}$ atom $\qquad\qquad n = N$

Fig. 3.26

number of mobile atoms is $(N - 1)$ (Fig. 3.26). If length of the chain be L and each atom separated by a distance a, then $a = L/N$. The general solution of the equation of motion of the nth particle

$$M\frac{d^2 X_n}{dt^2} = f(x_{n+1} + x_{n-1} - 2x_n) \tag{3.50}$$

can be written as the sum of two waves, one propagating to the right and the other to the left.

$$x_n(t) = \left\{ A_1 e^{i(kna + \beta_1)} + A_2 e^{i(-kna + \beta_2)} \right\} e^{i\omega t} \tag{3.51}$$

Here β_1 and β_2 are phase angles of two waves and A_1 and A_2 are their amplitudes. The boundary conditions are

$$x_n(t) = 0 \quad \text{for } n = 0 \quad \text{and} \quad n = N \text{ at all times.}$$

First boundary condition yields

$$A_1 = -A_2 \quad \text{and} \quad \beta_1 = \beta_2 = 0.$$

Since phase angles are equal, we choose them to be equal to zero.

Taking real part of Eq. (3.51), we have

$$x_n(t) = 2A_1 \sin nka \sin \omega t \tag{3.52}$$

which represents a standing wave and leads to the dispersion relation

$$M\omega^2 = 4f^2 \sin^2 \frac{ka}{2} \tag{3.53}$$

As that for, running wave solution, Eq. (3.45) with the difference that k is now limited to positive values ranging from 0 to π/a. The second boundary condition imposed $x_n(t) = 0$ for $N = n$ gives

$$\sin Nka \sin \omega t = 0$$

or $\quad \sin Nka = 0 = \sin \pi J \quad$ where J is an integer.

or $$k = \frac{\pi J}{Na} \tag{3.54}$$

Note that $J = 0$ must be excluded. Since this corresponds to $k = 0$, i.e., all the particles are at rest.

The maximum value of $k = \dfrac{\pi}{a}$, which gives $J_{\max} = N$. However, this value must be excluded for the same reason as $J = 0$. We then conclude that

$$J = 1, 2, \dots (N-1).$$

In other words, there are as many modes of vibration (k values) as there are the mobile atoms. Each value of k then corresponds a frequency ω_k. Hence, the frequency spectrum consists of $(N-1)$ discrete lines. For macroscopic chain lengths, the spacing of the lines is so close that we may call it a quasi-continuous spectrum.

(c) The Vibrational Mode of a Diatomic Linear Lattice

We now consider a linear lattice of two different kinds of atoms arranged alternatively as shown in Fig. 3.27. The particles are numbered in such a way that the even numbered have mass M and odd ones m $(M > m)$. The equation of motions of the two types of particles are different because of their different masses. In analogy with equation of motion of the mono-atomic lattice, we now have the following equations of motion assuming nearest neighbour interaction only.

$$M\frac{d^2 x_{2n+2}}{dt^2} = f(x_{2n+1} + x_{2n-1} - 2x_{2n}) \tag{3.55}$$

and $$m\frac{d^2 x_{2n+1}}{dt^2} = f(x_{2n+2} + x_{2n} - 2x_{2n+1}). \tag{3.55a}$$

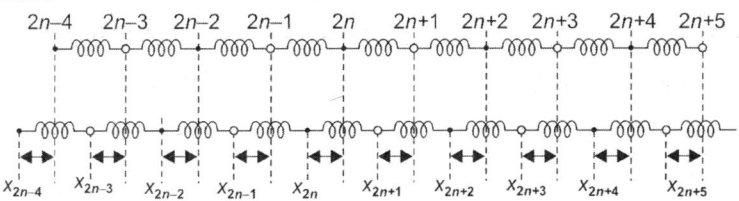

Fig. 3.27

Here x_n is the displacement of the nth particles from its equilibrium position.

Here we assume a single wave motion in which both types of particles participate; the displacement of the two masses may be different but they are both contributing to the same wave disturbance, hence for the motion of the two types of particles same ω and k is assumed. We assume a solution in the form of travelling wave with different amplitudes A and B.

$$x_{2n} = Ae^{i(\omega t + 2nka)} \tag{3.56}$$

and

$$x_{2n+1} = Be^{i[\omega t + (2n+1)ka]} \tag{3.56a}$$

Substituting in Eqs (3.55) and (3.55a), we have

$$(M\omega^2 - 2f)A + 2Bf \cos ka = 0 \tag{3.57}$$

and

$$(m\omega^2 - 2f)B + 2Af \cos ka = 0 \tag{3.57a}$$

These equations give

$$\frac{A}{B} = -\frac{2f \cos ka}{M\omega^2 - 2f} = -\frac{m\omega^2 - 2f}{2f \cos ka} \tag{3.58}$$

or

$$(M\omega^2 - 2f)(m\omega^2 - 2f) = 4f^2 \cos^2 ka$$

or

$$Mm\omega^4 - 2f(M+m)\omega^2 + 4f^2 \sin^2 ka = 0.$$

This gives the dispersion relation as

$$\omega^2 = f\left(\frac{1}{m} + \frac{1}{M}\right) \pm f\sqrt{\left(\frac{1}{m} + \frac{1}{M}\right)^2 - \frac{4\sin^2 ka}{Mm}}. \tag{3.59}$$

Since ω should be positive, each value of ω^2 leads to a single value for ω. Thus, in contrast to the monoatomic lattice, there are now two angular frequencies ω_+ and ω_- corresponding to a single value of the wave vector k. A plot of this result is shown in Fig. 3.28. There are two branches of ω versus k curve corresponding to whether the positive and negative sign is taken in Eq. (3.59). The upper branch $\omega_+(k)$ is called the *optical branch* (*optical mode of vibration*), while the lower one $\omega_-(k)$ is called the *acoustic branch* (*Acoustic mode of vibration*).

It is further observed that as in the monoatomic case, the frequency is a periodic function

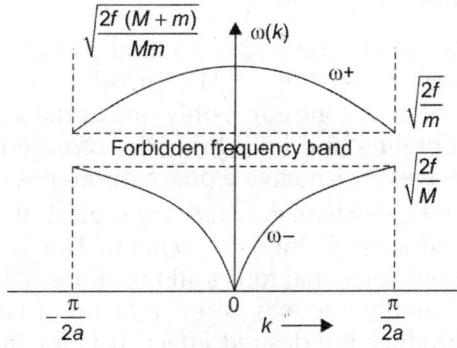

Fig. 3.28

of the wave vector k. The first zone, thus limits the values of k to the range between $-\dfrac{\pi}{2a}$ and $+\dfrac{\pi}{2a}$ as shown in Fig. 3.28 for $k = +\dfrac{\pi}{2a}$ the two frequencies are

$$\omega_+ = \sqrt{\frac{2f}{m}} \quad \text{and} \quad \omega_- = \sqrt{\frac{2f}{M}}.$$

The larger the ratio $\dfrac{m}{M}$, wider is the *frequency gap* between the two branches i.e., the width of the *forbidden band* depends on the difference of masses. If two masses are equal, the two branches become degenerate at $k = \dfrac{\pi}{2a}$.

It is of interest to investigate the physical difference between the acoustical and optical modes and why they are named so. To find out, we calculate $\dfrac{A}{B}$ as $k \to 0$ for both branches.

As $k \to 0$, $\cos ka \to +1$ and $\sin ka \to 0$.

$\therefore$
$$\omega_- = 0 \quad \text{and} \quad \omega_+ = \sqrt{2f\left(\frac{1}{m} + \frac{1}{M}\right)}.$$

For acoustic mode $A = B$ and for optical mode $-MA = mB$. For optical mode at $k = 0$, the vibrations of the two atoms are in opposite direction and amplitudes are inversely in the ratio of the masses, so that the centre of mass of the system remains fixed during the period of motion. For acoustical mode of vibrations, the two types of atoms move in the same directions with same amplitude as shown in Fig. 3.29.

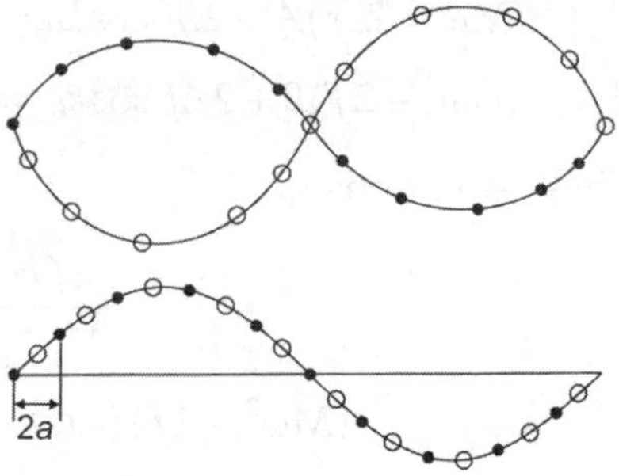

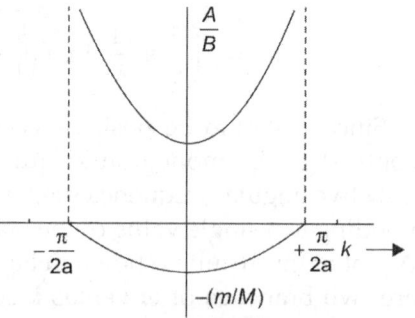

Fig. 3.29

The optical mode can have $k = 0$, but acoustic mode cannot. For other values of k, the ratio A/B may be calculated from Eq. (3.58). The results are shown in Fig. 3.30. It is further observed that at the edge of the Brillouin zone i.e., $k = +\dfrac{\pi}{2a}$.

$\qquad\qquad B = 0 \quad$ for acoustic mode
and $\qquad\qquad A = 0 \quad$ for optical mode

i.e. in acoustic mode, the light particles of mass m are at rest (Fig. 3.31). In other words at the Brillouin zone edge, only one of the sub-lattice is vibrating and both branches represent standing waves which have a phase difference of π.

Fig. 3.30

These characteristics are typical of the optical and acoustic mode in general. The acoustic mode vibrations can be excited by kind of some force that forces all the atoms in the lattice (crystal) to go in the same direction. For example, we may direct a beam of sound waves at the surface of a crystal. This will produce the desired effect. It is for this reason that such vibrations are called *acoustic vibrations* and constitute the acoustic branch in dispersion relation.

The optical mode of vibrations in ionic crystals, where the two types of atoms are oppositely charged, can be excited by an electric field, which tends to move the ions in opposite directions. Specially in ionic crystals, this mode can be excited by electric field associated with light waves from which, the term optical vibration or optical mode is derived.

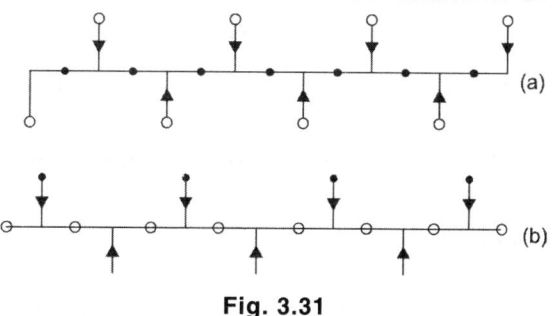

Fig. 3.31

If heavy mass $M \to \infty$, the acoustic branch disappears, the optical branch flattens in such a way that all k-values have the same frequency (Fig. 3.32a). Physically, this means that each atom is completely independent of its neighbour and each oscillates at its own natural uncoupled frequency. When $M \to \infty$, the mid-point between each atom is tied down isolating the atoms from one another. In the quantum theory of specific heat, Einstein used the same model with considerable success.

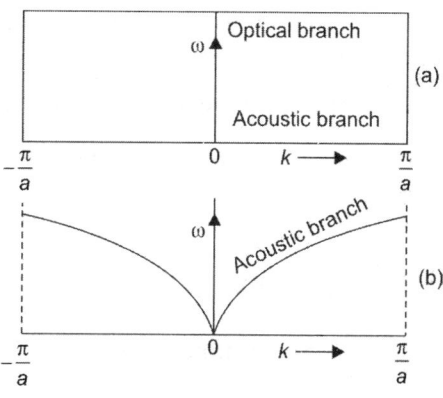

Fig. 3.32

If the light mass $m \to 0$, the optical branch disappears upwards. The acoustic branch is unchanged and we return to the monoatomic lattice. To make the correspondence exact, we would have to let force constant $f \to 2f$ to take into account the doubling the lattice constant.

The optical modes are essentially required for the natural frequency of the light atoms, perturbed by the heavy sub-lattice. At $k = \dfrac{\pi}{2a}$, the heavy sub-lattice is stationary (Fig. 3.31a) and the light atoms vibrate with natural frequency $\sqrt{\dfrac{2f}{m}}$. As k decreases, the perturbation increases until $k = 0$, the two sub-lattices are actually vibrating rigidly ($\lambda \to \infty$) against each other (Fig. 3.33). The increase in frequency as k decreases is due to an effectively large force constant $f \to f\left(1 + \dfrac{m}{M}\right)$ for the light atoms which results from the counter vibrating heavy sub-lattice.

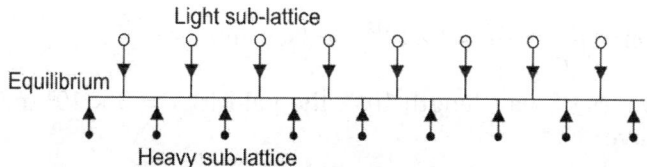

Fig. 3.33

Example 3.7: How many longitudinal modes are there for (a) linear monoatomic lattice, (b) a linear diatomic lattice of inter atomic spacing 10^{-10} m and length 1 m?

Solution: Since length is $L = 1$ m and inter atomic distance $a = 10^{-10}$ m. Hence no. of atoms $N = 10^{10}$.

Hence no. of modes for monoatomic fixed lattice is $N = 10^{10}$.

The no. of modes for periodic monoatomic fixed lattice is $\dfrac{N}{2} = 0.5 \times 10^{10}$.

The no. of modes for fixed diatomic lattice $\dfrac{N}{2} = 0.5 \times 10^{10}$ and the no. of modes for

periodic diatomic lattice $\dfrac{N}{4} = 0.25 \times 10^{10}$.

Example 3.8: If $v_0 = 10^3$ m/s compare the frequencies of sound waves of wavelength $\lambda = 10^{-9}$ m for (a) homogeneous line, (b) acoustic waves, (c) optical waves on a linear lattice containing two identical atoms per primitive cell of inter atomic spacing 2.5Å, and (d) light waves of the same wavelength.

Solution: (a) Frequency in case of homogeneous line is given by

$$\omega = v_0 k = 10^3 \times \dfrac{2\pi}{10^{-9}} = 2\pi \times 10^{12} \text{ rad/s}$$

(b) For acoustic waves in a diatomic lattice the frequency varies from $\omega = 0$ for $k = 0$

and $\omega = \sqrt{\dfrac{2f}{M}}$ for $k = \dfrac{\pi}{2a}$.

In case of diatomic lattice

$$v_0 = \sqrt{\dfrac{2f}{M}}a \quad \text{or} \quad \dfrac{v_0}{a} = \sqrt{\dfrac{2f}{M}} = \omega$$

where f is force constant, m is mass of particle and a, the linear atomic distance. Therefore,

$$\omega = \dfrac{v_0}{a} = \dfrac{10^3}{2.5 \times 10^{-10}} = 4 \times 10^{12} \text{ rad/s}$$

(c) For optical waves in diatomic lattice the frequency varies from

$$\omega = \sqrt{\dfrac{4f}{m}} \text{ for } k = 0 \text{ to } \omega = \sqrt{\dfrac{2f}{m}} \text{ for } k = \dfrac{\pi}{2a}$$

Since the lattice has two identical atoms primitive cell, hence, there will be no forbidden gap.

$$\omega = \sqrt{2}\,\dfrac{v_0}{a} = 4\sqrt{2} \times 10^{12} \text{ rad/sec} \quad \text{for} \quad k = 0$$

and

$$\omega = \sqrt{\dfrac{2f}{m}} = \dfrac{v_0}{a} = 4 \times 10^{12} \text{ rad/sec} \quad \text{for} \quad k = \dfrac{\pi}{2a}.$$

(d) For light waves of wavelength 10^{-9}, the velocity $c = 3 \times 10^8$ m/s, the velocity of light. Then, we have

$$\omega = 2\pi v = 2\pi \dfrac{c}{\lambda} = \dfrac{2\pi \times 3 \times 10^8}{10^{-9}} = 6\pi \times 10^{17} \text{ rad/s}.$$

Example 3.9: The unit cell of NaCl is cube of side 5.6 Å and Young's modulus in [100] direction is 5×10^{10} N/m². Estimate the wavelength at which electromagnetic radiation is strongly reflected by a sodium chloride crystal explaining any assumption you make (Atomic weight of Na and Cl are 23 and 37 respectively).

Solution: The frequency of radiation which is strongly reflected by ionic crystal is given by

$$\omega_0^2 = 2f\left(\frac{1}{m} + \frac{1}{M}\right)$$

where f is force constant while h, m and M are the masses of the atoms. Assuming that the extension in [100] direction produces negligible contraction in perpendicular direction. We can write

$$f = aE$$

Here a is the inter atomic distance and E, the Young's modulus

$$\therefore \qquad \omega_0^2 = 2aE\left[\frac{1}{m} + \frac{1}{M}\right]$$

$$= 2 \times 5.6 \times 10^{-10} \times 5 \times 10^{10}\left[\frac{1}{23} + \frac{1}{37}\right] \times \frac{1}{1.67 \times 10^{-27}} = \frac{56 \times 10^{27} \times 60}{23 \times 37 \times 1.67}$$

$$\omega_0 = 4.34 \times 10^{13} \text{ rad/s}$$

Hence the wavelength at which the electromagnetic radiation is strongly reflected is

$$\lambda = \frac{c}{\nu} = \frac{2\pi c}{\omega} = \frac{2\pi \times 3 \times 10^8}{4.34 \times 10^{13}}$$

$$= 4.324 \times 10^{-5} \text{ m} = 43.24 \text{ } \mu m.$$

Example 3.10: Show that the coupled vibrations of a periodically loaded string becomes waves in continuous medium.

Solution: Consider a flexible elastic string to which are attached $N + 1$ equal masses, equally spaced at a distanct a apart and the ends of the string are kept fixed. The particles are numbered 0 to N and that these particles undergo small displacements in transverse direction to the string. A constant tension T exists at all the points of the string and any increase in tension as particle oscillates is negligible. Assuming displacement of the pth particle to be y_p, the force on the pth particle is

$$F_p = -T \sin \theta_{p-1} + F \sin \theta_{p+1}$$

The values of sine from Fig. 3.34 are

$$\sin \theta_{p-1} \approx \tan \theta_{p-1} = \frac{y_p - y_{p-1}}{a}$$

and

$$\sin \theta_{p+1} \approx \tan \theta_{p+l} = \frac{y_{p+1} - y_p}{a}$$

$$\therefore \qquad F_p = -\frac{T}{a}(y_p - y_{p-1}) + \frac{T}{a}(y_{p+1} - y_p)$$

$$= \frac{T}{a}(y_{p+1} + y_{p-1} - 2y_p)$$

and equation of motion for p^{th} particle is

$$m\frac{d^2 y_p}{dt^2} = \frac{T}{a}(y_{p+1} + y_{p-1} - 2y_p)$$

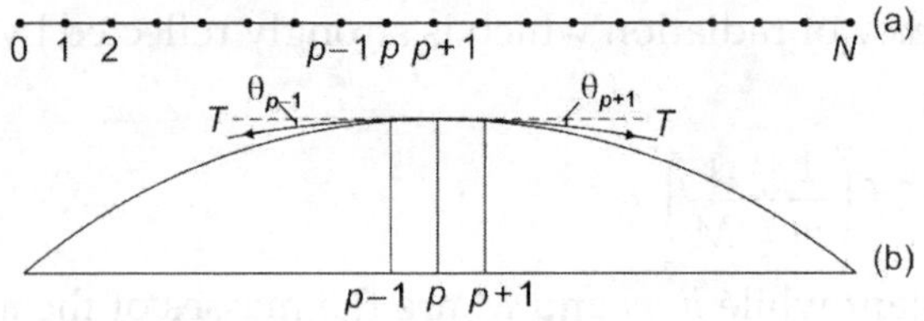

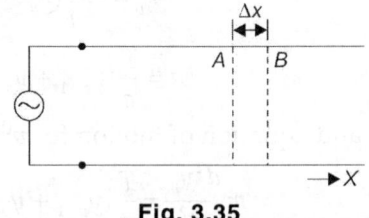

Fig. 3.34

The separation between the particles a may be put equal to (δx) and taking the limits as $(\delta x) \to 0$ i.e. as the masses merge into continuous string then

$$\frac{\partial^2 y_p}{\partial t^2} = \frac{T}{m}\left(\frac{y_{p+1} + y_{p-1} - 2y_p}{\delta x}\right)$$

$$= \frac{T}{m}\left(\frac{y_{p+1} - y_p}{\delta x} - \frac{y_p - y_{p-1}}{\delta x}\right)$$

$$= \frac{T}{m}\left\{\left(\frac{\delta y}{\delta x}\right)_{p+1} - \left(\frac{\delta y}{\delta x}\right)_p\right\}$$

But

$$\left(\frac{dy}{dx}\right)_{x+dx} - \left(\frac{dy}{dx}\right)_x = \frac{d^2 y}{dx^2}\cdot dx$$

Taking the limits as $(\delta x) \to 0$, we have

$$\frac{d^2 y_p}{dt^2} = \frac{T}{m}\cdot\frac{d^2 y}{dx^2}\cdot dx = \frac{T}{\rho}\cdot\frac{d^2 y}{dx^2}$$

where $\rho = \dfrac{m}{dx}$ the linear density of the string. This is the equation of wave motion.

3.11 ELECTRICAL TRANSMISSION LINES

A transmission line essentially consists of two long parallel conductors (wires) for carrying electrical energy from one point to the another. At one end of line, power is fed by AC generator, which causes vibrations in voltage and current which travel along the line in the form of an electrical wave. The generator provides the driving force for the the waves to travel in the transmission lines. We assume that the line is lossless i.e., we neglect the resistance of the conductors (wires) and any possible electrical leakage between them. Such a loss free line is called an *ideal transmission line*. When AC currents flow in the wires, they generate magnetic flux lines, which link the region between the wires, thus giving rise to self-inductance. The wires themselves constitute a capacitor. Let L and C be the inductance and capacitance per unit length of the conductors respectively.

Transmission Line Equations

Consider a two line transmission line. In this line, two parallel wires of identical cross-section are placed in X-direction (Fig. 3.35). The distance between the wires is very small in comparison to the wavelength of the waves propagating through the lines R, L, C and G

Fig. 3.35

respectively are the resistance, inductance, capacitance and conductance per unit length of the lines.

If current waves and voltage waves of angular frequency ω propagate through this line, then:

Series impedance per unit length of line $Z = R + i\omega L$ $(i = \sqrt{-1})$

and shunt admittance per unit length of line $Y = G + i\omega C$.

Consider a length dx of this line, then

$$\text{Potential drop } (dV) = -I(Z\,dx) \tag{3.60}$$

where I is the current at the position of element dx and Zdx is the series impedance.

The current loss at the same element

$$dI = -V(Y\,dx) \tag{3.60a}$$

$\therefore$
$$\frac{dV}{dx} = -IZ = -(R + i\omega L)I$$

and
$$\frac{dI}{dx} = -YV = -(G + i\omega C)V$$

Further differentiating these equations w.r.t. x, we have

$$\frac{d^2V}{dx^2} = -z\frac{dI}{dx} = -z(-y)V = yzV = r^2V \tag{3.61}$$

and
$$\frac{d^2I}{dx^2} = -y\frac{dV}{dx} = -y(-Iz)V = yzI = r^2I \tag{3.61a}$$

where,

$$yz = r^2 = (R + i\omega L)(G + i\omega c) \tag{3.62}$$

These Eqs (3.61 and 3.61a) are called *transmission line equations* and constant

$$r = \sqrt{Zy} = \sqrt{(R + i\omega L)(G + i\omega c)} \tag{3.63}$$

is called *propagation constant*. These transmission line equations depict the propagation of voltage and current waves. Their solutions can be written as

$$V = V_1e^{-rx} + V_2e^{rx} \tag{3.64}$$
$$I = I_1e^{-rx} + I_2e^{rx} \tag{3.64a}$$

V_1e^{-rx} depicts the voltage wave propagating in positive X-direction i.e., this represents the incident voltage wave. V_2e^{-rx} represents voltage wave propagating in negative X-direction i.e., reflected voltage wave. In this way incident and reflected waves both exist at every position and the resultant wave is the super-position of these two waves. Similarly, two terms of Eq. (3.64a) also represent incident and reflected current waves at any place. Now

$$\frac{dV}{dx} = -rV_1e^{-rx} + rV_2e^{rx} = -ZI$$

$\therefore$
$$I = -\frac{1}{Z}\frac{dV}{dx} = \frac{r}{Z}\left(V_1e^{-rx} - V_2e^{rx}\right)$$

$$= \frac{1}{Z_0}\left(V_1e^{-rx} - V_2e^{rx}\right) \tag{3.65}$$

where $Z_0 = \dfrac{Z}{r} = \dfrac{R + i\omega L}{\sqrt{(R + i\omega L)(G + i\omega c)}} = \sqrt{\dfrac{R + i\omega L}{G + i\omega c}}$

Here Z_0 is called the *characteristic impedance of the transmission line*. Then, we have

$$I_1 = \frac{V_1}{Z_0} \quad \text{and} \quad I_2 = \frac{V_2}{Z_0}$$

Hence, transmission line equations in terms of only two voltage constants are

$$V = V_1 e^{-rx} + V_2 e^{rx} \tag{3.66}$$

$$I = \frac{1}{Z_0}(V_1 e^{-rx} - V_2 e^{rx}) \tag{3.66a}$$

For determining complete solutions of transmission line equations, the conditions at the receiving end are used because in general, only this end is available for observation. Hence, the element under observation is taken from this end i.e., this end is treated as $x = 0$. If line has length l, then input end is taken as $x = -l$ (Fig. 3.36).

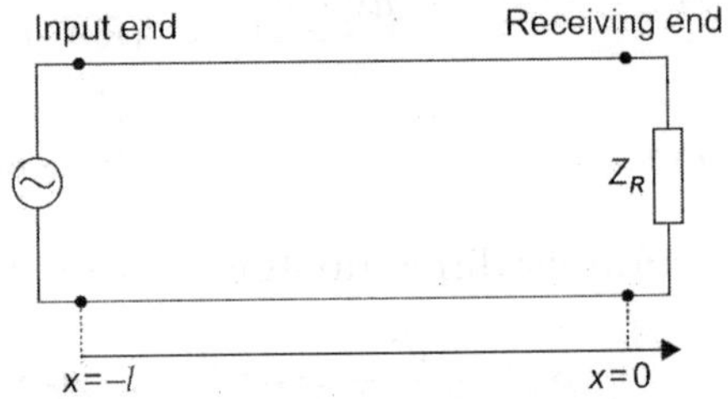

If a load Z_R is connected at the receiving end and their measured voltage is V_R and current I_R. Then at $x = 0$, $V = V_R$ and $I = I_R$ and $\dfrac{V}{I} = Z_R$.

Fig. 3.36

Using these values in Eqs (3.66) and (3.66a), we have

$$V_R = V_1 + V_2 \quad \text{and} \quad I_R = \frac{1}{Z_0}(V_2 - V_1)$$

$\therefore$
$$V_1 = \frac{V_R + Z_0 I_R}{2} = \frac{V_R}{2}\left[1 + Z_0 \frac{I_R}{V_R}\right]$$

$$= \frac{V_R}{2}\left[1 + \frac{Z_0}{Z_R}\right] \tag{3.67}$$

and
$$V_2 = \frac{V_R}{2}\left[1 - \frac{Z_0}{Z_R}\right] \tag{3.67a}$$

Using these values of V_1 and V_2 the Eqs (3.66 and 3.66a) become

$$V = \frac{V_R}{2}\left(1 + \frac{Z_0}{Z_R}\right)e^{-rx} + \frac{V_R}{2}\left(1 - \frac{Z_0}{Z_R}\right)e^{rx}$$

$$= \frac{V_R}{2}(e^{rx} + e^{-rx}) - \frac{V_R}{2}\frac{Z_0}{Z_R}(e^{rx} - e^{-rx})$$

$$= V_R \cos h\, rx - I_R \cdot Z_0 \sin h\, rx \tag{3.68}$$

In this way

$$I = I_R \cos h\, rx - \frac{V_R}{Z_0} \sin h\, rx \tag{3.68a}$$

and the impedance at this position is

$$Z = \frac{V}{I} = \frac{V_R \cos hrx - I_R Z_0 \sin hrx}{I_R \cos hrx - \frac{V_R}{Z_0} \sin hrx}$$

$$= \frac{I_R [Z_R \cos hrx - z_0 \sin hrx]}{\frac{I_R}{Z_0} [Z_0 \cos hrx - Z_R \sin hrx]}$$

or
$$Z = Z_0 \frac{[Z_R - Z_0 \tan hrx]}{[Z_0 - Z_R \tan hrx]} \qquad (3.69)$$

for input end $x = -l$, V_s = source voltage

$$V_s = V_R \cos hrl + I_R Z_0 \sin hrl$$

[Since $\cos h(-rl) = \cos hrl$ and $\sin h(-rl) = -\sin h rl$]

and source current I_s

$$I_s = I_R \cos hrl + \frac{V_R}{Z_0} \sin hrl$$

and source impedance

$$Z_S = \frac{V_s}{I_s} = Z_0 \frac{Z_R + Z_0 \tan hrl}{Z_0 + Z_R \tan hrl} \qquad (3.70)$$

Z_s is also called the *input impedance*.

In the above solutions we have only considered the dependence of V and I only on position. In case of dependence of these quantities on time also, the complete solution can be written as

$$V = (V_1 e^{-rx} + V_2 e^{rx}) e^{i\omega t}$$

and
$$I = \frac{1}{Z_0} \left(V_1 e^{-rx} + V_2 e^{rx} \right) e^{i\omega t}$$

3.12. PROPAGATION CONSTANTS OF TRANSMISSION LINE

The propagation constant is represented by r and its value is

$$r = \sqrt{(R + i\omega L)(G + i\omega c)}$$

This constant tells us about the propagation of the voltage and current waves on the transmission line. In general, it is a complex quantity and is written as

$$r = \alpha + i\beta.$$

Here, the real part (α) is called *attenuation constant* and the imaginary part (β) is called *phase constant*. α depicts the rate of decrease of amplitude. When the voltage and current waves propagate on the transmission line, there are a number of sources of power loss. Where the phase constant (β) depicts the rate of change of phase when the wave propagates on the transmission line. To obtain α and β in terms of distributed constants of transmission line. The expression is rationalised as under

$$r = \alpha + i\beta = \sqrt{(R + i\omega L)(G + i\omega c)}$$

Squaring both sides, we have

$$\alpha^2 - \beta^2 + 2i\alpha\beta = (R + i\omega L)(G + i\omega c)$$
$$= (RG - \omega^2 Lc) + i\omega(RC + LG)$$

Now, equating real and imaginary parts, we have

$$\alpha^2 - \beta^2 = (RG - \omega^2 Lc) \qquad \text{(i)}$$

and
$$2\alpha\beta = \omega(Rc + LG) \qquad \text{(ii)}$$

Substituting the value of β in (i), we have

$$\alpha^4 - \alpha^2(RG - \omega^2 Lc) - \frac{\omega^2}{4}(Rc + LG)^2 = 0$$

or
$$\alpha = \pm\left[\frac{1}{2}(RG - \omega^2 LC) + \frac{1}{2}\sqrt{(R^2 + \omega^2 L^2)(G^2 + \omega^2 C^2)}\right]^{\frac{1}{2}}$$

and now

$$\beta = \pm\left[\frac{1}{2}(\omega^2 LC - RG) + \frac{1}{2}\sqrt{(R^2 + \omega^2 L^2)(G^2 + \omega^2 C^2)}\right]^{\frac{1}{2}}$$

β measures the phase difference part per unit length of transmission line.

In the phase velocity of the propagating waves is v_p, angular frequency ω and wavelength λ, then the distribution of voltage or current wave online in distance λ will be equal to one complete cycle and the equivalent phase difference is 2π.

$$\therefore \qquad \beta = \frac{2\pi}{\lambda} \quad \text{or} \quad \lambda = \frac{2\pi}{\beta}$$

It v is wave frequency then phase velocity $v_p = v\lambda$ or

$$v_p = \frac{\omega}{2\pi} \times \frac{2\pi}{\beta} = \frac{\omega}{\beta}$$

or
$$\beta = \frac{\omega}{v_p} \qquad \text{(3.71)}$$

3.13 CHARACTERISTIC IMPEDANCE OF TRANSMISSION LINE

Let us consider a transmission line of infinite length. It is clear that voltage wave and current wave will not reach the other end in finite time. In such cases, there will be only incident wave propagating in positive X-direction. Hence in the equations for voltage and current on transmission line $V_2 = I_2 = 0$. Hence

$$V = V_1 e^{-rx} \quad \text{and} \quad I = I_1 e^{rx}$$

and
$$Z = \frac{V}{I} = \frac{V_1}{I_1} = Z_0$$

This means that, the ratio of voltage and current at each point on the transmission line of infinite length is constant and this value is called *characteristic impedance of transmission line*. It means that the characteristic impedance of an infinite transmission line is the impedance experienced by the generator at any point on the line. Z_0 mainly depends on R, L and C values of the transmission line. It also depends on the frequency of the wave propagating on the line. Z_0 is also a complex quantity and its value in terms of transmission line constants is

$$Z_0 = \sqrt{\frac{R + i\omega L}{G + i\omega c}}$$

For lossless line $R = G = 0$

$$Z_0 = \sqrt{\frac{L}{c}} \qquad\qquad (3.72)$$

For low loss lines $R << L$ and $G << C$. Hence, for such a line Z_0 is also given by

$$Z_0 = \sqrt{\frac{L}{c}}$$

3.14 ENERGY LOSS ON TRANSMISSION LINE

Main reasons for the loss of energy in transmission line are

- (i) Resistance of the conductors forming transmission line,
- (ii) Leakage of current in the dielectric medium between the parallel conductors,
- (iii) At high frequencies, the electromagnetic radiations due to accelerated charges on conductors.

Due to all these reasons, the amplitude of voltage and current waves decreases with the distance from the source end, resulting the decrease in the propagating power. For simplicity, let us consider an infinite line on which there is only incident wave but no reflected one. Then

$$V = V_1 e^{-rx} \quad \text{and} \quad I = I_1 e^{-rx} = \frac{V_1}{Z_0} e^{-rx}$$

But $\qquad\qquad r = \alpha + i\beta$

where α is attenuation constant and β is the phase constant.

$\therefore \qquad\qquad V = V_1 e^{-(\alpha + i\beta)x} = V_1 e^{-\alpha x} e^{-i\beta x}$

$$I = \left(\frac{V_1}{Z_0}\right) e^{-\alpha x} \cdot e^{-\beta x}$$

Thus, amplitude of the voltage and current waves decreases exponentially with distance x from the source (generator) end.

For two points on the transmission line unit distance apart, i.e., if for point A distance from source is x and for B, it is $x + 1$, then

$$|V_A| = V_1 e^{-\alpha x} \quad \text{and} \quad |V_B| = V_1 e^{-\alpha(x+1)}$$

$\therefore \qquad\qquad \left|\frac{V_A}{V_B}\right| = e^{\alpha}$

or $\qquad \alpha = \log_e \left|\frac{V_A}{V_B}\right| = 2.303 \log_{10} \left|\frac{A}{B}\right|$

The attenuation constant α when defined in this way has unit neper/metre. Now, powers at these two points are

$$P_A = |V_A||I_A| = \frac{|V_A|^2}{Z_0} \quad \text{where } [Z_0 \approx \sqrt{\frac{L}{C}} \text{ is a real quantity.}]$$

$$P_B = \frac{|V_B|^2}{Z_0}$$

$$\therefore \qquad \frac{P_A}{P_B} = \frac{|V_A|^2}{|V_B|^2}$$

and $\qquad \log_{10} \frac{P_A}{P_B} = 2\log_{10} \frac{|V_A|}{|V_B|}$

The quantity $10\log_{10} \dfrac{P_A}{P_B}$ is the measure of the power loss and its units are decibel (db).

Thus, if $P_B = \dfrac{1}{10}P_A$, i.e., in transmission from A to B, the power becomes $\dfrac{1}{10}$ of that of A, then

$$10\log_{10} \frac{P_B}{P_A} = 10\log_{10} 10 = 10 \text{ db}$$

Similarly, if $P_B = \dfrac{1}{2}P_A$, then power loss $= 10\log_{10} 2 = 3$ db. Thus, power loss in (db) is

$$\text{Power loss} = 10\log_{10} \frac{P_B}{P_A} = 20\log_{10} \left|\frac{V_A}{V_B}\right|$$

If the distance between A and B is l, then power loss is proportional to l.
Hence attenuation in (Neper)

$$= 2.303 \times l \log_{10} \frac{|V_A|}{|V_B|} \text{ and attenuation in decibel} = 20 \log_{10} \frac{|V_A|}{|V_B|}$$

$$1 \text{ Neper} = \frac{20}{2.303} = 8.686 \text{ db}$$

For lossless line $R = G = 0$.

$$\therefore \qquad r = \alpha + i\beta = \sqrt{(R + i\omega L)(G + i\omega C)} = i\omega\sqrt{LC}$$

Hence, $\alpha = 0$ and $\beta = \omega\sqrt{LC}$

$\therefore$ Wave velocity $v = \dfrac{\omega}{\beta} = \dfrac{1}{\sqrt{LC}}$

Characteristic impedence $= \sqrt{\dfrac{L}{C}}$

For small loss transmission line $R \ll \omega L$ and $G \ll \omega C$.

$$\therefore \qquad r = \sqrt{(R + i\omega L)(G + i\omega C)}$$

$$= \sqrt{RG + i\omega(LG + RC) - \omega^2 LC)}$$

$$= \sqrt{i\omega(LG + CR) - \omega^2 LC}$$

RG being smaller of all quantities

$\therefore$
$$r = (-\omega^2 LC)^{1/2} \sqrt{\left[1 - i\left(\frac{G}{\omega C} + \frac{R}{\omega L}\right)\right]}$$

$$= i\omega\sqrt{LC})\left[1 - \frac{i}{2}\left(\frac{G}{\omega C} + \frac{R}{\omega L}\right)\right]$$

$$= \frac{\sqrt{LC}}{2}\left[\frac{G}{C} + \frac{R}{L}\right] + i\omega\sqrt{LC}$$

$$= \frac{1}{2}\left[R\sqrt{\frac{C}{L}} + G\sqrt{\frac{L}{C}}\right] + i\omega\sqrt{LC}$$

$\therefore$
$$\alpha = \frac{1}{2}\left[R\sqrt{\frac{C}{L}} + G\sqrt{\frac{L}{C}}\right]$$

Under these conditions also $Z_0 \approx \sqrt{\frac{L}{C}}$

$\therefore$
$$\alpha = \left(\frac{R}{2Z_0} + \frac{GZ_0}{2}\right)$$

$$\beta = \omega\sqrt{LC}$$

and
$$v = \frac{\omega}{\beta} = \frac{1}{\sqrt{LC}}$$

3.15 STANDING WAVES ON TRANSMISSION LINE AND STANDING WAVE RATIO

When the transmission line is of finite length and has terminal load other than the characteristic load then both incident and reflected waves are always present. On putting Z_0 (the characteristic impedance) as terminal load, then, only incident waves are present there because in that state the load absorbs all the energy received by it. In this way the line behaves like an infinite line. Incident and reflected waves propagate in opposite direction and give rise to stationary waves.

At any point of the transmission line, the ratio of reflected voltage to the incident voltage is called *voltage reflection co-efficient* and is represented by r_c

$\therefore$
$$r_c = \frac{V_2 e^{rx}}{V_1 e^{-rx}} = \frac{V_2}{V_1} e^{2rx} \tag{3.73}$$

x is measured from the receiving end. Hence, for $x = -l$

$$r_c = \frac{V_2}{V_1} e^{-2rl}$$

and for $x = 0$

$$r_c = r_0 = \frac{V_2}{V_1} = \frac{\dfrac{V_R}{2}\left(1 - \dfrac{Z_0}{Z_R}\right)}{\dfrac{V_R}{2}\left(1 + \dfrac{Z_0}{Z_R}\right)} = \frac{Z_R - Z_0}{Z_R + Z_0}$$

when

$$\begin{aligned}
Z_R &= Z_0 & r_0 &= 0 & &\text{(Proper matching state)} \\
Z_R &= 0 & r_0 &= -1 & &\text{(Short-circuited line)} \\
Z_R &= \alpha & r_0 &= +1 & &\text{(Open circuit)}
\end{aligned}$$

In this way, due to reflection from load, the phase of the reflected wave is changed with respect to the incident wave. In case of short circuited line, the phase change is π (due to negative sign) and in case of open circuited line, there is no phase change. In general, if this phase difference is ϕ, then at $x = 0$

$$r_c = |r_0| e^{i\phi}$$

where $|r_0|$ is the magnitude of voltage reflection co-efficient.

Hence, for

$$x = -l$$

$$V = V_1 e^{rl} + V_2 e^{-rl}$$

$$= V_1 e^{rl}\left[1 + \frac{V_2}{V_1} e^{-2rl}\right]$$

$$= V_1 e^{rl}\left[1 + |r_0| e^{i\phi} e^{-2(\alpha + i\beta)l}\right]$$

For simplicity, if line is lossless, then $\alpha = 0$

and

$$V = V_1 e^{i\beta l}\left[1 + |r_0| e^{-i(2\beta l - \phi)}\right]$$

or

$$V = V_1 e^{i\beta l}\left[\{1 + |r_0|\cos(2\beta l - \phi)\} - i\{|r_0|\sin(2\beta l - \phi)\}\right]$$

$\therefore$

$$|V| = V_1\left[\{1 + |r_0|\cos(2\beta l - \phi)\}^2 + \{|r_0|\sin(2\beta l - \phi)\}\right]^{1/2}$$

$$= V_1\left[\{1 + |r_0|^2 + 2|r_0|\cos(2\beta l - \phi)\}\right]^{1/2}$$

In this way, the voltage depends upon the distance l from the receiving end. The maximum value of voltage is

$$|V_{max}|^2 = V_1[1 + |r_0|^2 + |2r_0|]^{1/2} \quad \text{when} \quad \cos(2\beta l - \phi) = +1$$

and

$$|V_{max}| = V_1(1 + |r_0|) = V_1\left(1 + \frac{V_2}{V_1}\right) = V_1 + V_2$$

The minimum value of resultant voltage is when

$$\cos(2\beta l - \phi) = -1$$

$\therefore$

$$|V_{min}| = V_1(1 - |r_0|) = (V_1 - V_2)$$

The points, where resultant voltage is maximum are called *voltage antinodal points* and the points of minimum voltage are called *voltage nodal points*.

The separation between neighbouring nodal or antinodal points is $\lambda/2$.

The ratio of maximum voltage and minimum voltage on transmission line is called *voltage standing wave ratio*.

$$\text{Voltage standing wave ratio } (\rho) = \frac{|V|_{max}}{|V|_{min}} = \frac{V_1 + V_2}{V_1 - V_2}$$

$$= \frac{1 + |r_0|}{1 - |r_0|} = \frac{|Z_R|}{|Z_0|}$$

3.16 EFFECT OF TERMINAL LOAD

The Eqs (3.68, 3.68a, 3.69 and 3.70) give the effect of terminal load on the behaviour of the transmission line. Here we will discuss some special cases.

(i) Matched Line, i.e. $Z_R = Z_0$ Terminal Load is Equal to the Characteristic Impedance

In this case $V_1 = V_R$ and $V_2 = 0$ [Eqs (3.67) and (3.67a)].

Hence

$$V = V_R e^{-rx}$$

and

$$I = \frac{V_R}{Z_0} e^{-rx}$$

and

$$Z_R = \frac{V}{I} = Z_0.$$

It means there is no reflected wave on line and only incident wave is present. Due to this, wave voltage and current values decreases exponentially in positive X-direction, i.e., from source end to receiving end. Absence of reflected wave means, whatever energy is received by receiving end is completely absorbed by the load. In this way the load is completely matched with the transmission line, i.e. the impedance at every point of line is equal to the characteristic impedance.

This situation is like an infinite line as such the line completed by characteristic impedance behaves like an infinite line. The variation of voltage on line is shown in Fig. 3.37. Same change will occur in current also. At source end $x = -l$, hence

$$V_s = V_R e^{rl}$$

and

$$I_s = \frac{V_R}{Z_0} e^{rl}$$

and input impedance $Z_R = 0$.

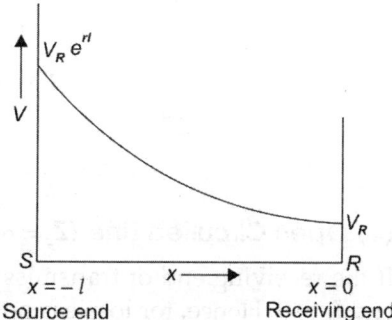

Fig. 3.37

(ii) Short-circuited Line ($Z_R = 0$)

In short circuited line, the impedance at receiving end is zero, hence, voltage will also be zero. For simplicity, let us assume the line be lossless then

$$r = \alpha + i\beta = i\beta$$

$$\therefore \qquad V = V_R \cos hrx - I_R Z_0 \sin hrx$$

$$= -I_R Z_0 \sinh(i\beta_x) = -il_R Z_0 \sin \beta x \tag{3.74}$$

and $\qquad I = I_R \cos hrx = I_R \cos(i\beta x) = I_R \cos \beta x \tag{3.74a}$

$$\therefore \qquad z = \frac{V}{I} = -iZ_0 \tan \beta x \tag{3.75}$$

Input impedance $(Z_s)_{sc}$ of short-circuited line

$$= \text{the value of } z \text{ at } x = -l$$

$$= -iZ_0 \tan(-\beta l)$$

$$= iZ_0 \tan(\beta l) \tag{3.76}$$

If length of line $l = \dfrac{\lambda}{2}, \lambda, \dfrac{3\lambda}{2}, \dots$ etc, then

$$\beta l = \frac{2\pi}{\lambda} l = \pi, 2\pi, 3\pi, \dots (n\pi)$$

and hence $\tan \beta l = 0$ as such, the input impedance of line will be zero. Such impedance is presented in series L–C resonance circuit which means that when line length is $\dfrac{n\lambda}{2}$, where $n = 1, 2$, etc. Under these circumstances, the line will behave like L-C series resonant circuit. When $l = \dfrac{\lambda}{4}, \dfrac{3\lambda}{4}, \dots,$

$$\beta l = \frac{2\pi}{\lambda} l = \frac{\pi}{2}, \frac{3\pi}{2}, \dots \quad \text{and} \quad \tan \beta l = \infty.$$

This means that input impedance of the line is infinity. Its behaviour is like L–C parallel resonance circuit. The changes in voltage and current is short circuited line of length λ are shown in Fig. 3.38.

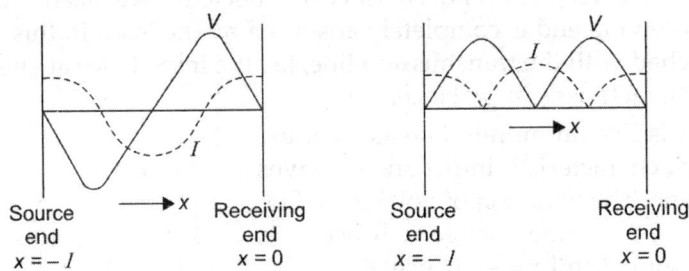

Source end $x = -l$ — Receiving end $x = 0$

Source end $x = -l$ — Receiving end $x = 0$

Fig. 3.38

(iii) Open Circuited Line ($Z_R = \infty$)

If the receiving end of transmission line is open, i.e., $Z_R = \infty$ as such the current at this end $I_R = 0$. Hence, for lossless open line

$$V = V_R \cos hrx = V_R \cos h(i\beta x) = V_R \cos \beta_x$$

and $\qquad I = -\dfrac{V_R}{Z_0} \sin hrx = -i\dfrac{V_R}{Z_0} \sin(\beta x)$

and $\qquad Z = iZ_0 \cot(\beta x)$

The input impedance of open line

$$(Z_S)_{oc} = \text{The value of } z \text{ at } l = -x$$
$$= iZ_0 \cot(-\beta l) = -iZ_0 \cot(\beta l)$$

When line is of length

$$l = \frac{\lambda}{4}, \frac{3\lambda}{4}, ..., \quad \beta l = \frac{\pi}{2}, \frac{3\pi}{2}, ...$$

and $\qquad \cot(\beta l) = 0 \quad \therefore \quad (Z_S)_{OC} = 0.$

Hence, at source end, the line will behave like at LC series resonance circuit.

When line length is

$$l = \frac{\lambda}{2}, \lambda, \frac{3\lambda}{2}, ..., \quad \beta l = \pi, 2\pi, 3\pi$$

and $\qquad \cot(\beta l) = \infty \qquad (Z_S)_{OC} = \infty$

As such the line behaves like an LC parallel resonance circuit. The voltage and current variation on an open line of length λ are shown in Fig. 3.39.

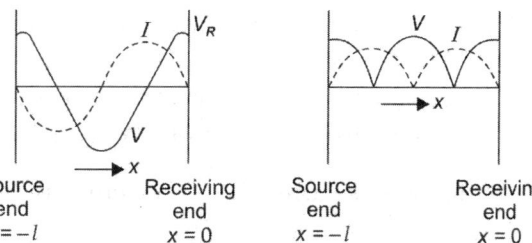

Source end $x = -l$ · Receiving end $x = 0$ · Source end $x = -l$ · Receiving end $x = 0$

Fig. 3.39

From above discussion, for a line of length l

$$(Z_S)_{SC} = iZ_0 \tan \beta l$$
$$(Z_S)_{OC} = -iZ_0 \cot \beta l$$
$$(Z_S)_{SC} = (Z_S)_{OC} = Z_0^2$$

or $\qquad Z_0 = \sqrt{(Z_S)_{SC}(Z_S)_{OC}} \qquad\qquad (3.77)$

(iv) Line of Length $\frac{\lambda}{4}$ and Terminal Load Z_R

For line of length $l = \frac{\lambda}{4}, \ \beta l = \frac{2\pi}{\lambda} \cdot \frac{\lambda}{4} = \frac{\pi}{2}$ and $\tan \beta l = \infty$

Hence input impedance

$$Z_S = Z_0 \frac{Z_R + iZ_0 \tan \beta l}{Z_0 + iZ_R \tan \beta l} = Z_0 \frac{\dfrac{Z_R}{\tan \beta l} + iZ_0}{\dfrac{Z_0}{\tan \beta l} + iZ_R}$$

$$= Z_0 \left(\frac{Z_0}{Z_R} \right)$$

$$= \frac{Z_0^2}{Z_R}$$

$\therefore \qquad\qquad Z_S \propto \dfrac{1}{Z_R}$

Such a line behaves like impedance transformer. If Z_R is large, the input impedance Z_S is small and if Z_R is small, Z_S is large.

(v) $\lambda/2$ Line of Terminal load Z_R

For line of length $\dfrac{\lambda}{2}$, $\beta l = \pi$ and $\tan \beta l = 0$.

Hence source end impedance (input impedance)

$$Z_s = Z_0 \frac{Z_R + iZ_0 \tan \beta l}{Z_0 + iZ_R \tan \beta l}$$

$$= Z_0 \left(\frac{Z_R}{Z_0} \right) = Z_R$$

Such a line behaves as (1:1) impedance transformer.

Example 3.11: An infinite lossless line is connected to a battery of voltage V_0 at $t = 0$. Show the voltage and current distribution on the line at a latter time t. What happens to the energy supplied by the battery during this time?

Solution: Only a forward wave $V(\omega t - \beta x)$ will be present. The conditions at the point $x = 0$ is $V(t, 0) = V_0$, $t > 0$.

There a constant voltage V_0 will appear on the line up to the point $x = vt_1$ in time t_1, where v is the propagation velocity.

The associated current is $\dfrac{V_0}{Z_0}$, where Z_0 is the characteristic impedance. Since the battery supplies a constant current $\dfrac{V_0}{Z_0}$, the energy supplied by the battery during the time t_1, is

$$E = V_0 \left(\frac{V_0}{Z_0} \right) t_1 = \frac{V_0^2}{Z_0} t_1 \text{ Joules}$$

The capacitance of the line length $x = vt_1$ has been charged to voltage V_0. The energy stored in the electric field is $\dfrac{1}{2}(Cx)V_0^2$. The magnetic field of inductance xL stores energy $\dfrac{1}{2}(xL)I^2$. Hence energy stored in electric and magnetic field is

$$= \frac{1}{2}CxV_0^2 + \frac{1}{2}xL \left(\frac{V_0}{Z_0} \right)^2 = \frac{1}{2}\frac{V_0^2}{Z_0} \left(Z_0 C + \frac{L}{Z_0} \right) x$$

$$= \frac{1}{2}\frac{V_0^2}{Z_0} \left[\sqrt{\frac{L}{C}} \cdot C + \frac{L}{\sqrt{\frac{L}{C}}} \right] x \qquad \because Z_0 = \sqrt{\frac{L}{C}}$$

$$= \frac{1}{2}\frac{V_0^2}{Z_0} [\sqrt{LC} + \sqrt{LC}]x$$

$$= \frac{V_0^2}{\sqrt{\frac{L}{C}}} \cdot \sqrt{LC}\, vt_1 \qquad \because v = \frac{1}{\sqrt{LC}}$$

$$= V_0^2 \sqrt{\frac{C}{L}} t_1 = \frac{V_0^2}{Z_0} t_1 = \text{Energy supplied by the battery.}$$

Thus, the energy supplied by the battery is stored in the form of electric and magnetic fields.

Example 3.12: A source of voltage $V(t) = at$, $0 < t < t_1$ and $V(t) = at_1$ for $t > t_1$ is switched on to a lossless line of characteristic impedance Z_0. What is the voltage and current distribution in the line? What is the energy supplied by source?

Solution: The voltage on the line for $t < t_1$ and for $t > t_1$ is shown in Fig. 3.40 for $t \le t_1$, the energy supplied by the battery is given by

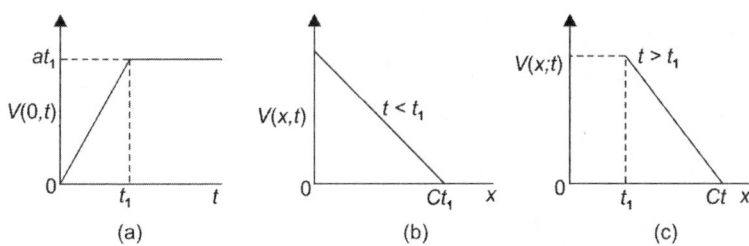

Fig. 3.40

$$E_1 = \int_0^{t_1} V(0,t) I(0,t) \, dt$$

$$= \int_0^{t_1} at \cdot \frac{at}{Z_0} \, dt = \int_0^{t_1} \frac{a^2}{Z_0} t^2 \, dt = \frac{a^2}{3Z_0} t_1^3$$

for $t \ge t_1$, the energy supplied is

$$E_2 = \int_0^t V(xt) \cdot I(xt) \, dt$$

$$= \int_0^{t_1} \frac{a^2 t^2}{Z_0} \, dt + \int_{t_1}^t \frac{(at_1)^2}{Z_0} \, dt$$

$$= \frac{a^2 t_1^3}{3Z_0} + \frac{a^2 t_1^2}{Z_0}(t - t_1)$$

$$= \frac{a^2 t_1^2}{Z_0}\left[\frac{t_1}{3} + t - t_1\right]$$

or $$E_2 = \frac{a^2 t_1^2}{Z_0}\left[t - \frac{2}{3} t_1\right].$$

Example 3.13: Calculate the impedance offered by a stretched string to the transverse waves propagating through it.

Solution: Let us consider a progressive wave on a string which is generated at one end by an impressed oscillating force $F = F_0 e^{i\omega t}$. The force is confined to the plane of paper and is in a direction transverse to the string (Fig. 3.41). If the

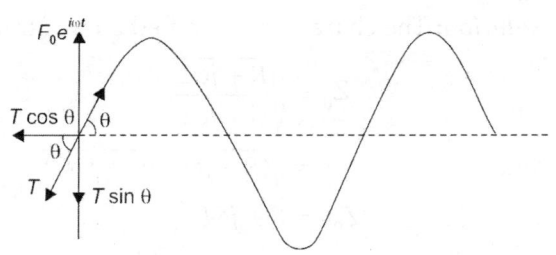

Fig. 3.41

string has a constant tension T, then at $x = 0$ end of the string, the applied force and tension component $T \sin \theta$ always balance each other, i.e.

$$F_0 e^{i\omega t} = -T \sin \phi \approx -T \tan \theta = -T \frac{\partial y}{\partial x} \text{ for small } \theta.$$

If Y represents the displacement of the progressive wave, then

$$Y = a e^{i(\omega t - kx)}$$

where the amplitude a may be complex due to phase relation with oscillating force. At $x = 0$

$$F_0 e^{i\omega t} = -T\left(\frac{\partial y}{\partial x}\right)_{x=0} = iT k a e^{i\omega t}$$

or

$$a = \frac{F_0}{ikT} = \frac{F_0 \omega}{i\omega kT} = -i\frac{F_0}{\omega}\left(\frac{c}{T}\right)$$

$$= \frac{F_0}{\omega}\left(\frac{c}{T}\right) e^{-\frac{i\pi}{2}}$$

where c is the velocity of propagation and the displacement is lagging behind the force in phase by $\frac{\pi}{2}$.

$\therefore$

$$Y = -i\frac{F_0}{\omega}\frac{c}{T} e^{i(\omega t - kx)}$$

and the transverse velocity of the particle is

$$v = \dot{y} = F_0 \frac{c}{T} e^{i(\omega t - kx)}$$

In mechanical vibrations, the characteristic impedance is defined as the ratio of transverse force and transverse velocity, i.e.

$$Z = \frac{\text{Transverse force}}{\text{Transverse velocity}} = \frac{F}{v}$$

$$= \left| \frac{F_0 e^{i\omega t}}{F_0 \frac{c}{T} e^{i(\omega t)}} \right| = \frac{T}{c}$$

Hence the characteristic impedance $= \frac{T}{c}$.

Example 3.14: Characteristic impedance and propagation constant of a 1200 Hz line are $Z_0 = (650 - j150)$ and $r = (0.005 - j0.007)$. Calculate distribution parameters of the line.

Solution: The characteristic impedance of transmission line is

$$Z_0 = \sqrt{\frac{R + j\omega L}{G + j\omega L}} \tag{1}$$

and

$$r = \sqrt{(R + j\omega L)(G + j\omega C)} \tag{2}$$

$\therefore$

$$Z_0 r = R + j\omega L \tag{3}$$

and

$$\frac{r}{Z_0} = G + j\omega L \tag{4}$$

$\therefore$ $\qquad R + j\omega L = (650 - j150)(0.005 - j0.007)$
$$= 4.3 + j3.8\,\Omega \cdot \mathrm{km}^{-1}$$

Comparing real and imaginary roots, we have
$$R = 4.3\ \Omega/\mathrm{km}$$
and $\qquad \omega L = 3.8\ \Omega/\mathrm{km}$

$\therefore$ $\qquad L = \dfrac{3.8}{\omega} = \dfrac{3.8}{2\pi \times 1200} = 0.5\ \mathrm{mH/km}.$

From equation (4), we have
$$G + j\omega C = \frac{(0.005 - J0.007)}{650 - J150} = \frac{5 + 7j}{6.5 + j1.5} \times 10^{-5}$$
$$= \{5 + j(11.9)\} \times 10^{-6}$$

Again equating real and imaginary roots, we have
$$G = 5 \times 10^{-6}\ \mathrm{mhos/km}$$
and $\qquad \omega C = 11.9 \times 10^{-6}\ \mathrm{mhos/km}$

$\therefore$ $\qquad C = \dfrac{11.9 \times 10^{-6}}{2\pi \times 1200} = 1.58 \times 10^{-3}\ \mu\mathrm{F/km}$

Example 3.15: For an open transmission line, the constants given are
$$R = 0.4\,\Omega/\mathrm{km},\ L = 2.5\ \mathrm{mH/km},\ \text{and}\ C = 0.01\ \mu\mathrm{F/km}$$

and G is negligible. If the frequency of the wave propagating on line is 10 kHz. Then, find (a) characteristic impedance, (b) damping constant α, (c) phase constant β, (d) phase velocity.

Solution: Characteristic impedance of transmission line is
$$Z_0 = \sqrt{\frac{R + j\omega L}{G + j\omega C}} \qquad \because G = 0 \quad \text{and} \quad R \ll \omega$$
$$= \sqrt{\frac{L}{C}} = \sqrt{\frac{2.5 \times 10^{-3}}{0.01 \times 10^{-6}}} = 500\,\Omega$$

Propagation constant
$$r = \alpha + i\beta = \sqrt{(R) + j\omega L)(G + j\omega C)}$$
$$= j\omega\sqrt{LC}\left(1 + \frac{R}{j\omega L}\right)^{1/2}$$
$$\approx j\omega\sqrt{LC}\left(1 + \frac{R}{2j\omega L}\right) \quad \text{(Using Binomial theorem)}$$
$$\approx \frac{R}{2}\sqrt{\frac{C}{L}} + j\omega\sqrt{LC}$$
$$= \frac{R}{2Z_0} + j\omega\sqrt{LC}$$

(b) $\alpha = \dfrac{R}{2Z_0} = \dfrac{0.4}{2 \times 500} = 4 \times 10^{-4}\ \mathrm{Neper^*/km}$

----

* Neper = 8.686 db

(c) $\beta = \omega\sqrt{LC} = 6.28 \times 10^4 (2.5 \times 10^{-3} \times 10^{-8})^{1/2}$

$\qquad = 31.4 \times 10^{-2}\,\text{rad/km}$

(d) $v = \dfrac{\omega}{\beta} = (2.5 \times 10^{-3} \times 10^{-8})^{1/2} = 2 \times 10^5\,\text{km/s}$

Example 3.16: For lossless transmission line following constants are given

$\qquad L = 0.6\,\mu\text{H/m}, C = 240\,\mu\mu\text{f/m}$, and $\omega = 2\pi \times 10^8\,\text{rad/s}$

Then find (i) β and wavelength (λ), (ii) Z_0, (iii) if length of line is $l = \dfrac{\lambda}{4}$ and terminal load impedance $Z_R = -i\,100\Omega$. Then, find input impedance (Z_S).

Solution: For lossless line

$$\beta = \omega\sqrt{LC} = 2\pi \times 10^8 \sqrt{6 \times 10^{-7} \times 240 \times 10^{-12}}$$

$$= 2.4\pi = 7.536\,\text{rad/m}$$

and wavelength

$$\lambda = \frac{2\pi}{\beta} = \frac{2\pi}{2.4 \times \pi} = \frac{1}{1.2} = 0.833\,\text{m}$$

(ii) For lossless transmission line

$$Z_0 = \sqrt{\frac{L}{C}} = \left(\frac{6 \times 10^{-7}}{240 \times 10^{-12}}\right)^{1/2} = 50\Omega$$

(iii) $Z_s = Z_0 \left[\dfrac{Z_R \cos h\,rl + Z_0 \sin h\,rl}{Z_0 \cos h\,rl + iZ_R \sin h\,rl}\right]$

for lossless transmission line $\alpha = 0,.\ r = i\beta$, then

$$\cos h\,rl = \cos\beta l \text{ and } \sin h\,rl = i\sin\beta l$$

$\therefore \qquad Z_S = Z_0 \left[\dfrac{Z_R \cos\beta l + iZ_0 \sin\beta l}{Z_0 \cos\beta l + iZ_R \sin\beta l}\right]$

But $Z_R = -i100\Omega$ $\beta l = \dfrac{2\pi}{\lambda} \cdot \dfrac{\lambda}{4} = \dfrac{\pi}{2}$

$\therefore \cos\beta l = 0$ and $\sin\beta l = 1$

$\therefore \qquad Z_S = Z_0 \left(\dfrac{iZ_0}{iZ_R}\right) = \dfrac{Z_0^2}{Z_R}$

$$= \frac{(50)^2}{-i100} = i25\Omega$$

REVIEW QUESTIONS AND PROBLEMS

1. Obtain resonant frequencies in an inductive coupled LC circuit. What is the effect of weak and strong coupling on resonant curve?
2. What is meant by normal mode of vibration of a coupled oscillator? Find frequencies of normal modes of vibration of a coupled oscillator.
3. Explain vibrations in one dimensional monoatomic lattice. Derive dispersion relation for this lattice.

4. What is transmission line? Define its characteristic impedance and propagation constant. Deduce a relation between voltage and current on a lossless transmission line.

5. Discuss the energy losses in transmission line.

6. Define and explain the following terms:
   (i) Characteristic impedance
   (ii) Infinite line
   (iii) Voltage standing wave ratio.

7. Discuss the dynamics of one dimensional monoatomic lattice. Explain wave velocity and group velocity, obtain expressions for them.

8. Discuss the electrically coupled circuit in detail.

9. What are normal co-ordinates and normal modes of a coupled system? What is their significance? Explain with the help of an example. How can you determine the frequencies of the normal modes experimentally?

10. What is reflected impedance? Show that for a perfect transformer, the impedance seen by the generator consists of the primary impedance in parallel with an impedance $\left(\dfrac{N_p}{N_s}\right)^2 Z_R$, where N_p and N_s are the number of primary and secondary transformer coil turns and Z_R is the load impedance.

11. What is the difference between group velocity and wave velocity? Two simple harmonic waves $y_1 = a\sin(\omega_1 t - k_1 x)$ and $y_2 = a\sin(\omega_2 t - k_2 x)$ superimpose over each other. Calculate group velocity. [Symbols have their usual meaning.]

12. Obtain dispersion relation for one dimensional monoatomic lattice. Show that such a lattice acts as a low pass filter.

13. Obtain frequencies of the normal modes of oscillations of two coupled LC circuits if coupling is capacitive.

14. Two identical LC circuits are capacitively coupled. For each circuit $L = 5$ mH and $C = 2\ \mu F$. Calculate normal mode frequencies (ω_1 and ω_2).

$$\left[Ans.\ \omega_1 = \sqrt{\dfrac{1}{LC}} = 10^4\ \text{rad/s}\quad \omega_2 = \sqrt{\dfrac{3}{LC}} = \sqrt{3}\times 10^4 \right]$$

15. Wave velocity in water is $\sqrt{\dfrac{g\lambda}{2\pi}}$. Prove that group velocity will be half of the velocity.

   g is acceleration due to gravity and λ is the wavelength of waves in water.

$$\left[Hint.\ V = \dfrac{\omega}{k} = \sqrt{\dfrac{g}{k}}\quad \text{and}\quad V_g = \dfrac{\partial \omega}{\partial k} = \dfrac{1}{2}V \right]$$

16. Define the following terms used for coupled circuits: (i) Natural frequency, (ii) Normal modes, (iii) Reflected impedance. Obtain the expression for the frequencies of normal modes of the two inductively coupled identical LC circuits. Show the effect of coupling with the help of the frequency response curve.

17. What is meant by dispersion relation? Write equation of motion for a one dimensional diatomic lattice. Show that the dispersion curve for such a lattice and explain acoustic and optical branches and forbidden gap.

18. According to de Broglie's hypothesis a moving particle is associated with matter waves. Prove that the group velocity of these waves is equal to the velocity of the particles.

19. Prove that for lossless line

   (i) The velocity of current and voltage waves is $\dfrac{1}{\sqrt{LC}}$, where L and C are respectively the inductance and capacitance per unit length.

   (ii) The input impedance of $\dfrac{\lambda}{4}$ line is inversely proportional to the normal impedance.

20. The input impedance of an open circuited line is $(-j\,188)\Omega$. When it is short-circuited, the input impedance is $(+j\,53.2)\Omega$. Calculate its characteristic impedance.

21. Obtain resonant frequencies in an inductance coupled LC circuit. What is the effect of weak and strong coupling on resonant curve?

22. A coupled oscillator is excited in mixed mode. Write down its equation of motion. How the displacement of a component oscillator will change with time?

23. Explain the transient behaviour of a coupled oscillator. If the damping forces are acting, what effect will be on transient behaviour?

24. Prove that the total energy of a coupled system is constant.

25. Describe various methods of terminating a transmission line. Find out input impedance for transmission line terminating with infinite load.

26. Characteristic impedance and propagation constant of a 1200 Hz line are $Z_0 = (650 - j\,150)$ and $r = (0.005 - j\,0.007)$. Calculate distribution parameters of the line.

27. Explain the motion of two coupled simple harmonic oscillators and obtain the equation of motion of two coupled oscillators. Prove that the total energy of a coupled system remains constant.

28. Discuss the vibrations of a one dimensional diatomic lattice and derive the dispersion relation for this lattice.

29. Prove that the motion of two types of atoms of a one dimensional diatomic lattice in optical branch are in opposite direction to each other, whereas in acoustic branch, they are in the same direction.

30. Derive the voltage and current wave equation of a general transmission line and express the solution of these equations in terms of characteristic impedance and propagation constant.

31. Terminal load on a lossless transmission line of length $\left(\dfrac{\lambda}{8}\right)$ is $(25 + j\,50)$ ohm. If characteristic impedance of the line is $100\,\Omega$, then calculate its input impedance.

32. Define coupled circuit and co-efficient of coupling. Prove that impedance of primary circuit changes when it is inductively coupled with a secondary circuit. In this arrangement also show that the primary circuit absorbs more energy from the alternating source.

33. Explain phase velocity and group velocity deriving their expression. Hence, differentiate, non-dispersive, normal dispersive and anomalous dispersive wave.

34. In a lossless ideal transmission line, find the velocity of propagation of current and voltage wave.

35. Define characteristic impedance of a transmission line. Prove that the characteristic impedance of an ideal transmission line is given by $Z = \sqrt{\dfrac{L}{C}}$. Here L is inductance perunit length and C is the capacitance per unit length.

36. Show that the characteristic impedance offered by a loss free gas to the sound waves travelling in it is given by $Z = \sqrt{\rho E}$, where ρ is the density of the gas and E is the bulk modulus.

37. A string of length $3l$ and negligible mass is attached to two fixed supports at its end. Tension in the string is T.
    (a) A particle of mass m is attached at a distance l from one end of the string. Set-up the equation for the small transverse oscillation of mass m and find its period.
    (b) An additional mass m is connected to the string dividing it into three equal segments each with tension T. Sketch the appearance of the string and masses in the two separate normal modes of transverse oscillations.
    (c) Calculate ω for the normal mode which has the higher frequency.

$$\left[\text{Ans. (a) } T = 2\pi\sqrt{\dfrac{2\,ml}{3T}}, \quad \text{(c) } \omega = \sqrt{\dfrac{3T}{ml}} \right]$$

SHORT ANSWER QUESTIONS

1. Two masses 0.1 kg and 0.5 kg are connected by a spring of force constant 400 N/m. If after stretching to a very small distance, these masses are left. Find the frequency of vibration of the system.

Ans. $v = \dfrac{1}{2\pi}\sqrt{\dfrac{k}{\mu}}, \mu = \dfrac{m_1 m_2}{m_1 + m_2} = \dfrac{0.1 \times 0.5}{0.6} = \dfrac{0.05}{0.6} = \dfrac{5}{60}$

$v = \dfrac{1}{2\pi}\sqrt{\dfrac{400 \times 60}{5.0}} = \dfrac{1}{2\pi}\sqrt{400 \times 12} = \dfrac{20 \times 2}{2\pi}\sqrt{3} = 11\text{Hz}$

2. In above problem, what is the ratio of the kinetic energies of the two masses?

Ans. $E = \dfrac{p^2}{2m}$ $\therefore$ $\dfrac{E_1}{E_2} = \dfrac{m_2}{m_1} = \dfrac{0.5}{0.1} = 5$

3. The input impedance of an open circuited line is $(-j188)$ ohm. When it is short circuited, the input impedance is $(+j53.2)$ ohm. Calculate its characteristic impedance.

Ans. For lossless line $Z_0 = \sqrt{(Z)_{OS}(Z)_{SC}}$

$= \sqrt{(-j188) \times (j53.2)} = 100\,\Omega$

4. The relation between input impedance and terminal impe-dance for $\lambda/4$ line is related as ...

Ans. For $\dfrac{\lambda}{4}$ line $\beta l = \dfrac{2\pi}{\lambda} \cdot \dfrac{\lambda}{4} = \dfrac{\pi}{2}$ and $\tan \beta l = \infty$.

Hence, input impedance $Z_S = Z_0 \dfrac{Z_R + jZ_0 \tan \beta l}{Z_0 + jZ_R \tan \beta l}$

$$\text{or } Z_S = \frac{\dfrac{Z_R}{\tan\beta l} + jZ_0}{\dfrac{Z_0}{\tan\beta l} + jZ_R} \cdot Z_0 = Z_0\left(\frac{Z_0}{Z_R}\right) = \frac{Z_0^2}{Z_R} \text{ or } Z_S \propto \frac{1}{Z_R}$$

5. Terminal load on a lossless transmission line of $(\lambda/8)$ is $25 + j50\,\Omega$. If characteristic impedance of the line is $100\,\Omega$, then calculate input impedance.

Ans. $Z_S = Z_0 \dfrac{Z_R + Z_0 \tan h\, rl}{Z_0 + Z_R \tan h\, rl}$ for lossless line $\alpha = 0, r = i\beta$

$$Z_S = Z_0 \frac{Z_R + iZ_0 \tan\beta l}{Z_0 + iZ_R \tan\beta l} \quad \beta l = \frac{\lambda}{8} \cdot \frac{2\pi}{\lambda} = \frac{\pi}{4} \text{ and } \tan\beta = 1$$

$$= 100 \times \frac{(25 + i50 + i100)}{100 + i(25) + i50} = (160 + i220)\,\Omega$$

6. Find the minimum length of lossless short circuited trans-mission line, when it acts as LC, series resonance circuit type at the input. Line is working at 50 kHz. Wave velocity is equivalent to velocity of light.

Ans. $(ZS)_{SC} = iZ_0 \tan\beta l = 0$ at resonance.

$\therefore \qquad \beta l = n\pi\ (n = 1, 2, 3, ...)$

for minimum length $n = 1$ and $\quad l = \dfrac{\pi}{\beta} = \dfrac{\pi\lambda}{2\pi} = \dfrac{\lambda}{2} = \dfrac{v}{2f}$

or $\qquad\qquad\qquad\qquad l = \dfrac{3 \times 10^8}{2 \times 50 \times 10^3} = 3 \times 10^3 \text{ m}$

7. If the phase velocity of component waves is same, then superposed wave moves with the same velocity $(v = v_0)$.

Ans. $v_p = \dfrac{\omega_1}{k_1} = \dfrac{\omega_2}{k_2} = v_0, \quad v_g = \dfrac{d\omega}{dk} = \dfrac{\omega_1 - \omega_2}{k_1 - k_2} = v_0$

8. Prove that on the basis of classical mechanics, the group velocity of matter waves is equal to particle velocity.

Ans. $\lambda = \dfrac{h}{p}$ and energy $E = \dfrac{p^2}{2m} = \dfrac{h^2 k^2}{8\pi^2 m} \quad \lambda = \dfrac{2\pi}{k}$

and $E = \dfrac{h\omega}{2\pi} = \dfrac{h^2 k^2}{8\pi^2 m}$ or $\omega = \dfrac{hk^2}{4\pi m} \quad \therefore \quad v_g = \dfrac{d\omega}{dk} = \dfrac{P}{m} = v$

9. In case of standing waves on transmission line, what is voltage standing wave ratio?

Ans. Voltage standing wave ratio $(\rho) = \dfrac{|v|_{max}}{|v|_{min}} = \dfrac{v_1 + v_2}{v_1 - v_2} = \dfrac{1 + |r_0|}{1 - |r_0|} = \dfrac{|Z_R|}{|Z_0|}$

10. Three equal masses m are equally spaced along a string of length $4a$. The tension in the string is T. Find the three normal mode frequencies.

Ans. $\omega_1 = \omega_0 \sqrt{1 - \dfrac{1}{\sqrt{2}}}, \quad \omega_2 = \omega_0, \omega_3 = \omega_0 \sqrt{1 + \dfrac{1}{\sqrt{2}}}$

where $\omega_0 = \sqrt{\dfrac{2T}{ma}}$

MULTIPLE CHOICE QUESTIONS

1. Wave velocity in water is $v = \sqrt{\dfrac{g\lambda}{2\pi}}$, where g is acceleration due to gravity and λ, the wavelength of waves in water. The group velocity of waves is:

    (a) v (b) $\dfrac{v}{2}$ (c) $2v$ (d) None of the above

2. According to de Broglie's hypothesis, a moving particle is associated with matter waves. The relation between v_g, the group velocity and v, the particle velocity is:

    (a) $v_g = v$ (b) $v_g = 2v$ (c) $v_g = \dfrac{v}{2}$ (d) There is no relation

3. The input impedance of $\dfrac{\lambda}{4}$ line, (Z_i) and terminal impedance Z_t are related as:

    (a) $Z_i \propto Z_t$ (b) $Z_i \propto \dfrac{1}{Z_t}$ (c) $Z_i^2 \propto Z_t$ (d) $Z_i \propto \dfrac{1}{Z_t^2}$

4. A short circuited transmission line of length $\dfrac{\lambda}{2}$ behaves like:

    (a) L-C parallel resonance circuit   (b) L-C series resonance circuit
    (c) Impedance transformer   (d) 1:1 impedance transformer

5. Two masses 1.0 kg and 5.0 kg are connected by a spring of forced constant 400 N/m and are placed on a frictionless table. The masses are slightly pulled apart and released. The frequency of the two-body system is:
    (a) 3.18 Hz   (b) 1.42 Hz   (c) 3.48 Hz   (d) 5 Hz

6. In above problem ratio of their kinetic energies is:
    (a) 5:1   (b) 1:5   (c) 4:1   (d) 1:4

ANSWERS

1. (b)   2. (a)   3. (a)   4. (b)   5. (c)   6. (a)

Wave Motion

4.1 INTRODUCTION

There are essentially two ways of energy transfer from one point to the other. The first involves the actual transportation of matter, i.e., the particles carrying energy more bodily from one place to the other whereas in the second mode, the energy is passed from one particle of the intervening medium to other and so on till the energy is conveyed to the last particle without actual transport of matter. It is the second mode in which we are interested and shall discuss in the present chapter. This mode of energy transfer is called *wave motion*.

The basic word in wave motion is disturbance or perturbation. The scope of this word need not be limited in a narrow sense to mean physical displacement of particles of a medium. In fact, a material medium is necessary. If some physical property of space can exist in vacuum, it then acts as a medium. We know that electric and magnetic fields can exist in vacuum. Disturbance in this case could be a perturbation (or change) in these fields. The disturbance travels in vacuum need no material medium for its propagation. Electromagnetic waves are an example of such a wave.

If we consider a system of particles in which every particle is tied to its neighbour by means of short-elastic strings, such that one cannot move without disturbing its neighbour. In this system if one particle is disturbed, it will also disturb its neighbour, which in turn disturbs its neighbour and so on. In this way the disturbance given to the first particle, disturbs the whole medium. This mode of energy transfer, we have defined is known as the *wave motion*.

From above example it is concluded that in order to set-up waves in a medium, the medium must possess certain definite properties which are:
1. The medium must possess good elastic nature so that one particle when distributed may have tendency to return to its original position, and also may be able to disturb its neighbour.
2. In order that the energy may be transferred efficiently the medium must possess some inertia to store energy which otherwise would dissipate, causing energy loss.
3. Every medium pose resistance of some kind which cause a loss of energy of oscillation of particles. If this resistance is high, the amplitude of disturbance will soon die out without propagation of energy to a large distance. Therefore, medium must possess low resistance.

The most common types of waves that we encounter are the harmonic waves in which the particles of the medium execute simple harmonic motion (oscillations) about their respective mean positions. If the particles vibrate in a plane perpendicular to the direction of wave propagation (i.e. the direction of the energy flow), the wave is said to be *transverse wave*. On the other hand, if the direction of vibration of the particles is in line with the direction of wave propagation, the waves are called *longitudinal waves*.

Further, if one particle executes the same motion as its preceeding one but lagging behind it, i.e. if there is a continuous change in phase from particle to particle, the waves are said to be *progressive waves*. On the other hand, if the motion does not progress further and the net transfer of energy through any point of the medium is zero, some of the particles are either thrown in a state of permanent rest or in permanent disturbance, the waves are called the *stationary waves*.

4.2 GENERAL EQUATION FOR A ONE-DIMENSIONAL PROGRESSIVE WAVE

We know that wave motion (transverse or longitudinal) is a result of oscillations of moving parts of continuous system. Figure 4.1 shows a graph of the displacement (disturbance) $\psi(x, t)$ of different particles against x at a given time. The length of the arrows represent the displacements of particles and the curve is the locus of displacement in continuous medium. This is what we observe as waves. The crests and troughs (in case of transverse waves) or compression and rarefractions (in case of longitudinal waves) move forward but the moving parts of the medium do not, they simply oscillate at their own location.

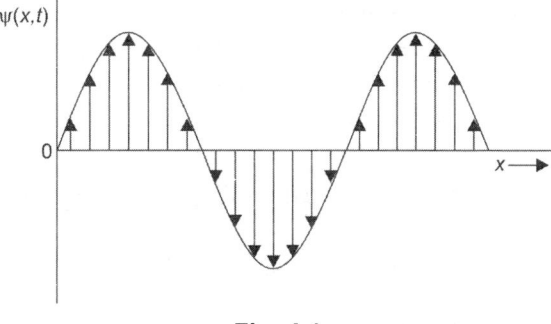

Fig. 4.1

What we observe is the state of motion of the moving parts which have their own motion which differs from others only in phase. We observe their phase relationships in the form of waves.

We shall deal with plane waves only which travel in one direction. Consider a plane perpendicular to the direction of wave propagation. In that plane all the particles have the same phase. In the next plane the phase of particles oscillations is different. The relation between the physical separation in space of these two planes and the phase difference between particle oscillations in these planes is called *wave equation*.

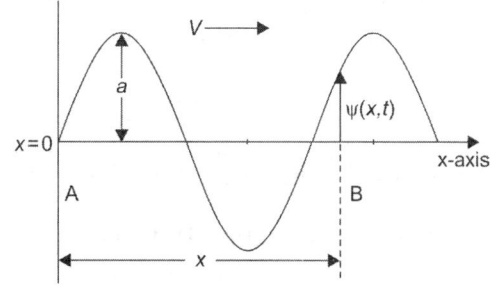

Fig. 4.2

Let us consider a harmonic wave propagating in medium with velocity v, i.e., velocity of propagation of disturbance through medium is v. Consider two planes A and B separated by a distance x (Fig. 4.2) and the disturbance is created at $x = 0$ (i.e. A) by giving oscillations of period T and amplitude a to the particles at this plane. The displacement of particles at plane A can be represented by equation

$$\psi(0, t) = a \sin \frac{2\pi}{T} t \qquad (4.1)$$

where $\psi(0, t)$ represents the displacement of particle located at position $x = 0$ and at time t. Since the disturbance has some finite velocity v, hence, the particles located at plane B at a distance x from the plane A will receive the disturbance after time x/v. Hence, the displacement of the particles at B will be same, which was that of the particles at A, x/v time earlier. Hence, the displacement of particles at x at time t is

$$= \text{displacement of particles at } x = 0 \text{ at time} \left(t - \frac{x}{v} \right)$$

$$= a \sin \left\{ \frac{2\pi}{T} \left(t - \frac{x}{v} \right) \right\}$$

or
$$\psi(x, t) = a \sin \left\{ \frac{2\pi}{T} \left(t - \frac{x}{v} \right) \right\} \tag{4.2}$$

This equation gives the displacement of the particles of the continuous medium as a function of x and t as a *harmonic wave* travels in the x direction with a velocity v. In writing the above equation, we have made two assumptions:

(i) The amplitude a of the particle oscillation does not change, as the wave propagates in the medium.

(ii) The medium is isotropic and homogenous, so that, the wave velocity v does not change from place to place.

In this way a wave travelling in the negative x-direction can be represented by equation

$$\psi(x, t) = a \sin \left\{ \frac{2\pi}{T} \left(t + \frac{x}{v} \right) \right\} \tag{4.2a}$$

Wavelength (λ), *the characteristics of wave motion is defined as the distance between two neighbouring particles vibrating in same phase.* In other words, the function $\psi(x, t)$ repeats itself after a distance λ. *Wavelength* λ is also called the *spacial periodicity of the wave.* As such, the wave is doubly periodic. It has temporal (in relation to time) periodicity T as well as spacial (in relation to space) periodicity λ.

The two characteristics of wave motion are related as:

$$v = \frac{\lambda}{T} = v\lambda \tag{4.3}$$

where $v = \frac{1}{T}$ and is called the *frequency*.

The Eq. (4.2) can be written as

$$\psi(x, t) = a \sin \left\{ 2\pi \frac{v}{\lambda} \left(t - \frac{x}{v} \right) \right\}$$

$$= a \sin \left\{ \frac{2\pi}{\lambda} (vt - x) \right\}$$

Here we define two more quantities called *wave vector* $K = \dfrac{2\pi}{\lambda}$ and the *angular frequency* $\omega = \dfrac{2\pi}{T} = 2\pi v$. In terms of ω and k, the wave equation is written as:

$$\psi(x, t) = a \sin (\omega t \mp kx) \tag{4.4}$$

Here negative (–) sign represents the harmonic wave propagating in positive X-direction and positive (+) sign represents the harmonic wave propagating in negative X-direction.

It may be mentioned here that the waves described by the above equations are equally well represented by a cosine function instead of a sine function.

4.3 DIFFERENTIAL FORM OF WAVE EQUATION

We will now derive the equation in differential form which governs the propagation of a one-dimensional harmonic wave. Partially differentiating Eq. (4.4) with respect to time twice, we have

$$\frac{\partial \psi(x,t)}{\partial t} = a\omega \cos(\omega t - kx)$$

and

$$\frac{\partial^2 \psi(x,t)}{\partial t^2} = -a\omega^2 \sin(\omega t - kx) = -\omega^2 \psi(x,t)$$

Now, partially differentiating Eq. (4.4) twice with respect to x, we have

$$\frac{\partial \psi(x,t)}{\partial x} = -ak \cos(\omega t - kx)$$

and

$$\frac{\partial^2 \psi(x,t)}{\partial x^2} = -ak^2 \sin(\omega t - kx) = -k^2 \psi(x,t)$$

From this differentiations, we have

$$\psi(x,t) = -\frac{1}{k^2}\frac{\partial^2 \psi}{\partial x^2}(x,t) = -\frac{1}{\omega^2}\frac{\partial^2 \psi(x,t)}{dt^2}$$

or

$$\frac{\partial^2}{\partial t^2}\psi(x,t) = \frac{\omega^2}{k^2}\frac{\partial^2 \omega(x,t)}{\partial x^2}$$

or

$$\frac{\partial^2}{\partial t^2}\psi(x,t) = v^2 \frac{\partial^2 \omega(x,t)}{\partial x^2} \tag{4.5}$$

where, $v^2 = \dfrac{\omega^2}{k^2}$.

Equation (4.5) is called *differential equation for wave motion* or *classical wave equation*. Two important conclusions can be drawn from Eq. (4.5). They are:

(i) Whenever the second order space derivative $\dfrac{\partial^2 \psi}{\partial x^2}$ of any physical quantity is related to the second order time derivative, a wave of some sort must propagate in the medium.

(ii) The propagation velocity of that wave is given by the square root of the co-efficient of the second order space derivative.

It must be remembered that the individual particles which make up the medium do not progress through the medium with the wave, they merely oscillate about their equilibrium positions. It is their phase relationships which we observe as waves. Therefore, the wave velocity is also called the *phase velocity*. It is the velocity with which planes of equal phase (crests or troughs in the case of transverse waves and compressions and rarefactions in case of longitudinal waves), travel through the medium. The phase velocity of a wave is given by

$$v = v\lambda = 2\pi v \frac{\lambda}{2\pi} = \frac{\omega}{k}$$

The phase of a wave travelling in the positive x-direction

$$\phi(x, t) = \omega t - kx$$

at a given space point x, the phase increases linearly with time (t). At a given time t, the phase decreases linearly with x. The decrease in phase is due to the fact that at great x, the phase corresponds to wave emitted at earlier times.

4.4 WAVE VELOCITIES IN CONTINUOUS SYSTEMS

(i) Transverse Vibrations of a String

For the study of transverse waves in a string, it is assumed that the string has the following properties:

(i) It is uniform in thickness and density, all over its length, to ensure a perfect shape of the wave,

(ii) It is perfectly flexible so that it may acquire any shape instantaneously during the passage of the wave;

(iii) It is inextensible, so that the tension applied by stretching the string may remain uniform along its length.

An ideal string is difficult to obtain in practice. However, a sufficiently long string of uniform cross-section, stretched between two fixed points, satisfies the conditions for the propagation of the transverse waves through it.

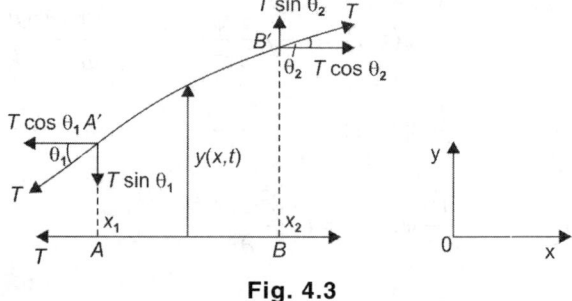

Fig. 4.3

Let the string be stretched by tension T along the X-axis. Its one end being at the origin and let the direction of displacement be Y-direction perpendicular to the length of the string (Fig. 4.3). Now, string is given a small displacement to one side perpendicular to its length and set free. Transverse vibrations will be set-up in the string.

Let us consider a small element AB of length Δx of this string at a distance x from the origin. We shall now obtain the equation of motion of this element. In equilibrium state, the forces acting on element AB are equal and opposite. When the element is displaced along the Y-axis to a new position $A'B'$, the element is no longer exactly straight, it has a slight curvature and the tension is tangential to the curve at A' and B'. Even in the deformed position, if the displacement is small, the tension remains unchanged.

Let $y(x, t)$ be the transverse displacement of the element located at x at time t. When the string is released, the displacement of the element will change with time. Thus, the displacement is a function of two variables x and t. For an extremely small element ($\Delta x \to 0$), the net force on the element along x and y directions is

$$F_x = T\{\cos \theta_2 - \cos \theta_1\}$$
$$F_y = T\{\sin \theta_2 - \sin \theta_1\}$$

where θ_1 and θ_2 are the directions of tangents to the string at the ends of the element, i.e. at $x = x_1$ and $x = x_2$ with $x_2 - x_1 = \Delta x$.

We have assumed the transverse displacement to be very small so that θ_1 and θ_2 are small angles as such, we can put

$$\cos\theta_1 \cong \cos\theta_2 \approx 1$$

and $\qquad \sin\theta_1 \approx \tan\theta_1 \quad \text{and} \quad \sin\theta_2 \approx \tan\theta_2$

under the assumption $f_x = 0$ and the transverse component of tension

$$F_y = T\{\tan\theta_2 - \tan\theta_1\}$$

$$= T\left\{\left(\frac{dy}{dx}\right)_{x_1} - \left(\frac{dy}{dx}\right)_{x_2}\right\}$$

$$= T[f(x_2) - f(x_1)]$$

where the function $f(x)$ stands for $\frac{dy}{dx}$. Using Taylor's series, we have

$$f(x_2) = f(x_1 + \Delta x)$$

$$= f(x_1) + \Delta x \left.\frac{\partial f}{\partial x}\right|_{x=x_1} + \frac{1}{2}(\Delta x)^2 \left.\frac{\partial^2 f}{\partial x^2}\right|_{x=x_1} + \ldots\ldots$$

Since Δx is very small, hence, the terms involving $(\Delta x)^2$ and higher order may be neglected. Hence, we can write as

$$f(x_2) - f(x_1) = \Delta x \left.\frac{\partial f}{\partial x}\right|_{x=x_1}$$

$$= \Delta x \frac{\partial}{\partial x}\left(\frac{\partial y}{\partial x}\right) = \Delta x \frac{\partial^2 y}{\partial x^2}$$

Here we have dropped the subscript x_1 because having ignored the higher derivatives, it does not matter now, where the interval Δx is, we evaluate the x-derivative. Since the displacement now is function of both x and t, $y(x, t)$ as such the space derivative is written as a partial derivative. Hence,

$$F_y = T\{f(x_2) - f(x_1)\} = T\Delta x \frac{\partial^2 y(x,t)}{\partial x^2} \qquad (4.7a)$$

The equation of motion now can be written as per Newton's second law of motion, i.e., force = mass × acceleration. If μ is the linear density (mass per unit length) of the string; hence, mass of the element Δx of the string is $\mu\Delta x$. Hence,

$$\mu\Delta x \frac{\partial^2 y(x,t)}{\partial t^2} = F_y = T\Delta x \frac{\partial^2 y(x,t)}{\partial x^2}$$

where $\dfrac{\partial^2 y(x,t)}{\partial t^2}$ is the acceleration of the element.

$$\therefore \qquad \frac{\partial^2 y(x,t)}{\partial t^2} = \frac{T}{\mu}\frac{\partial^2 y(x,t)}{\partial x^2}$$

$$= v^2 \frac{\partial^2 y(x,t)}{\partial x^2} \qquad (4.8)$$

where $v = \sqrt{\dfrac{T}{\mu}}$ has the dimensions of velocity. Equation (4.8) is *second order linear partial differential equation*, which we have defined as *classical wave equation*. v in Eq. (4.8) represents the *velocity* of transverse waves on a string.

Normal modes: We shall now obtain the normal modes (transverse) of a uniform string fixed at both ends and stretched with tension.

In case of strings stretched between two fixed points when it is plucked at any point, a wave starts in either direction and gets reflected at fixed ends and thus, two similar waves travelling in opposite directions are produced. Since, the two ends are rigidly fixed, there can be no displacement at these ends. In other words, the boundary conditions are:

$$y(0, t) = y(L, t) = 0 \qquad \text{for all the values of } t$$

where L is the total length of the string of linear density μ. The Eq. (4.8) describes the motion of any part of the string lying between $x = 0$ and $x = L$. In order to find normal modes, we assume the existence of a normal mode at angular frequency ω and phase constant ϕ. This means that every particle of the string executes simple harmonic motion (SHM) of angular frequency ω and phase constant ϕ. Thus, for a normal mode, we have

$$y(x, t) = A(x) \cos(\omega t + \phi) \tag{4.9}$$

There are an infinite number of equations, one for each particle characterised by its x value in the range o and L. The variable $y(x, t)$ is the displacement at time t of a particle located at x, and $A(x)$ is the amplitude of its motion. The amplitudes of all particles of the string will determine the shape or configuration of the mode. Differentiating partially twice with respect to x and t, we have

$$\frac{\partial^2 y(x,t)}{\partial x^2} = \frac{\partial^2 A(x)}{\partial x^2} \cos(\omega t + \phi)$$

and

$$\frac{\partial^2 y(x,t)}{\partial t^2} = -A(x)\omega^2 \cos(\omega t + \phi)$$

Notice that $A(x)$ is by definition a function of x only, we can write the total derivative $\dfrac{d^2 A(x)}{dx^2}$ instead of a partial derivative. Substituting these values in Eq. (6.8), we have

$$-\omega^2 A(x) \cos(\omega t + \phi) = v^2 \frac{d^2 A(x)}{dx^2} \cos(\omega t + \phi)$$

or

$$\frac{d^2 A(x)}{dx^2} = \frac{\omega^2}{v^2} A(x) = -k^2 A(x) \tag{4.10}$$

where $k = \dfrac{\omega}{v}$. The parameter k will be identified to be the wave vector. At the moment k just stands for $\dfrac{\omega}{v}$, where $v = \sqrt{\dfrac{T}{\mu}}$.

Equation (4.10) governs the shape of the mode. This equation is differential equation of SHM except that it represents oscillations in space (x) rather in time (t). The general solution of Eq. (4.10) can thus be written as

$$A(x) = A \sin kx + B \cos kx$$

where A and B are constant to be determined from initial conditions. The general solution for the displacement $y(x, t)$ of the string, in a given mode, can then be written as

$$y(x, t) = (A \sin kx + B \cos kx) \cos(\omega t + \phi) \tag{4.11}$$

Using the first boundary condition

$$y(0, t) = 0 \text{ for all values of } t$$

we have, $B = 0$.

Thus for the string fixed at $x = 0$, the equation reduces to

$$y(x, t) = A \sin kx \cos(\omega t + \phi) \tag{4.12}$$

Normal mode frequencies: The frequencies of the normal modes of transverse vibrations of the string can be obtained by using the second boundary condition namely $y(L, t) = 0$ for all values of t. This gives

$$A \sin kL = 0.$$

Here, A cannot be choosen as zero because this corresponds to a situation when string is permanently at rest. Hence, we have

$$\sin kL = 0$$

or $\qquad kL = n\pi$

where $n = 1, 2, 3, \ldots \ldots$ etc. Thus

$$k = \frac{n\pi}{L} \tag{4.13}$$

Here also, we have excluded $n = 0$ as this situation again gives the string, permanently at rest. Equation (4.3) reveals that only definite values of ω are allowed as ω and k are related as

$$\omega = vk \quad \text{where} \quad v = \sqrt{\frac{T}{\mu}}$$

$$\therefore \qquad \omega_n = \frac{n\pi}{L} \sqrt{\frac{T}{\mu}} \tag{4.14}$$

Here, we have used subscript n to indicate the value of ω for a particular integral value of n. Equation (4.14) gives the angular frequencies of the normal modes for transverse vibrations of a string fixed at both ends. The corresponding frequencies (in Hertz) of the modes are given by

$$v_n = \frac{n}{2L} \sqrt{\frac{T}{\mu}} \tag{4.15}$$

The mode with $n = 1$ is called *fundamental mode* and its frequency is

$$v_1 = \frac{1}{2L} \sqrt{\frac{T}{\mu}}$$

and is called *fundamental frequency*.

The modes with $n = 2, 3, 4, \ldots$, etc. are harmonics of the fundamental frequencies v_1, which follows from the fact that

$$v_2 = 2v_1, \quad v_3 = 3v_1, \quad v_4 = 4v_1, \ldots \text{ etc.}$$

Thus, the string has an infinite number of possible frequencies of vibration which are harmonic of the fundamental frequency v_1. It is the result of our assumption that the string is perfectly uniform and flexible. Strings in real practice (such as the strings of piano or violin, etc.) do not strictly obey this simple sequence of frequencies because they are not perfectly uniform and flexible.

Normal mode shapes: Equation (4.12) gives the displacement of the particles of the string in a normal mode. For n^{th} mode, this equation is

$$y_n(x, t) = A_n \sin k_n x \cos(\omega_n t + \phi_n)$$

where $k_n = \dfrac{n\pi}{L}$ and $\omega_n = k_n v = k_n \sqrt{\dfrac{T}{\mu}}$. The constants A_n and ϕ_n are to be determined from initial conditions. In order to know the shape of the first few modes, we proceed as follows:

For the fundamental mode ($n = 1$) which is also called the *first harmonic*, the particle displacements are given by

$$y_1(x, t) = A_1 \sin(k_1 x) \cos(\omega_1 t + \phi)$$
$$= A_1 \sin \frac{\pi x}{L} \cos(\omega_1 t + \phi_1)$$

This equation gives the displacement of all the particles of the string (i.e. all x-values) as a function of time t. Notice $y = 0$ at $x = 0, L$. There is no other value of x in the range 0 to L, where y can vanish. The displacement y_1 is maximum at $x = L/2$ for a given value of t. Figure 4.4(a) shows a plot of $y_1(x, t)$ against x for a particular value of t. This is the shape of the *fundamental mode*. In this mode, the string is vibrating in one segment. Its frequency of

vibration is $v_1 = \dfrac{1}{2L} \sqrt{\dfrac{T}{\mu}}$.

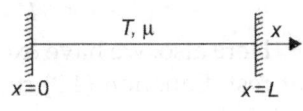

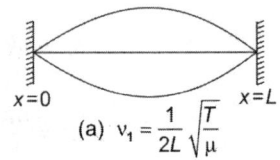

(a) $v_1 = \dfrac{1}{2L} \sqrt{\dfrac{T}{\mu}}$

In second harmonic ($n = 2$) also called as the *first overtone*, the displacements of the various particles of the string as a function of time are given by

$$y_2(x,t) = A_2 \sin\left(\frac{2\pi x}{L}\right) \cos(\omega_2 t + \phi_2)$$

In this case y_2 is zero at $x = 0$, $x = L/2$ and $x = L$. Figure 4.4(b) gives the shape of this mode. The string now vibrates in two segments at frequency v_2 (= $2v_1$). Similarly, the Fig. 4.4(c) and (d) shows the next two harmonics. These figures reveal that there are certain points on the string which are permanently at rest. The number of these points depends upon the number the mode under study. These points are called *nodes*. The points where the displacement is maximum (at a given time) are called *anti-nodes*.

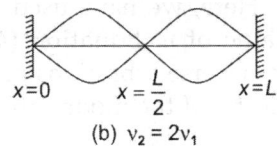

(b) $v_2 = 2v_1$

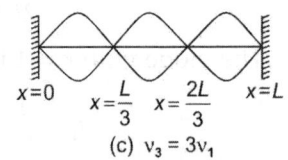

(c) $v_3 = 3v_1$

Equation (4.15) giving frequencies of different modes gives that the frequencies of all *overtones* of such a string are integral multiples of the fundamental frequency. *Overtones* bearing this simple relation to the fundamental are called *harmonics*. The fundamental is called *first harmonic*, the first overtone as *second harmonic* (twice the frequency) and so on.

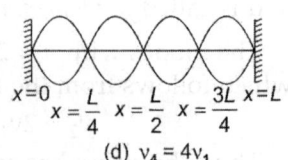

(d) $v_4 = 4v_1$

Fig. 4.4

The actual vibrating systems do not have exactly harmonic overtones due to non-uniformities in the string and the supports at its ends being not perfectly rigid. Very few vibrating systems have nearly harmonic overtones. These systems form the basis of most of the musical instruments. The reason is that, when the overtones are harmonic, the tonal quality of the sound is considerably improved.

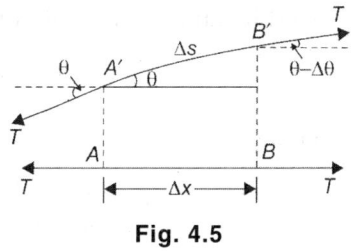

Fig. 4.5

Energy of a vibrating string: A vibrating string possesses kinetic as well as potential energy. Figure 4.5 shows the displaced position of an element AB of the string at particular moment of time when its displacement from the equilibrium position is $y(x, t)$, which henceforth for simplicity, we will write as y, which in fact is the function of both x and t.

The kinetic energy is due to the velocity of the element at that instant of time and the potential energy is the work done by the tension T is extending the element AB of the length Δx to the new length Δs when the string is vibrating. The length of the curved element $A'B'$ is[1]

$$\Delta s = \left[1 + \left(\frac{dy}{dx}\right)^2\right]^{1/2} \Delta x$$

Since $\frac{dy}{dx}$ is very small comparison to 1, hence, expanding bino-mially, we have (leaving higher powers)

$$\Delta s = \left[1 + \frac{1}{2}\left(\frac{dy}{dx}\right)^2\right] \Delta x$$

Now, the change in the length of the element when the string vibrates is

$$\Delta s - \Delta x = \frac{1}{2}\left(\frac{dy}{dx}\right)^2 \Delta x$$

Hence, the instantaneous kinetic energy of the element is

$$KE = \frac{1}{2}\mu\Delta s\left(\frac{dy}{dt}\right)^2$$

where μ is the linear density of string. Since y and its derivatives are assumed to be small, hence, here we can write $\Delta s \approx \Delta x$. Hence, total kinetic energy of the whole string of length L is

$$E_x = \int_0^L \frac{1}{2}\mu\Delta x\left(\frac{dy}{dt}\right)^2$$

$$= \frac{1}{2}\mu\int_0^L \left(\frac{dy}{dt}\right)^2 dx$$

---

1. $\frac{\Delta x}{\Delta s} = \cos\theta$ (Fig. 4.5) $\therefore$ $\sec^2\theta = \left(\frac{\Delta s}{\Delta x}\right)^2$

$\left(\frac{\Delta s}{\Delta x}\right)^2 = 1 + \tan^2\theta = 1 + \left(\frac{dy}{dx}\right)^2$ $\therefore$ $\Delta s = \left[1 + \left(\frac{dy}{dx}\right)^2\right]^{1/2}\Delta x.$

And instantaneous potential energy of the element is

$$E_p = T(\Delta s - \Delta x) = \frac{1}{2}T\left(\frac{\partial y}{\partial x}\right)^2 \Delta x$$

Hence, total potential energy of the string is

$$E_p = \int_0^L \frac{T}{2}\left(\frac{\partial y}{\partial x}\right)^2 dx = \frac{T}{2}\int_0^L \left(\frac{\partial y}{\partial x}\right)^2 dx$$

Total energy of the string in different modes is different and is sum of the kinetic and potential energy of the string. Thus, the kinetic energy in n^{th} mode is

$$(E_k)_n = \frac{1}{2}\mu\int_0^L \left(\frac{\partial y_n}{\partial t}\right)^2 dx$$

and potential energy is

$$(E_p)_n = \frac{1}{2}T\int_0^L \left(\frac{\partial y_n}{\partial x}\right)^2 dx$$

But the displacement in the nth mode is given by

$$y_n = A_n \sin k_n x \cos(\omega_n t + \phi_n)$$

$$\therefore \qquad \frac{\partial y_n}{\partial t} = -A_n \omega_n \sin k_n x \sin(\omega_n t + \phi_n)$$

$$\therefore \qquad (E_k)_n = \frac{1}{2}\mu A_n^2 \omega_n^2 \sin^2(\omega_n t + \phi_n)\int_0^L \sin^2 k_n x\, dx$$

But

$$\int_0^L \sin^2 k_n x\, dx = \frac{1}{2}\int_0^L (1 - \cos 2k_n x)\, dx$$

$$= \frac{1}{2} \qquad \because \quad \int_0^L \cos 2k_n x\, dx = 0$$

$$\therefore \qquad (E_k)_n = \frac{1}{4}\mu A_n^2 \omega_n^2 \sin^2(\omega_n t + \phi_n)$$

Similarly, potential energy

$$(E_p)_n = \frac{1}{2}T\int_0^L A_n^2 k_n^2 \cos^2 k_n x \cos^2(\omega_n t + \phi_n)\, dx$$

$$= \frac{1}{2}T A_n^2 k_n^2 \cos^2(\omega_n t + \phi_n)\int_0^L \cos^2 k_n x\, dx$$

But

$$\int_0^L \cos^2 k_n x\, dx = \frac{1}{2}\int_0^L (1 + \cos 2k_n x)\, dx = \frac{1}{2}$$

$$\because \qquad \int_0^L \cos^2 k_n x\, dx = 0$$

$$\therefore \qquad (E_p)_n = \frac{L}{4}T A_n^2 k_n^2 \cos^2(\omega_n + \phi_n)$$

Total energy

$$E_n = (E_k)_n + (E_p)_n$$

$$= \frac{L}{4}\mu A_n^2 \omega_n^2 \sin^2(\omega_n t + \phi_n) + \frac{L}{4}T A_n^2 k_n^2 \cos^2(\omega_n t + \phi_n)$$

But

$$k_n = \frac{\omega_n}{v} = \omega_n\sqrt{\frac{\mu}{T}}$$

$\therefore$

$$k_n^2 = \omega_n^2 \frac{\mu}{T}$$

$\therefore$

$$E_n = \frac{L}{4}\mu A_n^2 \omega_n^2 \sin^2(\omega_n t + \phi_n) + \frac{L}{4}T A_n^2 \frac{\omega_n^2 \mu}{T}\cos^2(\omega_n t + \phi_n)$$

$$= \frac{\mu L A_n^2 \omega_n^2}{4}[\sin^2(\omega_n t + \phi_n) + \cos^2(\omega_n t + \phi_n)]$$

$\therefore$

$$E_n = \frac{1}{4}m A_n^2 \omega_n^2 \qquad (4.16)$$

where $m = \mu L$, the mass of the string.

Putting

$$\omega_n = 2\pi v_n$$

$$E_n = \frac{1}{4}m 4\pi^2 v_n^2 A_n^2 = m\pi^2 v_n^2 A_n^2 \qquad (4.16a)$$

The total energy of the vibrating string is given by the sum of the energies associated with each of its normal mode

$\therefore$

$$E_{\text{total}} = \sum_n E_n = \sum_{n=1}^{\infty} m\omega_n^2 A_n^2$$

$$= \frac{1}{4}m \sum_{n=1}^{\infty} \omega_n^2 A_n^2 \qquad (4.17)$$

Characteristic Impedance of a String to Transverse Waves

When the wave propagates through medium, the medium will offer an opposition called *impedance*. If the medium is lossless, i.e. it does not have any resistance or dissipative component, the impedance is determined solely by its inertia and elasticity. These two properties are also responsible for storing and propagating energy in the medium in the form of waves.

Consider travelling waves propagating on a string which are generated at one end $x = 0$ by applying a transverse harmonic force at this end (Fig. 4.6). As a result of the application of force transverse waves travel in say the positive X-direction, in which the particle displacements are given by

$$y(x, t) = A\sin\left\{\frac{2\pi}{\lambda}(vt - x)\right\}$$

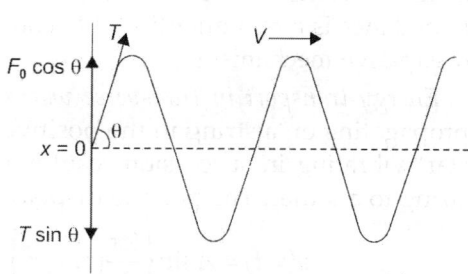

Fig. 4.6

Because in steady state the frequency of particle oscillations equals that of the driving force.

Differentiating partially the above Eq., we have

$$\frac{\partial y}{\partial x} = -A\left(\frac{2\pi}{\lambda}\right)\cos\left\{\frac{2\pi}{\lambda}(vt - x)\right\}$$

and

$$\frac{dy}{dt} = A\left(\frac{2\pi v}{\lambda}\right)\cos\left\{\frac{2\pi}{\lambda}(vt - x)\right\}$$

Comparing these equations, we have

$$\frac{\partial y}{\partial x} = -\frac{1}{v}\frac{\partial y}{\partial t}.$$

From the Fig. 4.6, it is apparent that transverse force exerted at the end $x = 0$ must be equal of and opposite to $T \sin\theta$, the transverse component of tension, i.e.

$$F = F_0 \cos\omega t = -T\sin\theta \approx -T\tan\theta = -T\left(\frac{\partial y}{\partial x}\right)_{x=0}$$

$$= \frac{T}{v}\left(\frac{\partial y}{\partial t}\right)_{x=0}$$

Here, we have assumed θ to be small and used assumption $\sin\theta \approx \tan\theta$. But

$$\left(\frac{\partial y}{\partial t}\right)_{x=0} = A\frac{2\pi v}{\lambda}\cos\frac{2\pi vt}{\lambda} = v_0\cos\omega t$$

where $v_0 = A\frac{2\pi v}{\lambda} = A\omega$ is the velocity amplitude. Thus, we have

$$F_0 \cos\omega t = v_0(\cos\omega t)\frac{T}{v}$$

or

$$F_0 = \frac{Tv_0}{v}$$

The characteristic impedance of the string is defined as

$$Z = \frac{\text{Transverse force amplitude}}{\text{Transverse velocity amplitude}} = \frac{F_0}{v_0}$$

or

$$Z = \frac{T}{v} = \sqrt{\mu T} = \mu v \quad \left(\because \ v = \sqrt{\frac{T}{\mu}}\right) \tag{4.18}$$

Since velocity is determined by the inertia and elasticity, it is clear that the impedance is also governed by these two properties of the medium. For a lossless medium, the impedance is a real quantity. It becomes a complex quantity, if the medium contains a dissipative mechanism.

Energy transport in transverse waves on a string: Let us consider a transverse wave propagating on a string in the positive X-direction. The different particles on the string start vibrating in succession resulting in the transfer of energy, from one part of the string to another. The particle displacements are given by

$$y(x,t) = A\sin\left\{\frac{2\pi}{\lambda}(vt - x)\right\}$$

where $v = \sqrt{\dfrac{T}{\mu}}$, T being tension and μ is linear mass density. Differentiating, we have

$$\frac{\partial y}{\partial x} = -A \frac{2\pi}{\lambda} \cos\left\{\frac{2\pi}{\lambda}(vt - x)\right\} \tag{4.19}$$

and

$$\frac{\partial y}{\partial t} = -A \frac{2\pi v}{\lambda} \cos\left\{\frac{2\pi}{\lambda}(vt - x)\right\} \tag{4.20}$$

In order to calculate energy transfer along the string, we first of all calculate the energy density, i.e. the energy per unit length of the string.

Let us consider the element of length dx of the string. We have already seen that potential energy of this element is

$$du = \frac{1}{2}T \left(\frac{\partial y}{\partial x}\right)^2 dx$$

and kinetic energy dk of the element is given by

$$dk = \frac{1}{2}\mu \left(\frac{\partial y}{\partial x}\right)^2 dx$$

Hence, the potential energy density and kinetic energy density are

$$\frac{du}{dx} = \frac{1}{2}T \left(\frac{\partial y}{\partial x}\right)^2$$

and

$$\frac{dk}{dx} = \frac{1}{2}\mu \left(\frac{\partial y}{\partial x}\right)^2$$

Using Eqs (4.19 and 4.20), the energy densities at time $t = 0$ are

$$\frac{du}{dx} = \frac{1}{2}T A^2 \left(\frac{2\pi}{\lambda}\right)^2 \cos^2\left(\frac{2\pi x}{\lambda}\right)$$

and

$$\frac{dk}{dx} = \frac{1}{2}\mu A^2 \left(\frac{2\pi v}{\lambda}\right)^2 \cos^2\left(\frac{2\pi x}{\lambda}\right)$$

Hence, the total energy associated with one complete wavelength of the sinusoidal wave on the string can be obtained by integrating these expressions over one wavelength, i.e., $x = 0$ to $x = \lambda$.

Thus,

$$u = \int_0^\lambda \frac{du}{dx}dx = \frac{1}{2}TA^2 \left(\frac{2\pi}{\lambda}\right)^2 \int_0^\lambda \cos^2\left(\frac{2\pi x}{\lambda}\right)dx$$

$$= \frac{1}{4}TA^2 \left(\frac{2\pi}{\lambda}\right)^2 \int_0^\lambda \left\{1 + \cos\left(\frac{4\pi x}{\lambda}\right)\right\} dx$$

$$= \frac{1}{4}TA^2\lambda \left(\frac{2\pi}{\lambda}\right)^2$$

$$\therefore \quad \int_0^\lambda \cos\left(\frac{4\pi x}{\lambda}\right)dx = 0$$

$$\therefore \qquad u = \frac{\pi^2 T A^2}{\lambda} = \pi^2 A^2 \mu v^2 \lambda \qquad (4.21)$$

$$\therefore \qquad v^2 = \frac{T}{\mu} \quad \text{and} \quad v = v\lambda$$

Similarly,

$$k = \int_0^\lambda \frac{dk}{dx} = \frac{1}{2} k A^2 \left(\frac{2\pi v}{\lambda}\right)^2 \int_0^\lambda \cos^2\left(\frac{2\pi x}{\lambda}\right) dx$$

$$= \pi^2 A^2 \mu v^2 \lambda \qquad (4.21a)$$

Notice that potential and kinetic energies over a length of the string equal to a wavelength are equal in magnitude. Hence, the total energy per wavelength

$$E = u + k = 2\pi^2 A^2 v^2 \lambda \qquad (4.22)$$

We will now calculate the rate of transfer of energy in string. Let a harmonic force F is applied at the end $x = 0$ of the string. We have seen already that this force must be equal and opposite to the transverse component of tension, i.e.

$$F = -T \sin\theta = -T \tan\theta = -T \left(\frac{\partial y}{\partial x}\right)_{x=0}$$

and from Eq. (4.19)

$$\left(\frac{\partial y}{\partial x}\right)_{x=0} = -A\left(\frac{2\pi}{\lambda}\right)\cos\frac{2\pi vt}{\lambda}$$

Therefore,

$$F = AT\left(\frac{2\pi}{\lambda}\right)\cos\left(\frac{2\pi vt}{\lambda}\right)$$

The rate at which energy is supplied to the string at $x = 0$ or the input power P at a time t, by definition is

$$P(t) = \text{Force} \times \text{velocity}$$

$$= F\left(\frac{\partial y}{\partial t}\right)_{x=0}$$

But $$\left(\frac{\partial y}{\partial t}\right)_{x=0} = A\left(\frac{2\pi v}{\lambda}\right)\cos\left(\frac{2\pi vt}{\lambda}\right)$$

Hence, substituting the values, we have

$$P(t) = AT\left(\frac{2\pi}{\lambda}\right)\cos\left(\frac{2\pi vt}{\lambda}\right) \times A\left(\frac{2\pi v}{\lambda}\right)\cos\left(\frac{2\pi vt}{\lambda}\right)$$

$$= A^2 Tv\left(\frac{2\pi}{\lambda}\right)^2 \cos^2\left(\frac{2\pi vt}{\lambda}\right)$$

But $\dfrac{v}{\lambda} = \dfrac{1}{\tau}$, where τ is time period.

$$\therefore \qquad P(t) = A^2 Tv\left(\frac{2\pi}{\lambda}\right)^2 \cos^2\left(\frac{2\pi t}{\tau}\right)$$

hence, average power input $\langle P(t) \rangle$ for one period is obtained by integrating over one period

$$\therefore \quad \langle P(t) \rangle = A^2 T v \left(\frac{2\pi}{\lambda}\right)^2 \frac{\int_0^\tau \cos^2\left(\frac{2\pi t}{\tau}\right) dt}{\int_0^\tau dt}$$

$$= \frac{1}{2} \frac{A^2 T v}{\tau} \int_0^\tau \left(\frac{2\pi}{\lambda}\right)^2 \left\{1 + \cos\left(\frac{4\pi t}{\tau}\right)\right\} dt$$

$$= \frac{1}{2} A^2 T v \left(\frac{2\pi}{\lambda}\right)^2$$

or

$$\langle P(t) \rangle = 2\pi^2 A^2 \mu v^2 v, \quad v = \frac{1}{2}\mu v \omega^2 A^2 \tag{4.23}$$

$$\therefore \quad \tau = \frac{1}{v}, \quad v = \sqrt{\frac{T}{\mu}} \quad \text{and} \quad v = v\lambda.$$

From Eq. (4.22), we have that $2\pi^2 A^2 \mu v^2$ is the total energy per unit length of the string. Thus, we conclude that the time averaged input power is equal to the total energy per unit length times the wave velocity. This also implies that the energy does not stay at the deriving end but it flows along the string which thus, acts as a medium for transport of energy from one point to the another, the speed of transport being equal to the wave velocity. Equation (4.23) also tells that energy crossing unit area in unit time, the area being taken perpendicular to the direction of propagation of energy is also $2\pi^2 A^2 v^2 \mu v$ which by definition is called *intensity*. Hence

$$I = 2\pi^2 A^2 v^2 \mu v \tag{4.24}$$

This shows that the intensity of the wave at any point is proportional to the square of amplitude of the wave at that point.

Example 4.1: A simple harmonic wave travelling in the positive x-direction has velocity 330 m/s and amplitude 2.5 cm. If the frequency of wave is 660 Hz. Determine: (i) The equation of wave; (ii) The particle velocity; (iii) The particle acceleration taking origin at $x = 0$.

Solution: The wave equation is given by

$$y = A \sin(\omega t - kx)$$

$$v = \frac{\omega}{k}, \quad \omega = 2\pi f \quad \text{and} \quad k = \frac{2\pi}{\lambda}$$

(i) Wave equation $y = A \sin\omega\left(t - \frac{k}{\omega}x\right) = A \sin\omega\left(t - \frac{x}{v}\right)$

$$= 2.5 \times 10^{-2} \sin(2\pi \times 660)\left\{t - \frac{x}{330}\right\}$$

$$= 2.5 \times 10^{-2} \sin(1320\pi\left\{t - \frac{x}{330}\right\}$$

(ii) Particle velocity is $\dfrac{dy}{dt} = 2.5 \times 10^{-2} \times 1320\pi \cos(1320\pi)\left(t - \dfrac{x}{330}\right)$

$$= 33\pi \cos(1320\pi)\left(t - \dfrac{x}{330}\right) \text{m/s}$$

(iii) Particle acceleration $\dfrac{d^2x}{dt^2} = -33\pi \times 1320\pi \sin(1320\pi)\left(t - \dfrac{x}{330}\right)$

$$= -4.356 \times 10^4 \pi^2 \sin(1320\pi)\left(t - \dfrac{x}{330}\right)$$

Example 4.2: A transverse wave of amplitude 0.01 m is generated at one end ($x = 0$) of a long string by a tuning fork of frequency 500 Hz. At a given instant of time, the displacement of the particle at $x = 0.1$ m is –0.005 m and that of particle at $x = 0.2$ m is 0.005 m. Calculate the wavelength and wave velocity. Obtain the equation of the wave assuming that the wave is travelling along the x-direction and that the end $x = 0$ is at the equilibrium position at $t = 0$.

Solution: The general equation of the wave propagating along the positive x-direction with displacement at the end $x = 0$ at $t = 0$ being zero is given by

$$y(x, t) = A \sin(\omega t - kx)$$

where A the amplitude is 0.01 m.

When $x_1 = 0.1$ m, $y = -0.005$ m

∴ $\qquad -0.005 = 0.01 \sin(\omega t - kx_1)$

or $\quad \sin(\omega t - kx_1) = -0.5$

or $\qquad \omega t - kx_1 = \dfrac{7\pi}{6}$ $\qquad\qquad$ (i)

When $\qquad x_2 = 0.2$ m, $y = 0.005$ m

∴ $\qquad 0.005 = 0.01 \sin(\omega t - kx_2)$

or $\qquad \sin(\omega t - kx_2) = 0.5$

∴ $\qquad \omega t - kx_2 = \dfrac{\pi}{6}$ $\qquad\qquad$ (ii)

From (i) and (ii), we have:

$$k(x_2 - x_1) = \pi$$

or $\quad \dfrac{2\pi}{\lambda}(0.2 - 0.1) = \pi$

or $\quad \lambda = 2 \times 0.1 = 0.2$ m

Frequency of wave = Frequency of the tuning fork = 500 Hz

∴ $\qquad$ Wave velocity $v = v\lambda = 0.2 \times 500 = 100$ m/s

∴ $\quad$ Wave equation is $y(x, t) = 0.01 \sin\left\{2\pi \times 500t - \dfrac{2\pi}{0.2}x\right\}$

$$= 0.01 \sin 10\pi\{100t - x\}$$

Example 4.3: A long uniform string of linear density 0.1 kg/m is stretched with tension of 40 N. One end of the string ($x = 0$) is given transverse sinusoidal oscillations with amplitude 0.02 m and period 0.1 s, so that transverse waves are set-up in the x-direction:

   (a) Find the velocity and wavelength of the wave.
   (b) If at the driving end ($x = 0$), the displacement (y) at time $t = 0$ is 0.01 m but the velocity is negative, what is the equation of the travelling waves?
   (c) What is the velocity of the string at a point $x = 10$ m at time $t = 1$ s.

Solution: (a) Velocity $u = \sqrt{\dfrac{T}{\mu}} = \sqrt{\dfrac{40}{0.1}} = 20$ m/s

Frequency $v = \dfrac{1}{T} = \dfrac{1}{0.1} = 10$ Hz

   (b) Since displacement at the end ($x = 0$) at $t = 0$ is finite and velocity negative, also given that waves are sinusoidal and they travel in the x-direction. These conditions are satisfied, if we take the wave equation as

$$y(x, t) = A \sin\{(kx - \omega t) + \theta\}$$

where θ is the initial phase to be determined.

Now, at $t = 0$ and $x = 0$

$$y = 0.01 \text{ m, with } A = 0.02 \text{ m}$$

∴ $$0.01 = 0.02 \sin \theta$$

or $\sin \theta = \dfrac{1}{2}$ or $\theta = \dfrac{\pi}{6}$

∴ Wave equation is

$$y(x, t) = 0.02 \sin\left[\left(\frac{2\pi}{2}x - \frac{2\pi}{0.1}t\right) + \frac{\pi}{6}\right]$$

$$= 0.02 \sin \pi\left(x - 20t + \frac{1}{6}\right)$$

   (c) The velocity of the string is given by

$$\frac{dy}{dt} = 0.02(-20\pi) \cos \pi\left(x - 20t + \frac{1}{6}\right)$$

$$= -0.4\pi \cos \pi\left(x - 20t + \frac{1}{6}\right)$$

∴ Velocity at $x = 10$ m and $t = 1$ s is

$$\frac{dy}{dt} = -0.4\pi \cos \pi\left(10 - 20 + \frac{1}{6}\right)$$

$$= -0.4\pi \cos\left(10\pi - \frac{\pi}{6}\right) \quad \because \cos(-\theta) = \cos t\,(+\,\theta)$$

$$= -0.4\pi \cos\left(-\frac{\pi}{6}\right)$$

$$= -0.4\pi \frac{\sqrt{3}}{2} = -1.088 \text{ m/s}$$

Hence, velocity at $x = 10$ m and $t = 1$ s is negative and 1.088 m/s.

Example 4.4: A progressive harmonic wave travelling in a string is given by $y = 10 \cos(0.1x - 4.0t)$, where x and y are expressed in centimetres and time in seconds. Calculate the following parameters of the wave motion:

(i) The amplitude, frequency, velocity and wavelength of the waves.

(ii) The maximum transverse speed and acceleration of a particle in string.

(iii) Energy flux of the wave, the linear density of the string is 1.25 gm/cm.

Solution: (i) Comparing the given equation with the standard wave equation:

$$y = A \cos(kx - \omega t) \quad \text{and} \quad y = 10 \cos(0.1x - 4.0t)$$

We have   Amplitude $A = 10$ cm

$$\text{Wave vector } k = \frac{2\pi}{\lambda} = 0.1 \quad \text{or} \quad \lambda = 20\pi = 62.8 \text{ cm}$$

$$\text{Angular frequency } \omega = 2\pi\nu = 4 \quad \text{or} \quad \nu = \frac{4}{2\pi} = 0.637 \text{ Hz}$$

$$\text{Velocity } v = \frac{\omega}{k} = \frac{4}{0.1} = 40 \text{ cm/s}$$

(ii) Particle velocity is $\frac{dy}{dt}$. Hence, differentiating, we have

$$\frac{dy}{dt} = 10 \times 4 \sin(0.1x - 4.0t)$$

Since maximum value of $\sin\theta = \pm 1$

Hence, maximum particle velocity $\frac{dy}{dt} = 40$ cm/s

Particle acceleration is $\frac{d^2y}{dt^2} = -10 \times 4^2 \cos(0.1x - 4.0t)$

Hence maximum particle acceleration $= -16 \times 10 = -160$ cm/s²

(iii) Energy flux of the wave is given by

$$I = 2\pi^2 A^2 v^2 \mu \times v$$

$$= 2\pi^2 (10)^2 \left(\frac{2}{\pi}\right)^2 \times (1.25) \times 40$$

$$= 4 \times 10^4 \text{ ergs/cm}^2 \text{ s} = 40 \text{J/m}^2\text{s} = 40 \text{ watt/m}^2$$

Example 4.5: A string of length l and mass m hangs freely from a fixed point. Calculate

(i) The velocity of transverse wave along the string at any position.

(ii) Time taken by a traverse pulse to travel the string.

Solution: (i) Consider a point on the string at a distance x from the free end. Tension at that point = mass of x length of string × g.

$$\therefore \qquad T = \frac{m}{l} x g$$

$\therefore$ Velocity of transverse waves along the string

$$V = \sqrt{\frac{\text{Tension}}{\text{Mass per unit length}}} = \sqrt{\frac{\dfrac{m}{l} x g}{\dfrac{m}{l}}} = \sqrt{xg}$$

This shows that the velocity depends on the position of the point.

(ii) We have

$$v = \frac{dx}{dt} = \sqrt{xg}$$

$\therefore$

$$dt = \frac{1}{\sqrt{g}} \frac{dx}{\sqrt{x}}$$

$\therefore$ Time required for pulse to propagate through the string L is

$$t = \int_0^l \frac{1}{\sqrt{g}} \frac{dx}{\sqrt{x}} = \frac{1}{\sqrt{g}} \left[\frac{x^{\frac{1}{2}}}{\frac{1}{2}} \right]_0^l$$

$$= \frac{2}{\sqrt{g}}[\sqrt{l} - 0] = 2\sqrt{\frac{l}{g}}$$

Example 4.6: Two strings A and B of the same material, cross-sectional area and length are fixed at their ends and subjected to tension in the ratio of 2.89:1 respectively when the strings are vibrated, 8 beats per second are heard between the third harmonic of string A and the fifth harmonic of string B. Calculate fundamental frequencies of each string.

Solution: If T is the tension in string B then tension in A is 2.89T.

Hence frequency of third harmonic of string A is

$$v_a = \frac{3}{2l} \sqrt{\frac{2.89T}{\mu}} = \frac{5.1}{2l} \sqrt{\frac{T}{\mu}}$$

and frequency of the fifth harmonic of B is

$$v_b = \frac{5}{2l} \sqrt{\frac{T}{\mu}}$$

$\therefore$ $\qquad v_a - v_b = 8$

or $\qquad \dfrac{5.1}{2l} \sqrt{\dfrac{T}{\mu}} - \dfrac{5}{2l} \sqrt{\dfrac{T}{\mu}} = 8$

or $\qquad \dfrac{0.1}{2l} \sqrt{\dfrac{T}{\mu}} = 8$

or $\qquad \dfrac{1}{2l} \sqrt{\dfrac{T}{\mu}} = \dfrac{8}{0.1} = 80$

Hence, fundamental frequency of B is 80 Hz.

The fundamental frequency of A is

$$v_1 = \frac{1}{2l} \sqrt{\frac{2.89T}{\mu}} = 1.7 \times 80 = 136 \text{ Hz}$$

4.5 LONGITUDINAL VIBRATIONS OF A ROD

The longitudinal waves can be produced in a rod by striking it lengthwise (or by rubbing at its end along the length of the rod). The vibrations are in audible range of frequencies. In order to find the frequencies of the normal modes of vibration of the rod, we will first obtain the equation of motion of the particles of the rod.

Consider a rod of uniform cross-sectional area lying along the X-axis. Let us divide the rod into a large number of small slices each of length Δx. Consider one such slice AB (Fig. 4.7a). The end A is at a distance $x_1 = x$ from some origin and the end B is at a distance $x_2 = x + \Delta x$. When the rod is struck lengthwise, the particles of the rod are displaced along the X-axis, i.e. along the length of the rod. Hence, the particle displacements are longitudinal rather than transverse. We shall give the symbol ξ to denote the displacement along the X-axis of the particle in a plane at $x = x_1$ perpendicular to the axis of the rod. We shall now obtain the equation of motion of a thin slice AB of the rod which in the undisturbed state is contained between x and $x + \Delta x$ as shown in Fig. 4.7(a).

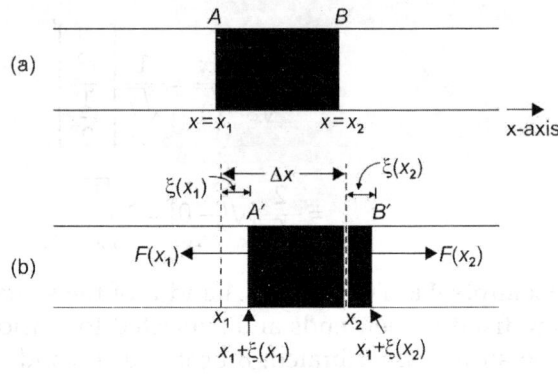

Fig. 4.7

Figure 4.7(b) shows the displaced position of the slice. Let the x co-ordinates of the end A of the slice in displaced position be $x_1 + \xi(x_1)$, so that $\xi(x_1)$ represents the displacement of the particles at the plane x_1. Similarly, let $x_2 + \xi(x_2)$ is the displacement of the particles originally located at a plane $x = x_2$, so that $\xi(x_2)$ is the displacement of the

particles at the plane x_2. Therefore, change in length of the slice $= \xi(x_2) - \xi(x_1) = \left(\dfrac{\partial \xi}{\partial x}\right)_{x_1} \Delta x$

Here we have used Taylor's expansions

$$\xi(x_2) = \xi(x_1) + \left(\frac{\partial \xi}{\partial x}\right)_{x_1} \Delta x + \ldots\ldots$$

and retained up to order (Δx), as in the case of string. Since the original length of the slice is Δx, the longitudinal strain at x_1 is given by

$$\xi(x_1) = \frac{\text{Change in length}}{\text{Original length}} = \frac{\left(\dfrac{\partial \xi}{\partial x}\right)_{x_1} \Delta x}{\Delta x} = \left(\frac{\partial \xi}{\partial x}\right)_{x_1}$$

If Y is the Young's modulus of the material of the rod, the stress S at x_1 is given by

$$S(x_1) = Y\left(\frac{\partial \xi}{\partial x}\right)_{x_1}$$

Similarly, the stress at x_2 is given by

$$S(x_2) = Y\left(\frac{\partial \xi}{\partial x}\right)_{x_2}$$

Hence, net stress on the element AB is

$$S(x_2) - S(x_1) = Y[f(x_2) - f(x_1)]$$

when, $f(x) = \left(\dfrac{\partial \xi}{\partial x}\right)$. Since $f(x_2) - f(x_1) = \dfrac{\partial \xi}{\partial x} \Delta x$, we have

$$S(x_2) - S(x_1) = Y \dfrac{\partial f}{\partial x} \Delta x$$

$$= Y \dfrac{\partial}{\partial x} \left\{\dfrac{\partial \xi}{\partial x}\right\} \Delta x = Y \dfrac{\partial^2 \xi}{dx^2} \Delta x$$

If α is the area of cross-section of the rod, the net longitudinal force on the element is given by

$$F = F(x_2) - F(x_1)$$

$$= \text{Stress} \times \text{area} = Y \alpha \Delta x \dfrac{\partial^2 \xi}{\partial x^2}$$

In order to obtain equation of motion, we use Newton's laws of motion. If ρ is the density of the material of the rod, then mass of the slice is $\alpha \Delta x \rho$. Therefore, we have

$$F' = \rho \alpha \Delta x \dfrac{\partial^2 \xi}{dx^2}$$

In dynamic equilibrium, the two forces F and F' must balance giving

$$\rho \alpha \Delta x \dfrac{\partial^2 \xi}{\partial t^2} = Y \alpha \Delta x \dfrac{\partial^2 \xi}{\partial x^2}$$

or

$$\dfrac{\partial^2 \xi}{\partial t^2} = \dfrac{Y}{\rho} \dfrac{\partial^2 \xi}{\partial x^2}$$

$$= v^2 \dfrac{\partial^2 \xi}{\partial x^2} \tag{4.24}$$

with

$$v = \sqrt{\dfrac{Y}{\rho}} \tag{4.25}$$

Hence v is the velocity of the longitudinal (sound) waves in the rod.

Normal modes: We will now find the possible normal modes of longitudinal vibrations of the rod under the given boundary conditions. Let us assume that there exists a normal mode at angular frequency ω and phase constant ϕ. This means in that mode, every particle of the rod executes SHM of angular frequency ω and phase constant ϕ. Thus, for normal mode, we have

$$\xi(x, t) = A(x) \cos(\omega t + \phi).$$

Differentiating partially this equation with respect to x and t then substituting in Eq. (4.24), we have

$$\dfrac{\partial^2 A(x)}{\partial x^2} = -k^2 A(x)$$

where $k = \dfrac{\omega}{v}$ with v given by Eq. (4.25). The general solution of the equation is

$$A(x) = A \sin kx + B \cos kx$$

where A and B are constants to be determined from the boundary conditions. The general displacements $\xi(x, t)$ of the rod, in a given mode is then expressed by

$$\xi(x, t) = (A \sin kx + B \cos kx) \cos(\omega t + \phi) \tag{4.26}$$

Now, we shall determine frequencies and shapes of the normal modes of longitudinal vibrations of the rod under following boundary conditions.

Rod clamped at one end: Let a uniform rod of length L be rigidly clamped at $x = 0$ and $x = L$ be the free end. Since the end $x = 0$ is rigidly fixed, hence, there can be no particle displacement at this end, i.e.

$$\xi(x, t) = 0$$

The condition at $x = L$ must express the fact that this end is free, i.e., the particles at this end are free to vibrate and under no stress. In other words, stress at

$$x(x = L) = Y\left(\frac{\partial \xi}{\partial x}\right)_{x=L} = 0.$$

Using the boundary condition $\xi(0, t) = 0$ for all values of t, we have

$$B = 0$$

The general displacement equation now reduces to

$$\xi(x, t) = A \sin kx \cos(\omega t + \phi)$$

differentiating this equation with respect to x, we have

$$\frac{\partial \xi(x,t)}{\partial x} = Ak \cos kx \cos(\omega t + \phi) \tag{4.27}$$

The frequencies of the normal modes are obtained by using the second boundary condition, i.e. $\left(\frac{\partial \xi}{\partial x}\right)_{x=L} = 0$ for all values of t. This gives

$$A k \cos kL = 0.$$

When

$$kL = \frac{\pi}{2}, \frac{3\pi}{2}, \frac{5\pi}{2}, \text{......}$$

$$= (2n - 1)\frac{\pi}{2}$$

where n is an integer having values 1, 2, 3, ... ∞.

$$\therefore \qquad \omega_n = \frac{(2n - 1)\pi v}{2L} \quad \text{since } v = \frac{\omega}{k} \text{ or } k = \frac{\omega}{v}$$

or

$$\omega_n = (2n - 1)\frac{\pi}{2L}\sqrt{\frac{Y}{\rho}} \tag{4.28}$$

The Eq. (4.28) gives the angular frequencies of the normal modes for longitudinal vibrations of a rod fixed at one end and free at the other. The corresponding frequencies are

$$v_n = \frac{(2n - 1)}{4L}\sqrt{\frac{y}{\rho}} \tag{4.28a}$$

In n^{th} mode, the displacement of the particles of the rod is given by

$$\xi_n(x,t) = A_n \sin\left\{(2n-1)\frac{\pi x}{2L}\right\} \cos(\omega_n t + \phi_n) \tag{4.29}$$

The constants A_n and ϕ_n are determined from initial conditions. The lowest mode ($n = 1$) or the fundamental mode has the frequency

$$v_1 = \frac{1}{4L}\sqrt{\frac{Y}{\rho}} \tag{4.30}$$

and particle displacement is given by

$$\xi_1(x,t) = A_1 \sin\left(\frac{\pi x}{2L}\right) \cos(\omega_1 t + \phi_1)$$

We observe that $\xi_1(x, t)$ is zero at $x = 0$ and maximum at $x = L$. There is no other value of x in the range 0 and L where ξ_1 can be zero. This mode is shown in Fig. 4.8(a) which is graph between ξ_1 and x. Remember the displacement ξ_1 are longitudinal and not the transverse.

The next mode ($n = 2$) has the frequency given by

$$v_2 = \frac{3}{4L}\sqrt{\frac{Y}{\rho}} = 3v_1$$

which is three times the frequency of the fundamental mode. The particle displacements in this mode are given by

$$\xi_2(x,t) = A_2 \sin\left(\frac{3\pi x}{2L}\right) \cos(\omega_2 t + \phi_2)$$

Notice that ξ_2 is zero at $x = 0$ as well as at $x = \frac{2L}{3}$. At $x = L$, ξ is maximum but negative.

The shape of the mode with frequency $v_3 = 5v_1$. There are two nodes at $x = \frac{2L}{5}$ and $x = \frac{4L}{5}$ in addition to one at $x = 0$. Thus, a rod fixed at one end and free at the other has only *odd harmonics* of frequencies 3, 5, 7, ... times the fundamental frequency. All the even harmonics are absent.

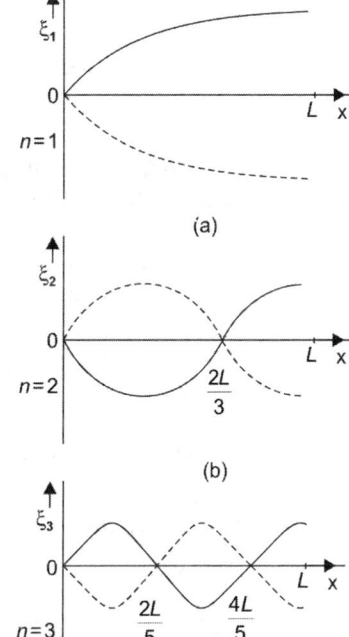

Fig. 4.8

Rod clamped in the middle: Consider a uniform rod which is free at its ends, i.e., at $x = 0$ and $x = L$ but is clamped at $x = \frac{L}{2}$. The boundary conditions now, are

$$\left(\frac{\partial \xi}{\partial x}\right)_{x=0} = \left(\frac{\partial \xi}{\partial x}\right)_{x=L} = 0$$

Equation (4.26) gives the displacement of the particles of the rod in a mode of angular frequency ω and phase constant ϕ. Differentiating with respect to x, we have

$$\frac{\partial \xi}{\partial x} = v(A\cos kx - B\sin kx)\cos(\omega t + \phi)$$

Using the condition $\dfrac{\partial \xi}{\partial x} = 0$ at $x = 0$, we have

$$A = 0$$

Hence

$$\frac{\partial \xi}{\partial x} = -Bk \sin kx \cos(\omega t + \phi)$$

The condition $\dfrac{\partial \xi}{\partial x} = 0$ at $x = L$ is satisfied if

$$\sin KL = 0 \quad \text{or} \quad KL = n\pi$$

or

$$\frac{\omega L}{v} = n\pi$$

where n has integral value $n = 1, 2, 3, \ldots$ etc. Thus,

$$\omega_n = \frac{n\pi v}{L} = \frac{n\pi}{L}\sqrt{\frac{Y}{\rho}} \tag{4.31}$$

This equation gives the angular frequencies of normal modes of longitudinal vibrations of a rod which is free at both ends. The corresponding frequencies are

$$v_n = \frac{\omega_n}{2\pi} = \frac{n}{2L}\sqrt{\frac{Y}{\rho}}$$

In nth mode the particle displacements are given by equation

$$\xi(x,t) = B_n \cos\left(\frac{n\pi x}{L}\right)\cos(\omega_n t + \phi) \tag{4.32}$$

Since the rod is clamped in middle, i.e., $\xi\left(\dfrac{L}{2}, t\right) = 0$ for all times, this gives

$$\cos\left(\frac{n\pi L}{2L}\right) = \cos\left(\frac{n\pi}{2}\right) = 0$$

To safisfy this condition, the value of $n = 2, n = 4, n = 6,$ ... etc., are not allowed[2] in Eq. (4.32). Hence, the allowed values of n are 1, 3, 5, ... etc.

The fundamental mode ($n = 1$) has a frequency given by

$$v_1 = \frac{1}{2L}\sqrt{\frac{Y}{\rho}}$$

The higher mode frequencies are $3v_1, 5v_1, 7v_1, \ldots$ etc. Figure 4.9 shows the shapes of the first three modes of rod clamped in the middle.

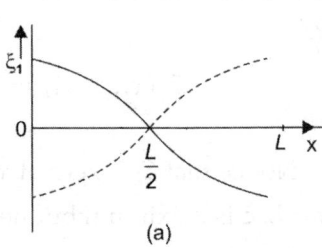

(a)

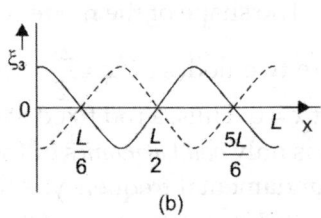

(b)

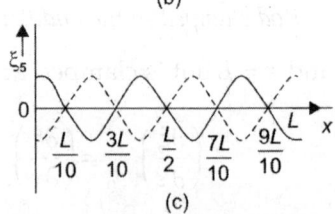

(c)

Fig. 4.9

---

2. Because if they are allowed, then $\cos\dfrac{n\pi}{2} = \pm 1$ for these values of n; hence $x = \dfrac{L}{2}$ will have some displacement, which is not admissible.

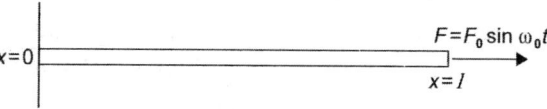

Example 4.7: A uniform rod of length L fixed at one end ($x = 0$) and subjected to sinusoidal force $F = F_0 \sin \omega_0 t$ at the free end ($x = L$) as shown in Fig. 4.10. Show that the steady state displacement for forced oscillations of the rod is given by

$$\xi(x, t) = \frac{F_0 v}{\alpha Y \omega_0} \sec\left(\frac{\omega_0 l}{v}\right)$$

$$\sin\left(\frac{\omega_0 x}{v}\right) \sin \omega_0 t$$

Fig. 4.10

where α is cross-sectional area and Y the Young's modulus of the rod and $v = \sqrt{\dfrac{Y}{\rho}}$, where ρ is the density of the material of the rod.

Solution: The equation of motion of the rod in longitudinal vibration is

$$\frac{\partial^2 \xi}{\partial t^2} = v^2 \frac{\partial^2 \xi}{\partial x^2} \tag{i}$$

and the steady state displacement of the forced vibrations will be sinusoidal of frequency ω_0. Hence, steady state vibrations can be written as

$$\xi(x, t) = A(x) \sin \omega_0 t$$

$$\therefore \quad \frac{\partial^2 \xi}{\partial t^2} = -\omega_0^2 A(x) \sin(\omega_0 t) = -\omega_0^2 \xi(x, t) \tag{ii}$$

Substituting in (i), we have

$$\frac{\partial^2 \xi}{\partial x^2} + \frac{\omega_0^2 \xi}{v^2} = 0 \tag{iii}$$

The solution of (iii) will give $A(x)$ and can be written as

$$A(x) = A \cos kx + B \sin kx \quad \text{where } k = \frac{\omega_0}{v}$$

$$\therefore \quad \xi(x, t) = \{A \cos kx + B \sin kx\} \sin \omega_0 t$$

Since the rod is fixed at $x = 0$, i.e. $\xi(0, t) = 0$ for all the times which gives $A = 0$.

$$\therefore \quad \xi(x, t) = B \sin kx \sin \omega_0 t$$

At the free end $x = l$, the strain due to the longitudinal vibrations is $\dfrac{\partial \xi}{\partial x}$. Hence, the internal force at this end is

$$F_i = \text{stress} \times \alpha = Y \text{ strain} \times \alpha = Y\alpha \frac{\partial \xi}{\partial x}$$

and this must be equal to the externally applied force $F_0 \sin \omega_0 t$, i.e.

$$Y\alpha\left(\frac{\partial \xi}{\partial x}\right) = F_0 \sin \omega_0 t$$

$$\therefore \quad Y\alpha B k \cos kx \sin \omega_0 t = F_0 \sin \omega_0 t$$

$$\therefore \qquad B = \frac{F_0}{Y\alpha k \cos kl} = \frac{F_0 v}{Y\alpha\omega_0}\sec\left(\frac{\omega_0 l}{v}\right)$$

$$\therefore \qquad v = \frac{\omega_0}{k}$$

$$\therefore \qquad \xi(x,\,t) = \frac{F_0 v}{Y\alpha\omega_0}\sec\left(\frac{\omega_0 l}{v}\right)\sin\left(\frac{\omega_0}{v}x\right)\sin\omega_0 t$$

Example 4.8: A bar of length l and mass M is fixed at one end and carries a mass m at the other end as shown in Fig. 4.11. Derive the equation which determines the normal mode frequencies of the system. Show that if $m \le M$, the frequency equation reduces to the equation of normal mode frequencies of a rod fixed at one end and free at the other.

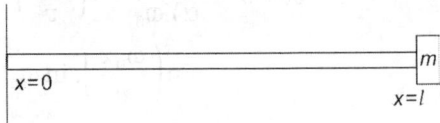

Fig. 4.11

Solution: The displacement in the rod in the normal mode is given by

$$\xi(x,y) = (A_n \sin k_n x + B_n \cos k_n x)\cos(\omega_n t + \phi_n)$$

where $\qquad k_n = \dfrac{\omega_n}{v}$ and $v = \sqrt{\dfrac{Y}{\rho}}$.

Since the rod is fixed at $x = 0$, i.e. $\xi(0, t) = 0$ which gives $B_n = 0$.

$$\therefore \qquad \xi(x,t) = A_n \sin k_n x \cos(\omega_n t + \phi_n)$$

At the free end $x = l$, the force $\alpha Y\dfrac{\partial \xi}{\partial x}$ due to vibrations must be equal and opposite to the Newton's force, i.e., mass × acceleration.

$$\therefore \qquad \alpha Y\left(\frac{\partial \xi}{\partial x}\right)_{x=l} = -m\left\{\frac{\partial^2 \xi}{\partial t^2}\right\}_{x=l}$$

where Y is Young's modulus of the material of the rod and α, the area of the cross-section of the rod.

$$\frac{\partial \xi}{\partial x} = A_n k_n \cos k_n x \cos(\omega_n t + \phi_n)$$

$$\therefore \qquad \left(\frac{\partial \xi}{\partial x}\right)_{x=l} = A_n k_n \cos k_n l \cos(\omega_n t + \phi_n)$$

and $\qquad \left(\dfrac{\partial^2 \xi}{\partial t^2}\right)_{x=l} = -A_n \omega_n^2 \sin k_n l \cos(\omega_n t + \phi_n)$

$$\therefore \qquad \alpha Y A_n k_n \cos k_n l \cos(\omega_n t + \phi_n) = mA_n \omega_n^2 \sin k_n l \cos(\omega_n t + \phi_n)$$

or $\qquad \tan k_n l = \dfrac{Y\alpha k_n}{m\omega_n^2} = \dfrac{Y\alpha\omega_n}{mv\omega_n^2} = \dfrac{YM}{\rho lmv\omega_n}$

$$\because \qquad k_n = \frac{\omega_n}{v} \text{ and } \alpha l\rho = M \text{ or } \alpha = \frac{M}{\rho l}$$

$$\therefore \qquad \tan k_n l = \frac{M}{m}\frac{v}{l\omega_n} \qquad \because v^2 = \frac{Y}{\rho}$$

$$\therefore \qquad \frac{M}{m} = \frac{l\omega_n}{v}\tan\left(\frac{\omega_n l}{v}\right)$$

This is the required equation which determines the normal mode frequencies of the system.

If $m \leq M$, $\tan\left(\dfrac{\omega_n l}{v}\right) \to \infty$

or $\dfrac{\omega_n l}{v} = (2n+1)\dfrac{\pi}{2}$ where $n = 0, 1, 2, ...$etc.

or $\omega_n = \dfrac{(2n+1)v\pi}{2l}$

or $v_n = \dfrac{\omega_n}{2\pi} = \dfrac{(2n+1)v}{4l} = \dfrac{(2n+1)}{4l}\sqrt{\dfrac{Y}{\rho}}$

which agrees with Eq. (4.28a) for the frequencies of the normal modes of rod, one end of which is free and other is clamped.

Example 4.9: A uniform rod of length l is free at one end ($x = l$) and is subjected to sinusoidal displacement $a_0 \sin \omega_0 t$ at the other end ($x = 0$). Show that the steady state displacement of the rod is given by

$$\xi(x,t) = a_0 \cos\left(\dfrac{\omega_0 x}{v}\right)\left\{1 + \tan\left(\dfrac{\omega_0 t}{v}\right)\tan\left(\dfrac{\omega_0 x}{v}\right)\right\}\sin \omega_0 t$$

Hence, show that resonance will occur if $\omega_0 = (2n+1)\dfrac{\pi v}{2l}$, where $n = 0, 1, 2, ...$ etc.

Solution: The equation of the motion of longitudinal vibrations of the rod is given by

$$\dfrac{\partial^2 \xi}{\partial t^2} = v^2 \dfrac{\partial^2 \xi}{\partial x^2}$$

where $\xi(x, t)$ is the displacement at any cross-section of the bar. In steady state the longitudinal vibrations will have the same frequency as that of the displacement of end $x = 0$. Hence, the general solution will be of the form

$$\xi(x, t) = A(x) \sin(\omega_0 t) = (A \sin kx + B \cos kx) \sin \omega_0 t$$

where $k = \dfrac{\omega_0}{v}$

The steady boundary conditions are

$$\xi(0, t) = a_0 \sin \omega_0 t$$

and $\left(\dfrac{\partial \xi}{\partial x}\right)_{x=l} = 0$

The first boundary condition yields

$$a_0 \sin \omega_0 t = B \sin \omega_0 t \quad \text{or} \quad B = a_0$$

The second boundary condition yields

$$\left(\dfrac{\partial \xi}{\partial x}\right)_{x=l} = k\{A \cos kl - B \sin kl\} \sin \omega_0 t = 0$$

or $\tan kl = \dfrac{A}{B}$

$\therefore$ $A = B \tan kl = a_0 \tan kl$

Hence, the steady state displacement is given by

$$\xi(x, t) = \{a_0 \tan kl \sin kx + a_0 \cos kx\} \sin \omega_0 t$$

$$= a_0 \cos kx\{1 + \tan(kl) \tan kx\} \sin \omega_0 t$$

But

$$k = \frac{\omega_0}{v}$$

$\therefore$

$$\xi(x, t) = a_0 \cos\left(\frac{\omega_0 x}{v}\right)\left\{1 + \tan\left(\frac{\omega_0 l}{v}\right) \tan\left(\frac{\omega_0 x}{v}\right)\right\} \sin \omega_0 t$$

Resonance occur when displacement ξ becomes very large, i.e., theoretically $\to \infty$.

$$\tan\left(\frac{\omega_0 l}{v}\right) = \infty$$

or

$$\frac{\omega_0 l}{v} = (2n + 1)\frac{\pi}{2} \quad \text{where} \quad n = 0, 1, 2, \dots \text{etc.}$$

or

$$\omega_0 = (2n + 1)\frac{\pi v}{2l}$$

4.6 PRESSURE WAVES IN GAS COLUMNS

Air or any other gas can be considered as a continuous medium at ordinary pressure. The average distance between the molecules is very small (molecular density of air at normal pressure is of the order of $10^{19} \sim 10^{20}$ molecules per cc) and as an approximation a column of gas then can be considered to behave as if it were a continuous medium. As such, a column of gas represents a system almost equivalent to a solid rod (both are continuous systems and have internal elasticity) with the difference, that in air columns, the relevant modulus of elasticity is the bulk modulus or volume elasticity of the gas.

When a phase progressive longitudinal wave travels through a gas, the particles of the gas execute *simple harmonic motion* along the direction of propagation of the wave with continuous varying phase. The distance between the particles so alters that at any instant particles are alternatively crowded and spread out. Hence, pressure varies from particle to particle inside the gas, thus sets up a pressure wave propagating through the column.

Consider a gas contained in a cylindrical tube of cross-sectional area α and ρ be the density of gas. Now, let us divide the column of gas into small elements each of length Δx, Figure 4.13(a) shows one such element AB in the undisturbed state. Let P_0 be the equilibrium pressure and V the volume of the element. Then

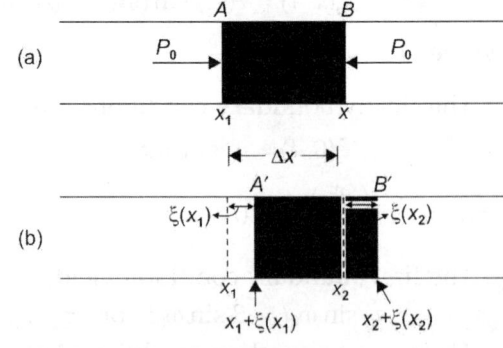

Fig. 4.13

$$V = \alpha \Delta x$$

The gas enclosed inside an element is like a compressed spring which would like to extend itself as such, the gas exerts an outward pressure on the walls of the element. Let us analyse the motion of this element of the gas when it is disturbed along x-axis. This can be done by creating disturbance at the mouth of the tube by placing a vibrating tuning fork at the open end of the tube. The particles at the plane A move to the right and oscillate about A. The particles at plane B also oscillate about B.

Let the particles at A be apart x_1 from some origin and at end B be apart x_2. So that the length of element AB is $\Delta x = x_2 - x_1$. Figure 4.13(b) shows the disturbed position of the element AB at an instant of time when A is shifted to A' and B to B'. Now, let the co-ordinates of A' and B' be $x_1 + \xi(x_1)$ and $x_2 + \xi(x_2)$ respectively, where $\xi(x_1)$ and $\xi(x_2)$ are the displacements of the particles originally at x_1 and x_2 consecutively. If at this moment $\xi(x_2) > \xi(x_1)$, there is an increase in length of element and hence, increase in its volume. Let increase in volume be ΔV, then

$$\Delta V = \alpha\{\xi(x_2) - \xi(x_1)\} = \alpha\frac{\partial\xi(x)}{\partial x}\Delta x$$

$\therefore$ Volume strain $= \dfrac{\Delta V}{V} = \dfrac{\partial\xi(x)}{\partial x}$ $\because V = \alpha\Delta x$

This increase in volume is due to decrease in pressure. Let new pressure at A' be $P(x_1) = P_0 - \Delta P$ while that at B' is $P(x_2)$. Here Δp is decrease in pressure at A.

In case of gases, the volume elasticity, i.e., bulk modulus E is defined as

$$E = -\frac{\Delta P}{\Delta V/V} = -\frac{V\Delta P}{\Delta V}$$

i.e., the pressure difference for a fractional change in volume. Since increase in volume is due to decrease in pressure as such negative sign is inserted to keep E positive. Hence, change in pressure is given by

$$\Delta P = -E\frac{\Delta V}{V} = -E\frac{\partial\xi}{\partial x} \qquad\qquad (4.31a)$$

Now, the pressure difference across the ends of the elements $A'B'$ is given by

$$P(x_2) - P(x_1) = \frac{\partial P(x)}{\partial x}\Delta x$$

$$= \frac{\partial}{\partial x}[P_0 - \Delta P]\Delta x$$

$$= -\frac{\partial}{\partial x}(\Delta P)\Delta x$$

Substituting the values of ΔP, we have

Excess pressure $= -\dfrac{\partial}{\partial x}\left(-E\dfrac{\Delta V}{V}\right)\Delta x$

$$= E\frac{\partial}{\partial x}\left(\frac{\partial\xi}{\partial x}\right)\Delta x$$

Hence, the net force acting on the element is

$$F = \text{Cross-sectional area} \times \text{Excess pressure}$$

$$= \alpha \cdot E\Delta x\frac{\partial^2\xi}{\partial x^2}$$

If ρ is the density of the gas, the mass of the element of length Δx is equal to $\alpha\rho\Delta x$.

Hence, the Newton's force on element is given by $\rho\alpha\Delta x\dfrac{\partial^2\xi}{\partial t^2}$.

Equating the two forces, we get

$$\alpha \cdot \rho \Delta x \frac{\partial^2 \xi}{\partial t^2} = \alpha \Delta x E \frac{\partial^2 \xi}{\partial x^2}$$

or

$$\frac{\partial^2 \xi}{\partial t^2} = \frac{E}{\rho} \frac{\partial^2 \xi}{\partial x^2} = v^2 \frac{\partial^2 \xi}{\partial x^2} \tag{4.32}$$

where $v = \sqrt{\dfrac{E}{\rho}}$ is the velocity of the longitudinal waves in gas.

Normal modes: We have derived the equation of motion of the particles of the gas and now are in a position to obtain the possible normal modes of the longitudinal vibrations of the column of a gas under specified boundary conditions. The particle displacements in a normal mode of angular frequency ω and phase constant f are given by

$$\xi(x,t) = (A \sin kx + B \cos kx) \cos(\omega t + \phi) \tag{4.33}$$

where $k = \dfrac{\omega}{v}$ and $v = \sqrt{\dfrac{E}{\rho}}$. Hence in that mode, the vibrations in pressure along the tube are given by

$$\Delta P = -E \frac{\partial \xi}{\partial x}$$

But

$$\frac{\partial \xi}{\partial x} = k(A \cos kx - B \sin kx) \cos(\omega t + \phi)$$

and hence

$$\Delta P = -E \frac{\partial \xi}{\partial x} = kE(B \sin kx - A \cos kx) \cos(\omega t + \phi) \tag{4.33a}$$

We shall now determine the frequencies and shapes of the gas enclosed under the following boundary conditions:

(a) **Tube closed at one end and open at the other:** Consider a gas enclosed in a uniform cylindrical tube of length L lying along x-axis. The end $x = 0$ is open while the end $x = L$ is closed. An open end represents a condition of zero pressure variation during the oscillation and maximum movement of gas particles. This implies that at the open end

$$\Delta P = 0 \quad \text{or} \quad \frac{\partial \xi}{\partial x} = 0$$

In other words, the particles at the open end are under no strain and therefore free to move. The closed end of the other hand, is the place of zero movement and maximum pressure variation. Thus, at the closed end $\xi = 0$. So, the boundary conditions are (for all t values)

$$(\Delta P) = 0 \quad \text{or} \quad \left(\frac{\partial \xi}{\partial x} \right)_{x=0} = 0.$$

and

$$\xi(L, t) = 0$$

Using the first boundary condition yields $A = 0$, and hence, Eq. (4.33) reduces to

$$\xi(x, t) = B \cos kx \cos(\omega t + \phi) \tag{4.34}$$

Using second boundary condition, we will obtain the normal mode frequencies. This condition yields

$$\xi(L, t) = B \cos Lk \cos(\omega t + \phi) = 0.$$

which gives

$$kL = (2n - 1)\frac{\pi}{2}$$

or

$$\omega_n = vk_n = (2n - 1)\frac{\pi}{2L}\sqrt{\frac{E}{\rho}} \qquad (4.33(a))$$

where $n = 1, 2, 3, \dots \infty$. The corresponding frequencies in hertz are

$$\nu_n = \frac{\omega_n}{2\pi} = \frac{(2n - 1)}{4L}\sqrt{\frac{E}{\rho}} \qquad (4.35)$$

Figure 4.14 shows the first three modes of the tube. Remember that the particle displacements are longitudinal. Notice that a tube closed at one end has only odd harmonics of frequency $3, 5, 7, \dots$ etc. times the fundamental frequency ν_1. All even harmonics of frequencies $2\nu_1, 4\nu_1, 6\nu_1, \dots$ etc. are absent. The frequency of the fundamental mode is given by

$$\nu_1 = \frac{1}{4L}\sqrt{\frac{E}{\rho}}$$

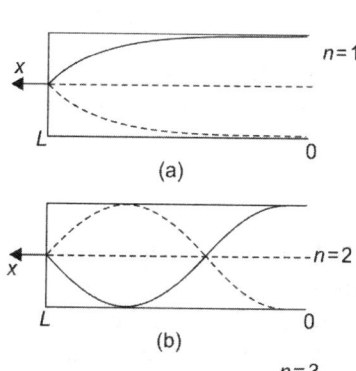

(b) **Tube open at both ends:** For tube open at both ends called the *open tube*, the boundary conditions are

$$\left(\frac{\partial \xi}{\partial x}\right) = 0 \quad \text{and} \quad \left(\frac{\partial \xi}{\partial x}\right)_{x=L} = 0$$

Using Eq. (4.33a), we find that these conditions are satisfied if $A = 0$ and

$$\sin kL = 0$$

or

$$kL = n\pi$$

given

$$\omega_n = vk_n = \frac{n\pi}{L}v = \sqrt{\frac{E}{\rho}}$$

and

$$\nu_n = \frac{\omega_n}{2\pi} = \frac{n}{2L}\sqrt{\frac{E}{\rho}}$$

where $n = 1, 2, 3, \dots$ etc. Figure 4.15 shows the first three modes of the tube open at both ends. Here in open tube, all the harmonics, i.e., even as well as odd harmonics are present. The positions where the particles displacement are maximum are called *anti-nodes*. The positions of the particles which are permanently at rest are called *nodes*. It may be noted that a tube with both ends closed has the same set of natural frequencies as the one with both ends open but the positions of nodes and anti-nodes are interchanged.

Fig. 4.14

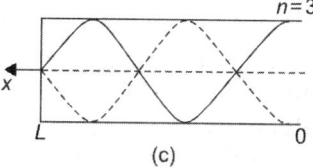

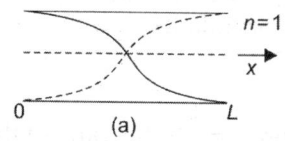

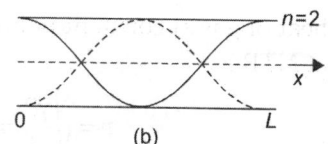

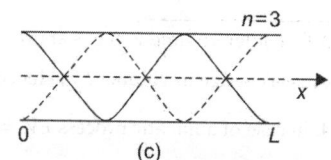

Fig. 4.15

4.7 NEWTON'S FORMULA

First of all, Newton purely on theoretical basis gave an expression for the velocity of sound waves in a gas. He assumed that when sound wave travels through a gaseous medium, the temperature variation in the regions of compression and rarefaction are negligible. The conditions, therefore, are isothermal and Boyle's law can be applied. Hence, Newton assumed that in expression for the velocity of sound in gases.

$$v = \sqrt{\frac{E}{\rho}}$$

E is isothermal volume elasticity, which is equal to the pressure of the gas P.[3]
Hence under isothermal conditions

$$v = \sqrt{\frac{P}{\rho}}$$

For air at NTP, density $\rho = 1.29$ kg/m^3 and pressure $P = 1.01 \times 10^5$ N/m^2. Using these values, the velocity of sound in air at NTP is

$$v = \sqrt{\frac{1.01 \times 10^5}{1.29}} = 280 \text{ m/s}$$

The various experiments performed for the determination of the velocity of sound in air give the value 332 m/s which is about 16% higher than theoretical value predicted by Newton. Newton could not give satisfactory explanation for this discrepancy. It was Laplace who gave the satisfactory explanation and modified the formula which gave the results very close to the experimental values.

Laplace's correction: According to Laplace as sound wave travels in a gas, the region of compression is heated and the region of rarefaction is cooled. Since the thermal conductivity of gas is small and these thermal changes occur so rapidly that the heat developed in compression and cooling produced in rarefaction is not transferred out or in, to achieve thermal equilibrium with surroundings in short time, i.e. the time required by the sound wave to travel from compression to rarefaction, i.e. there is region of unequal temperature. Thus, Laplace was of the opinion that the changes are adiabatic. Hence, E in the formula for velocity of sound should be adiabatic bulk modulus. Its value is[4]

$$E_a = \gamma P$$

where $\gamma = \dfrac{C_p}{C_v}$, the ratio of the specific heat of gas at constant pressure (C_p) and the specific heat of gas at constant volume (C_v). For air $\gamma = 1.4$ and hence, the velocity of sound in air at NTP is

$$v = \sqrt{\frac{\gamma P}{d}} = 331.6 \text{ m/s}.$$

---

3. Consider volume V of gas at pressure P. When a sound wave propagates through it, the change in pressure (ΔP) causes change in volume ΔV. Hence $E = -\dfrac{\Delta P}{\Delta V/V}$ and using Boyle's law $PV = $ const., we have $E_i = -\dfrac{\Delta PV}{\Delta V} = P$

4. In case of adiabatic process $PV^r = $ constant.

$\therefore PrV^{r-1}\Delta V + V^r \Delta P = 0.$ $\therefore E_a = -\dfrac{(\Delta P)V}{\Delta V} = rP$

which is in close agreement with experimental value and hence, proves the correctness of Laplace explanation. If we assume gas to be ideal, then using gas equation, we have

$$PV = \frac{m}{M}RT$$

where R is the universal gas constant and T is the absolute temperature of the gas, m is mass of gas and M molecular weight.

$$\therefore \qquad \frac{P}{\rho} = \frac{RT}{M}$$

$$\therefore \qquad v = \sqrt{\frac{\gamma RT}{M}} \qquad\qquad (4.36)$$

This equation reveals that velocity of sound in a gas:

(a) is independent of pressure or density,

(b) is proportional to the square root of absolute temperature,

(c) is inversely proportional to the square root of the molecular weight, and

(d) depends on the value of γ of the gas, i.e. the atomicity of the gas.

4.8 ACOUSTIC IMPEDANCE

The acoustic impedance Z offered by an elastic medium to the propagation of sound wave through it is defined as

$$Z = \frac{\text{Pressure due to wave}}{\text{Particle velocity}}$$

The particle displacement for a wave travelling in the positive X-direction is given by

$$\xi(x,t) = A\sin\left\{\frac{2\pi}{\lambda}(vt - x)\right\}$$

where $v = \sqrt{\dfrac{E}{\rho}}$, A is amplitude (displacement) and λ is the wavelength.

Differentiating, we have:

$$\text{Volume strain} = \frac{\partial \xi}{\partial x} = -A\left(\frac{2\pi}{\lambda}\right)\cos\left\{\frac{2\pi}{\lambda}(vt - x)\right\} \text{ and}$$

$$\text{Particle velocity} = \frac{\partial \xi}{\partial t} = A\left(\frac{2\pi v}{\lambda}\right)\cos\left\{\frac{2\pi}{\lambda}(vt - x)\right\}$$

But we know that pressure due to wave is

$$\Delta P = -E\frac{\partial \xi}{\partial x} = EA\left(\frac{2\pi}{\lambda}\right)\cos\left\{\frac{2\pi}{\lambda}(vt - x)\right\}$$

Hence, acoustic impedance Z is given by

$$Z = \frac{\Delta P}{\dfrac{\partial \xi}{\partial t}} = \frac{E}{v} = \rho v \qquad \because \ v = \sqrt{\frac{E}{\rho}}$$

Since the wave velocity v depends upon the inertial (mass, density) and elastic (Bulk modulus E) properties of the medium, the characteristic impedance is also governed by these two properties of the medium. We have assumed that the medium is lossless and possesses no dissipative mechanism in which case Z is real.

4.9 ENERGY TRANSPORT IN PROGRESSIVE SOUND WAVES IN A GAS

Suppose a sound wave is travelling in a gas. Consider an infinitesimal small volume ΔV of the medium whose dimensions are small compared to the wavelength of the wave, hence, the amplitude and phase of oscillations of all the particles within this volume element can be considered to be the same. Let n be number of particle in this volume element ΔV and each has mass m. Hence energy of each particle is equal to $\dfrac{1}{2}mA^2\omega^2$ where ω is the angular frequency of oscillation and A, its amplitude.

Hence, energy of oscillations of these n particles $= \dfrac{1}{2}nmA^2\omega^2$

$$= \frac{1}{2}(\rho \Delta V)A^2\omega^2$$

where $\rho = \dfrac{mn}{\Delta V}$ is the density of the medium. Hence energy density, i.e. energy permit volume is

$$U = \frac{1}{2}\rho A^2\omega^2 = \frac{1}{2}\rho A^2(2\pi v)^2 = 2\pi^2\rho A^2 v^2$$

where $\omega = 2\pi v$ and v is frequency.

Intensity is defined as the energy crossing unit area per second, area taken perpendicular to the direction of propagation of wave. Now, consider a plane wave propagation along X-direction. Let $ABCD$ be a plane wave front of area s at any time t (Fig. 4.16). In a small interval of time Δt, let the wave front is displaced through a distance $\Delta x = v\Delta t$ in the direction of propagation, i.e. x-direction. The new position of wave front is $A'B'C'D'$. Because of this, the particles of the medium in volume $s\Delta x$ are set into motion

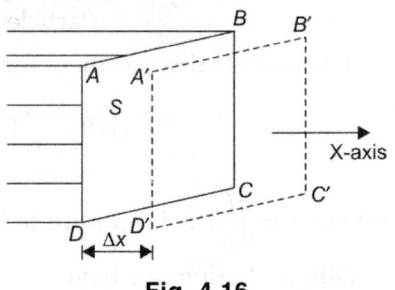

Fig. 4.16

(oscillations) since U is the energy density of the medium, then in time Δt, the medium of volume $S\Delta x$ receives energy equal to $u(\Delta xS)$. Thus, the energy propagated through area s in time $\Delta t = u\Delta xS$.

Hence, energy flux or intensity

$$I = \frac{u \times \Delta x \times S}{S \times \Delta t} = u\frac{\Delta x}{\Delta t} = u.v.$$

where v is the wave velocity $\left(= \dfrac{\Delta x}{\Delta t}\right)$. Thus, the energy flux or intensity is equal to **energy** density times velocity

$$\therefore \qquad I = 2\pi^2\rho A^2 v^2 v \qquad\qquad (4.37)$$

This is the same expression as for transverse waves on a string.

Example 4.10: If the velocity of sound in hydrogen gas at 0°C is 1260 m/s then find the velocity of sound at 0°C in mixture of H_2 and O_2 in volume ratio of 2:1. The ratio of densities of O_2 and H_2 is 16:1.

Solution: If volume of oxygen is $V_0 = V$, then volume of hydrogen is $V_H = 2V$. Hence the density of the mixture is

$$\rho = \frac{\rho_0 V_0 + \rho_H V_H}{V_0 + V_H} = \frac{16\rho_H V + \rho_H \times 2V}{V + 2V}$$

where ρ_H is density of hydrogen $\rho_0 = 16\rho_H$

$\therefore$ $$\rho = 6\rho_H$$

Velocity of sound in hydrogen is

$$V_H = \sqrt{\frac{\gamma P}{\rho_H}}$$

and velocity of sound in mixture

$$V = \sqrt{\frac{\gamma P}{\rho}}$$

$\therefore$ $$\frac{V}{V_H} = \sqrt{\frac{\rho_H}{6\rho_H}} = \frac{1}{\sqrt{6}}$$

$\therefore$ $$V = \frac{V_H}{\sqrt{6}} = \frac{1260}{\sqrt{6}} = 514.4 \text{ m/s}$$

Example 4.11: Find the ratio of the velocity of sound in helium and hydrogen at the same temperature. $\gamma = 5/3$ for monoatomic gas and $7/5$ for diatomic gas.

Solution: Velocity of sound in gas is given by

$$V = \sqrt{\frac{\gamma R T}{M}}$$

where M is molecular weight of the gas.

$\therefore$ $$V_H = \sqrt{\frac{\gamma_H R T}{M_H}}$$

and $$V_{He} = \sqrt{\frac{\gamma_{He} R T}{M_{He}}}$$

$\therefore$ $$\frac{V_{He}}{V_H} = \sqrt{\frac{\gamma_{HE} \cdot H_H}{\gamma_H M_{He}}} = \sqrt{\frac{\frac{5}{3} \times 2}{\frac{7}{5} \times 4}}$$

$$= \sqrt{\frac{25}{42}} = 0.77$$

4.10 FOURIER METHOD: GENERAL MOTION OF A CONTINUOUS SYSTEM

In our treatment so far, we have been discussing only harmonic waves in which the source or the particles of the medium perform simple harmonic motion. But let us consider the situation when several pure tones are heard simultaneously. For example, vibrations from string of a musical instrument or from our vocal cords, one gets a fundamental tone accompanied by a number of overtones (harmonics) of frequencies integer times the frequency of the fundamental. Such vibrations are complex periodic vibrations and not simple harmonic oscillations. Such complex periodic vibrations can be analysed into a set of single harmonic vibrations having frequencies which are multiples of the frequency of the complex periodic vibration. It is done using Fourier theorem given by J.B.J. Fourier in 1822 and the process of analysing complex vibrations is called *Fourier analysis*. The various component motions with their relative amplitudes and phases, when added together according to the principle of superposition will result into the given complex periodic motion.

(a) Fourier Theorem

This theorem states that any periodic wave form $y(t)$ in the interval of one time period can be expressed as a series of harmonics of repetition frequency (fundamental frequency). These harmonics will have varying amplitudes and phases depending upon the nature of the wave form. When amplitude of these components is plotted against frequency, the resulting curve is called the *frequency spectrum of the wave*. The theorem is represented mathematically as follows:

$$Y(t) = C_0 + C_1 \sin(\omega t + \phi_1) + C_2 \sin(2\omega t + \phi_2) + \dots \dots + C_n(n\omega t + \phi_n) + \dots \dots \quad (4.38)$$

The periodicity of $y(t)$ is given by $(\omega/2\pi)$ $C_0, C_1, C_2, \dots C_n, \dots$ etc. are the amplitudes of the constituent harmonics and $\phi_1, \phi_2, \phi_3, \dots, \phi_n$, etc. are the initial phase of constituent harmonics. Hence, we have taken the vibration to be represented as sine functions. These components can also be represented as cosine functions without altering the validity of representation. Using the relation

$$\sin(\alpha + \beta) = \sin\alpha\cos\beta + \cos\alpha\sin\beta$$

and representing the co-efficients of sine and cosine terms properly, we can represent $y(t)$ as

$$Y(t) = C_0 + A_1 \sin\omega t + A_2 \sin 2\omega t + \dots + A_n \sin n\omega t + \dots$$
$$+ B_1 \cos\omega t + B_2 \cos 2\omega t + \dots + B_n \cos n\omega t + \dots$$
$$= C_0 + \sum_{n=1}^{\infty} A_n \sin\omega t + \sum_{n=1}^{\infty} B_n \cos n\omega t \quad (4.39)$$

This series is known as *Fourier series* and $C_0, A_1, A_2, A_3, \dots A_n$ and $B_1, B_2, B_3, \dots, B_n$ etc. are known as *Fourier co-efficients* or *constants*. The values of these constants can be easily determined by the term integration series after multiplying it by a suitable factor.

(b) Limitations of Fourier Theorem

In using Fourier theorem for analysis of any complex period function, it should be ascertained that the given function satisfies the following conditions:

   (i) *It should be single valued*: It means that the displacement should proceed only in one direction so that at any time, there may not be two or more values of the displacements.
  (ii) *It should be continuous*: The function to which Fourier theorem is applied should be continuous in nature, i.e., it should have no discontinuity.
 (iii) *It should be finite*: It means that displacement during vibration should always remain finite, it should not assume infinite value at any instant.

In case of all mechanical vibrations (hence in case of sound vibrations), all the above three conditions are satisfied. As no particle can have two or more different displacements at one and the same time and in all cases of periodic motion, there cannot exists any discontinuity whatsoever. The condition three is also true as in all cases of sound vibrations, displacement never cross the limit of finiteness.

However, if a function $y(t)$ is to be analysed in the form of a Fourier series in a given interval (a, b), the properties possessed by the function in this interval should be known. The matter becomes still simple, if the conditions given by Dirichlet are follows:

(i) $y(t)$ is continuous in interval (a, b) except for a finite number of finite disconti-nuities.

(ii) $y(t)$ has a finite number of maxima and minima in this interval.

These are known as *Dirichlet's conditions* which are just sufficient for applying the theorem to a function but are not necessary.

(c) Evaluation of Fourier Co-efficients

(i) Determination of C_0

To determine C_0 multiply Eq. (4.39) by dt and integrate both sides for one complete cycle, i.e., from $t = 0$ to $t = T = \dfrac{2\pi}{\omega}$. We have

$$\int_0^T y(t)dt = \int_0^T C_0 dt + A_1 \int_0^T \sin \omega t\, dt + A_2 \int_0^T \sin 2\omega t + ... + A_n \int_0^T \sin \omega t\, dt + ...$$

$$+ B_1 \int_0^T \cos \omega t\, dt + B_2 \int_0^T \cos 2\omega t\, dt ... + B_n \int_0^T \cos n\omega t\, dt + ...$$

But we know that

$$\int_0^{2\pi} \sin x\, dx = \int_0^{2\pi} \cos x\, dx = 0$$

Thus, we get

$$\int_0^T Y(t)dt = C_0 \int_0^T dt = C_0 T$$

$$\therefore \qquad C_0 = \frac{1}{T}\int_0^T Y(t)\, dt \tag{4.40}$$

(ii) Determination of A_n

To determine the values of A_n, the coefficient of sin terms, multiply Eq. (4.39) by $\sin n\omega t\, dt$ and integrate for one time period. We get

$$\int_0^T y(t)\sin n\omega t\, dt = C_0 \int_0^T \sin n\omega t\, dt + A_1 \int_0^T \sin n\omega t \sin \omega t\, dt + ... + A_n \int_0^T \sin^2 \omega t\, dt + ...$$

$$+ B_1 \int_0^T \cos \omega t \sin n\omega t\, dt ... + B_n \int_0^T \sin n\omega t \cos n\omega t\, dt + ...$$

or $\quad \displaystyle\int_0^T Y(t)\sin \omega t\, dt = \frac{1}{2}A_n \int_0^T (1 - \cos 2n\omega t)\, dt$

Since all other integerals are zero.

$$\therefore \qquad \int_0^T Y(t)\sin n\omega t\, dt = \frac{1}{2}A_n T$$

or $$A_n = \frac{2}{T}\int_0^T Y(t)\sin n\omega t\, dt \tag{4.41}$$

(iii) Determination of B_n

To determine the value of B_n, we multiply Eq. (4.39) by $\cos n\omega t$ and integrate for one period. Thus

$$\int_0^T Y(t)\cos n\omega t\, dt = C_0 \int_0^T \cos n\omega t\, dt + A_1 \int_0^T \sin \omega t \cos n\omega t\, dt + \dots$$

$$+ A_n \int_0^T \sin n\omega t \cos n\omega t\, dt + \dots + B_1 \int_0^T \cos \omega t \cos n\omega t\, dt + \dots$$

$$+ B_n \int_0^T \cos^2 n\omega t\, dt + \dots$$

$$\therefore \quad \int_0^T Y(t)\cos n\omega\, dt = B_n \int_0^T \cos^2 n\omega t\, dt = \frac{1}{2} B_n \int_0^T (1 + \cos 2n\omega t)\, dt = \frac{1}{2} B_n T$$

Since all other integrals are zero.

$$\therefore \quad B_n = \frac{2}{T}\int_0^T Y(t)\cos n\omega t\, dt \qquad (4.42)$$

$$\therefore \quad Y(t) = \frac{1}{T}\int_0^T Y(t)\, dt + \sum_{n=1}^{\infty}\left[\frac{2\sin n\omega t}{T}\int_0^T Y(t)\sin n\omega t\, dt\right]$$

$$+ \sum_{n=1}^{\infty}\left[\frac{2}{T}\cos n\omega t \int_0^T Y(t)\cos n\omega t\, dt\right] \qquad (4.43)$$

and $\quad C_n = \sqrt{A_n^2 + B_n^2} \qquad (4.44)$

where C_n represents the amplitude of different constituent harmonics given by Eq. (4.38) and corresponding phase angles ϕ_n are given by

$$\phi_n = \tan^{-1}\frac{B_n}{A_n} \qquad (3.45)$$

If $A_n = 0$, $\phi_n = \dfrac{\pi}{2}$, and if $B_n = 0$, $\phi_n = 0$, i.e., in these cases, all constituents harmonic vibrations are in phase. If $C_0 = 0$, the axis of the curve coincides with the time axis.

The representation of complex periodic vibration as a function of amplitude, frequency and phase of various harmonic motions is called the *Fourier's analysis*.

4.11 TRANSVERSE MOTION OF A STRING FIXED AT BOTH ENDS

We have seen that a uniform flexible string has finite number of possible frequencies of vibration and if the string is rigidly fixed at its ends, these frequencies bear a simple relation with the fundamental frequency. The frequencies of the higher harmonics are multiples of the fundamental frequency. If such a string set into vibrations in just the right manner, it will vibrate with just one of these frequencies. But under arbitrary initial condition, the general motion of the string is given by the superposition of the normal modes. Thus, the general displacement of the string is a superposition of all the normal mode displacements

$$Y(x, t) = Y_1(x,t) + y_2(x,t) + \dots + Y_n(x,t)$$

$$= A_1 \sin k_1 x \cos(\omega_1 t + \phi_1) + A_2 \sin k_2 x \cos(\omega_2 t + \phi_2) + \dots A_n \sin k_n x \cos(\omega_n t + \phi_n)$$

or $\quad Y(x, t) = \displaystyle\sum_{n=1}^{\infty} A_n \sin k_n x \cos(\omega_n t + \phi_n)$

where $K_n = \dfrac{n\pi}{L}$ and $\omega_n = \dfrac{n\pi}{L}\sqrt{\dfrac{T}{\mu}}$

We can write the above equation as

$$y(x,t) = \sum_{n=1}^{\infty} \sin k_n x \{B_n \cos \omega_n t + C_n \sin \omega_n t\} \tag{4.46}$$

where the constants B_n and C_n are related to constant A_n and ϕ_n as

$$B_n = A_n \cos \phi_n \quad \text{and} \quad C_n = -A_n \sin \phi_n$$

and particle velocity is given by

$$V(x,t) = \frac{\partial y}{\partial t}(x,t) = \sum_{n=1}^{\infty} \omega_n \sin k_n x (-B_n \sin \omega_n t + C_n \cos \omega_n t) \tag{4.47}$$

The constants B_n and C_n in Eqs (4.46 and 4.47) are finite arbitrary constants which are determined from the initial conditions corresponding to the infinite number of points along the string whose displacements and velocities must all be specified at time $t = 0$. To evaluate these constants, we shall make use of the Fourier technique discussed in the previous section. Once these constants are determined, the general motion, i.e. $y(x, t)$ is determined, since the other constants namely k_n and ω_n are already known.

Here, we observe that all the overtones are harmonics, i.e. $\omega_n = n\omega_1$ as such by the time fundamental mode (frequency ω_1) has completed one cycle, the second harmonic (frequency $\omega_2 = 2\omega_1$) has completed two cycles and third harmonic (frequency $\omega_3 = 3\omega_1$) has completed three cycles and so on. Thus, during the second cycle of the fundamental, the motion is an exact repetition of the first cycle. In other words, the motion is periodic, but not the harmonic because the resultant of a superposition of harmonic oscillations of different frequencies is not a harmonic oscillation.

Let $Y(x,0) = Y_0(x)$ and $V(x,0) = \left(\dfrac{\partial y}{\partial t}\right)_{t=0} = V_0(x)$ respectively be the displacements and velocities of all the particles of the string (with x in the range 0 to L) at time $t = 0$. Then, Eqs (4.46 and 4.47) give $\left(\text{using } k_n = \dfrac{n\pi}{L}\right)$.

$$Y_0(x) = \sum_{n=1}^{\infty} B_n \sin k_n x = \sum_{n=1}^{\infty} B_n \sin\left(\frac{n\pi x}{L}\right) \tag{4.48}$$

and

$$V_0(x) = \sum_{n=1}^{\infty} C_n \omega_n \sin k_n x = \sum_{n=1}^{\infty} C_n \omega_n \sin\left(\frac{n\pi x}{L}\right) \tag{4.49}$$

These initial conditions must satisfy these two equations. The right hand side of these equations are called *Fourier series*. These series have spatial periodicity, i.e., function $Y_0(x)$ or $V_0(x)$ repeats itself in space. The special periodicity is $\lambda = 2L$. Since $Y_0(x)$ repeats if x changes from a value x to a value $x + \lambda$ where $\lambda = 2L$. This is apparent because

$$\sin\left(\frac{n\pi x}{L}\right) = \sin\frac{n\pi}{L}(x + 2L) = \sin\left\{\frac{n\pi x}{L} + 2n\pi\right\} = \sin\frac{n\pi x}{L}$$

It is worth to mention here that initial displacement $Y_0(x)$ and the velocity of the particle $V_0(x)$ are such function of x which vanish at $x = 0$ and $x = L$ when the string is rigidly fixed.

We can directly write down the co-efficient using the Fourier analysis

$$B_n = \frac{2}{L}\int_0^L Y_0(x)\sin\left(\frac{n\pi x}{L}\right)dx \tag{4.50}$$

$$C_n = \frac{2}{L\omega_n}\int_0^L V_0(x)\sin\left(\frac{n\pi x}{L}\right)dx \tag{4.51}$$

and

$$Y(x,\,t) = \sum_{n=1}^{\infty}\sin\frac{n\pi x}{L}(B_n\cos\omega_n t + C_n\sin\omega_n t) \tag{4.52}$$

where $\omega_n = \dfrac{n\pi v}{L}$ with $v = \sqrt{\dfrac{T}{\mu}}$, here T is the tension and μ is the linear density of the string and $n = 1, 2, 3, ...\infty$.

We shall now take some specific forms of $v_0(x)$ and $y_0(x)$ and determine the general motion of a string fixed at both ends.

Plucked string: A string is said to be plucked if it is given a finite displacement but zero initial velocity. Let us consider a simple example of string fixed at the ends and is plucked at the centre through distance h and released at $t = 0$ (Fig. 4.17). Thus at $t = 0$, the velocity of the string is zero, i.e., $v_0(x) = 0$ which gives

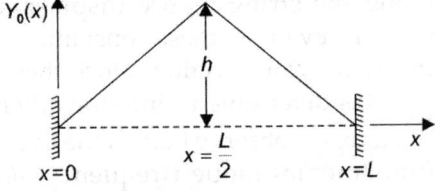

Fig. 4.17

$$C_n = 0.$$

The initial displacement of the string is given by

$$Y_0(x) = \begin{cases} \dfrac{2hx}{L}; & 0 \le x \le \dfrac{L}{2} \\[2mm] \dfrac{2h}{L}(L-x), & \dfrac{L}{2} \le x \le L \end{cases}$$

Using Eq. (4.50), we have

$$B_n = \frac{2}{L}\int_0^{L/2}\frac{2hx}{L}\sin\left(\frac{n\pi x}{L}\right)dx + \frac{2}{L}\int_{L/2}^{L}\frac{2h}{L}(L-x)\sin\left(\frac{n\pi x}{L}\right)dx$$

Integration by parts gives

$$B_n = \frac{8h}{n^2\pi^2}\sin\left(\frac{n\pi}{2}\right) = \begin{cases} 0,\ \text{if } n \text{ is an even integer.} \\[2mm] (-1)^{n-1/2}\,\dfrac{8h}{n^2\pi^2},\ \text{if } n \text{ is an odd integer.} \end{cases}$$

Therefore, the general motion of the string, using Eq. (4.52) is

$$Y(x,\,t) = \sum_{n=1}^{\infty} B_n\sin\left(\frac{n\pi x}{L}\right)\cos\omega_n t$$

$$= \frac{8h}{\pi^2}\sum_{n=1}^{\infty}\frac{1}{n^2}\sin\left(\frac{n\pi}{2}\right)\sin\left(\frac{n\pi x}{L}\right)\cos\omega_n t$$

$$\text{or } Y(x,\,t) = \frac{8h}{\pi^2}\left[\sin\left(\frac{\pi x}{L}\right)\cos\omega_1 t - \frac{1}{9}\sin\left(\frac{3\pi x}{L}\right)\cos\omega_3 t + \frac{1}{25}\sin\left(\frac{5\pi x}{L}\right)\cos\omega_5 t + ...\right] \tag{4.53}$$

where $\omega_1 = \dfrac{\pi}{L}\sqrt{\dfrac{T}{\mu}}$ and $\omega_3 = 3\omega_1$, $\omega_5 = 5\omega_1$, etc. are the normal mode frequencies of the transverse vibration of the string fixed to both ends. Since all the terms on right hand side of Eq. (6.48) are known, the general motion of the string is determined for all x and t.

Equation (4.53) gives us the following important information about the motion of the string:

(i) It tells us that all the even harmonics will be absent from the sound emitted by the string for they are not present.

(ii) The amplitude of the third harmonic is $1/9$ that of the fundamental, the amplitude of the fifth harmonic is $1/25$ that of the fundamental, etc. It means that the intensity of the third harmonic is $1/81$ of that of the fundamental, the intessity of the fifth harmonic is $1/625$ that of the fundamental and so on.

(iii) Only the first few terms in the series contribute to the displacement or energy of the string.

Which harmonics are absent depends on where the string is plucked. If the string is plucked at $x = L/4$, we can show that the fourth, eighth etc. harmonics will be absent. The general rule (which can be proved by computing the value of B_n) is that in motion of any plucked string, all those harmonics are absent which have node at the point where the string is plucked.

Total energy of the string plucked at the centre: We have already calculated the total energy of a vibrating string which is given by

$$E = \frac{1}{4}m\sum_{n=1}^{\infty}\omega_n^2 A_n^2 \qquad \text{(Eq. 4.17)}$$

$$= \frac{1}{4}m\sum_{n=1}^{\infty}\omega_n^2 (B_n^2 + C_n^2)$$

For plucked string $C_n = 0$ and $B_n = \dfrac{8h}{n^2\pi^2}\sin\left(\dfrac{n\pi}{2}\right)$.

Hence total energy of the plucked string is given by

$$E = \frac{16mh^2}{\pi^4}\sum_{n=1}^{\infty}\frac{\omega_n^2}{n^4}\sin^2\left(\frac{n\pi}{2}\right)$$

$$= \frac{16mh^2}{\pi^4}\frac{\pi^2 v^2}{L^2}\sum_{n=1}^{\infty}\frac{1}{n^2}\sin^2\left(\frac{n\pi}{2}\right)\quad \left(\because \omega_n = \frac{n\pi v}{L}\right)$$

$$= \frac{16mh^2 v^2}{\pi^2 L^2}\left[\frac{1}{1^2} + \frac{1}{3^2} + \frac{1}{5^2} + \dots\right].$$

But $\dfrac{1}{1^2} + \dfrac{1}{3^2} + \dfrac{1}{5^2} + \dots, \dots = \dfrac{\pi^2}{8}$

or $$E = \frac{16mh^2 v^2}{\pi^2 L^2}\frac{\pi^2}{8} = \frac{2h^2 v^2 m}{L^2} = \frac{2h^2 mT}{L^2\mu}\quad \because v = \sqrt{\frac{T}{\mu}}$$

or $$E = \frac{2Th^2}{L}\quad \because \mu L = m$$

Thus, *the vibrational energy of the string is directly proportional to the tension and the square of the distance through which the centre of the string is plucked, and inversely proportional to the length of the string.*

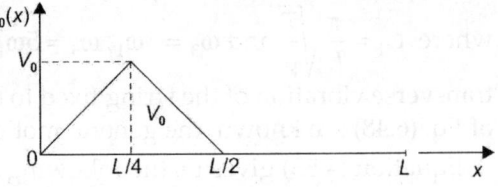

Fig. 4.18

Struck string: A string is said to be struck if it is given a finite velocity but zero initial displacement. Consider a uniform string stretched taut between its end $x = 0$ and $x = L$. Suppose the string is struck (with hammer) at $x = L/4$ keeping the portion between $x = L/2$ to $x = L$ at rest. Let at $t = 0$, the hammer imparts a velocity of v_0 to the string which increases linearly from zero at $x = 0$ to value v_0 at $x = L/4$ and decreases linearly to zero at $x = L/2$. The velocity profile is shown in Fig. 4.18. Thus, the initial displacement $Y_0(x) = 0$ giving $B_n = 0$ and the initial velocity v_0 is given by

$$V_0(x) = \begin{bmatrix} \dfrac{4v_0 x}{L} & \text{for} \quad 0 \le x \le \dfrac{L}{4} \\[3mm] \dfrac{4v_0}{L}\left\{\dfrac{L}{2} - x\right\} & \text{for} \quad \dfrac{L}{4} \le x \le \dfrac{L}{2} \\[3mm] 0 & \text{for} \quad \dfrac{L}{2} \le x \le L \end{bmatrix}$$

Substituting these values of $v_0(x)$ in (4.51), we have $\left(\because \omega_n = \dfrac{n\pi v}{L} \right)$

$$C_n = \frac{2}{n\pi v}\left[\int_0^{L/4} \frac{4v_0 x}{L} \sin\left(\frac{n\pi x}{L}\right) dx + \int_{L/4}^{L/2} \frac{4v_0}{L}\left(\frac{L}{2} - x\right)\sin\left(\frac{n\pi x}{L}\right) dx \right]$$

Now integrating by parts, we have

$$C_n = \frac{16 v_0 L}{v(n\pi)^3} \sin\left(\frac{n\pi}{4}\right)\left\{1 - \cos\left(\frac{n\pi}{4}\right)\right\}$$

Here $C_n = 0$ for $n = 4, 8, 12, \ldots$ etc. Hence, the general solution is given by

$$Y(x, t) = \frac{16 v_0 L}{v\pi^3} \sum_{n=1}^{\infty} \frac{1}{n^3} \sin\left(\frac{n\pi}{4}\right)\left\{1 - \cos\frac{n\pi}{4}\right\} \sin\left(\frac{n\pi x}{L}\right)\sin\left(\frac{n\pi vt}{L}\right)$$

$$= \frac{8 v_0 L}{v\pi^3}\left[(\sqrt{2} - 1)\sin\frac{\pi x}{L}\sin\frac{\pi vt}{L} + \frac{1}{4}\sin\frac{2\pi x}{L}\sin 2\pi vt \right.$$

$$\left. + \frac{1}{27}(\sqrt{2} + 1)\sin\frac{3\pi x}{L}\sin\frac{3\pi vt}{L} - \frac{1}{125}(\sqrt{2} + 1)\sin\frac{5\pi x}{L}\sin\frac{5\pi vt}{L} + \ldots \right]$$

we observe that the fourth, eighth, etc., harmonics (i.e., those having a node at $x = L/4$) are absent in this case.

The more general case in which both the displacement and velocity are finite at time $t = 0$ can also be discussed but it is much more complicated and is not of much interest.

Here, we have analysed the transverse vibrations of strings by making use of Fourier analysis. This method can quite generally be used to analyse any bounded uniform continuous system, e.g., longitudinal vibrations of rods and gas columns.

Example 4.12: A uniform rod of length L is fixed at both ends, the rod is set into longitudinal vibrations by giving it a constant velocity v_0 at time $t = 0$ at all points along the rod in the X-direction. Find the subsequent longitudinal vibrations of the rod.

Solution: Since rod is fixed at both ends, the general solution for displacement is

$$Y(x,t) = \sum_{n=1}^{\infty} \sin\left(\frac{n\pi x}{L}\right)\left[B_n \cos\left(\frac{n\pi vt}{L}\right) + C_n \sin\left(\frac{n\pi vt}{L}\right)\right]$$

where

$$B_n = \frac{2}{L}\int_0^L Y_0(x)\sin\left(\frac{n\pi x}{L}\right)dx$$

and

$$C_n = \frac{2}{\pi n v}\int_0^L v_0(x)\sin\left(\frac{n\pi x}{L}\right)dx$$

where $v = \sqrt{\dfrac{Y}{\rho}}$, Y is Young's modulus and ρ is density of the rod. Here $Y(x, t)$ is the displacement (longitudinal) of the particles at distance x from the fixed end at any time t.

Initial conditions are

$$Y(x, 0) = 0 \quad \text{for all values of } x \text{ giving } B_n = 0$$

and

$$v_0(x, 0) = v_0 \text{ for all } x.$$

$\therefore$

$$C_n = \frac{2}{n\pi v}\int_0^L v_0 \sin\left(\frac{n\pi x}{L}\right)dx = \frac{2v_0 L}{(n\pi)^2 v}(1 - \cos n\pi)$$

$$= \frac{4v_0 L}{(n\pi)^2 v} \quad \text{when} \quad n = 1, 3, 5, \ldots$$

$$= 0 \quad \text{when } n = 2, 4, 6, \ldots$$

Thus, the resulting motion of the bar is given by

$$Y(x, t) = \frac{4v_0 L}{\pi^2 v}\sum_{n=1}^{n}\frac{1}{n^2}\sin\left(\frac{n\pi x}{L}\right)\sin\left(\frac{n\pi vt}{L}\right) \quad \text{for } n = \text{odd}$$

$$= \frac{4v_0 L}{\pi^2 v}\left[\sin\left(\frac{\pi x}{L}\right)\sin\left(\frac{\pi vt}{L}\right) + \frac{1}{9}\sin\left(\frac{3\pi x}{L}\right)\sin\left(\frac{3\pi vt}{L}\right) + \ldots, \ldots\right]$$

Example 4.13: A string of length L is tightly stretched with a tension T between its two fixed ends $x = 0$ and $x = L$. A hammer blow is given to a small part of length of the string at a distance a from the end $x = 0$, as a result of which the part b of the string has an initial velocity V_0, the rest of the string is initially undisturbed. Show that the subsequent vibration of the string is given by

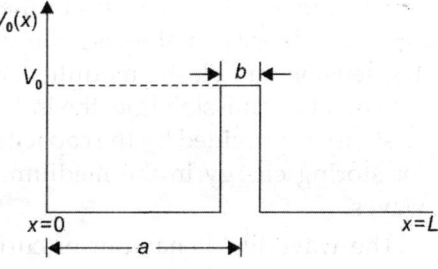

Fig. 4.19

$$Y(x,t) = \frac{4v_0 L}{\pi^2 v}\sum_{n=1}^{\infty}\frac{1}{n^2}\sin\left(\frac{n\pi a}{L}\right)\sin\left(\frac{n\pi b}{2L}\right)\sin\left(\frac{n\pi x}{L}\right)\sin\left(\frac{n\pi vt}{L}\right)$$

where $v = \sqrt{\dfrac{T}{\mu}}$, μ being the linear density of the string. Assume that the tension T remains constant.

Solution: The vibrations of the string are given by

$$Y(x, t) = \sum_{n=1}^{\infty} \sin\left(\frac{n\pi x}{L}\right)\left[B_n \cos\left(\frac{n\pi vt}{L}\right) + C_n \sin\left(\frac{n\pi vt}{L}\right)\right]$$

with

$$B_n = \frac{2}{L}\int_0^L Y_0(x)\sin\left(\frac{n\pi x}{L}\right)dx$$

and

$$C_n = \frac{2}{L}\int_0^T v_0(x)\sin\left(\frac{n\pi x}{L}\right)dx$$

The initial conditions are

$$Y_0(x) = 0 \quad \text{for all } x, \text{ giving } B_n = 0$$

$$v_0(x) = v_0 \text{ from } x = a - \frac{b}{2} \text{ to } x = a + \frac{b}{2}$$

Using these values of $v_0(x)$, we have

$$C_n = \frac{2}{n\pi v}\int_{a-\frac{b}{2}}^{a+\frac{b}{2}} v_0 \sin\left(\frac{n\pi x}{L}\right)dx$$

$$= \frac{2v_0 L}{(n\pi)^2 v}\left|-\cos\left(\frac{n\pi x}{L}\right)\right|_{a-\frac{b}{2}}^{a+\frac{b}{2}}$$

$$= \frac{4v_0 L}{(n\pi)^2 v}\sin\left(\frac{n\pi a}{L}\right)\sin\left(\frac{n\pi b}{L}\right)$$

Hence

$$Y(x, t) = \frac{4v_0 L}{(\pi)^2 v}\sum_{n=1}^{\infty}\frac{1}{n^2}\sin\left(\frac{n\pi a}{L}\right)\sin\left(\frac{n\pi b}{L}\right)\sin\left(\frac{n\pi x}{L}\right)\sin\left(\frac{n\pi vt}{L}\right)$$

4.12 ELECTROMAGNETIC WAVES IN SPACE

We have seen that the velocity of waves in a material medium is determined by the inertia and elastic properties of the medium. The linear density (in the case of strings) and mass density (in the case of rods and gases) are the measure of inertia of the system. The tension and elastic modulus are the measure of the elasticity of the system. In the case of a transmission line, the inductance of the line is measure of magnetic inertia and elasticity is provided by the capacitance of the line. These two properties are responsible for storing energy in the medium. This energy travels in the medium in the form of waves.

The wave like behaviour of current and voltage on a transmission line helps us to understand the propagation of electromagnetic disturbances in space. The existence of voltage across the transmission line implies the existence of an electric field in the space between the cables. Since the voltage varies in space and time, so the associated electric field must vary in space and time. The existence of space and time dependent magnetic field in the space around the cable. Since the current and voltage propagate in the line in the form of waves, the electric and magnetic fields must exhibit a similar wave like behaviour.

When electric and magnetic fields vary in space and time, they produce what we call the electromagnetic waves. It is clear that an oscillating charge will produce electromagnetic waves. An oscillating charge has an oscillating electric and magnetic fields around it and hence, it produces electromagnetic waves. Electrons falling from a higher to a lower energy orbit in an atom radiate electromagnetic waves of a particular wavelength and frequency. The motion of electrons in an antenna radiate electromagnetic waves by a process called *bremsstrahlung*.

The propagation of electromagnetic waves in a medium is also due to the inertia and elastic properties of the medium described by what we call the magnetic permeability μ of the medium and electric permittivity of medium ε. In the SI system, μ is expressed in henries per meter. This property provides the magnetic inertia of the medium. The elasticity of the medium is provided by the capacitive property called the *electrical permittivity ε of the medium*. In SI system ε is expressed in farad per metre. Permeability μ stores the magnetic energy and permittivity ε stores the electric field energy. The electromagnetic energy propagates in the medium in the form of electromagnetic waves.

Maxwell's Equations and Plane Electromagnetic Waves

The electric and magnetic fields are connected by Maxwell's equations, which in a region of space, where there is no charge or current are as follows:

$$\left.\begin{array}{ll} \text{(i) } \nabla \cdot \mathbf{E} = 0 & \text{(ii) } \nabla \cdot \mathbf{B} = 0 \\[2mm] \text{(iii) } \nabla \times \mathbf{E} = -\dfrac{\partial \mathbf{B}}{dt} & \text{(iv) } \nabla \times \mathbf{B} = \mu_0 \varepsilon_0 \dfrac{\partial \mathbf{E}}{dt} \end{array}\right\} \tag{4.53}$$

In these equations, the electric field $\mathbf{E}$ is expressed in volts per metre and B in tesla. Equation (iv) is a generalisation of Ampere's law in electricity which gives the magnetic field associated with a current carrying conductor and equation (iii) is the generalisation of Faraday's law in electromagnetism. Whereas equations (i) and (ii) are Gauss law in electrostatics and magnetostatics. They constitute a set of coupled, first order, partial differential equations.

Applying curl to (iii), we have[5]

$$\nabla \times \nabla \times \mathbf{E} = \nabla(\nabla \cdot \mathbf{E}) - \nabla^2 \mathbf{E}$$

From (i) $\nabla \cdot \mathbf{E} = 0$ and from (ii) $\nabla \times \mathbf{E} = -\dfrac{\partial \mathbf{B}}{\partial t}$

$$\therefore \qquad \nabla \times \left(-\frac{\partial \mathbf{B}}{dt}\right) = -\nabla^2 \mathbf{E}$$

$$\therefore \qquad \nabla^2 E = \frac{\partial}{\partial t}(\nabla \times B)$$

$$= \frac{\partial}{\partial t}\left(\mu_0 \, \varepsilon_0 \, \frac{\partial E}{dt}\right)$$

---

5. We have used the identity $\mathbf{A} \times \mathbf{B} \times \mathbf{C} = \mathbf{B}(\mathbf{A} \cdot \mathbf{C}) - \mathbf{C}(\mathbf{A} \cdot \mathbf{B})$.

Here, we have used Eq. (iii)

$$\therefore \qquad \nabla^2 E = \mu_0 \, \epsilon_0 \, \frac{\partial^2 E}{dt^2} \qquad\qquad (4.54)$$

Similarly applying curl to (iv), we have

$$\nabla \times (\nabla \times \mathbf{B}) = \nabla(\nabla \cdot \mathbf{B}) - \nabla^2 \mathbf{B} = \nabla \times \left(\varepsilon_0 \mu_0 \frac{\partial \mathbf{E}}{dt} \right)$$

But $\nabla \cdot \mathbf{B} = 0$

$$-\nabla^2 \mathbf{B} = \varepsilon_0 \mu_0 \frac{\partial}{\partial t} (\nabla \times \mathbf{E})$$

But $\nabla \times \mathbf{E} = \dfrac{\partial \mathbf{B}}{\partial t}$

$$\therefore \qquad -\nabla^2 \mathbf{B} = \varepsilon_0 \mu_0 \frac{\partial}{\partial t} \left(-\frac{\partial \mathbf{B}}{\partial t} \right) = - \varepsilon_0 \mu_0 \frac{\partial^2 \mathbf{B}}{\partial t^2}$$

or $\qquad\qquad \nabla^2 \mathbf{B} = \mu_0 \varepsilon_0 \dfrac{\partial^2 \mathbf{B}}{\partial t^2} \qquad\qquad (4.55)$

Hence in vacuum the components of **E** and **B** satisfy the equation

$$\nabla^2 \psi = \frac{1}{v^2} \frac{\partial^2 \psi}{\partial t^2}$$

which is called *wave equation*. Hence, Eq. (4.55) describes waves travelling with velocity v such that

$$v^2 = \frac{1}{\varepsilon_0 \mu_0}$$

or $\qquad\qquad v = \dfrac{1}{\sqrt{\varepsilon_0 \mu_0}} \qquad\qquad (4.56)$

According to Maxwell's equation, then empty space supports the propagation of electromagnetic waves at speed

$$v = \frac{1}{\sqrt{\varepsilon_0 \mu_0}}$$

Using the values

$$\mu_0 = 4\pi \times 10^{-7} \text{ Henry/m}$$

and $\qquad\qquad \varepsilon_0 = \dfrac{9 \times 10^9}{4\pi} \text{ Farad/m}$

Then $\qquad\qquad v = 3 \times 10^8 \text{ m/s}$

which is precisely the speed of light. The result is stunning. Perhaps light is an electromagnetic wave. This conclusion is an established fact and does not surprise any one today but imagine what a triumph it was in Maxwell's time.

We have seen that E and B in free space satisfy the three dimensional wave equation

with wave velocity $v = \dfrac{1}{\sqrt{\mu_0 \varepsilon_0}}$. Which is the speed of the light in vacuum c. We now confine our attention to sinusoidal waves of single frequency ω, i.e. to monochromatic waves. Suppose for the moment that the waves are travelling in the X-direction and have no Y or Z-dependence, these are called *plane waves* because the fields are uniform over every plane perpendicular to the direction of propagation (Fig. 4.20). In such a case

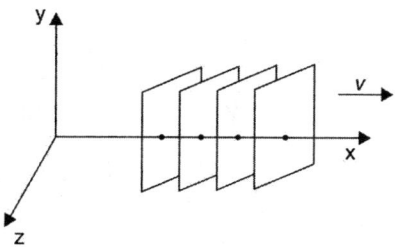

Fig. 4.20

$$E_y = E_0 \sin(kx - \omega t) \\ B_z = B_0 \sin(kx - \omega t)$$ (4.57)

where E_0 and B_0 are the amplitudes of the electric and magnetic fields.

Now, the wave Eqs (4.54 and 4.55) are derived from Maxwell's equations. However, whereas every solution to Maxwell's equation (in empty space) must obey the wave equation, the converse is not true. Maxwell's equations impose special constraints on E_0 and B_0. In particular since

$$\nabla \cdot E = 0 \quad \text{and} \quad \nabla \cdot B = 0$$

it follows that

$$(E_0)_x = (B_0)_x = 0$$ (4.58)

i.e., amplitude of E and B in X-direction is zero that is electromagnetic waves are transverse; the electric and magnetic fields are perpendicular to the direction of propagation. Moreover, Faraday's law

$$\nabla \times E = -\frac{\nabla B}{\partial t}$$

implies a relation between electric and magnetic field amplitudes, i.e.

$$\frac{\partial E}{\partial x} = kE_0 \cos(kx - \omega t)$$

and

$$\frac{\partial B}{\partial t} = \omega B_0 \cos(kx - \omega t)$$ (4.58a)

$\therefore$

$$E_0 = \frac{\omega B_0}{k} = cB_0 \quad \text{or} \quad B_0 = \frac{E_0}{c}$$ (4.58b)

where c is velocity given by $c = \dfrac{\omega}{k}$.

Equation (4.58) reveals that E and B fields are in phase reaching their zero and maximum values at the same time and are naturally perpendicular that is if **E** points in y-direction then **B** points in Z-direction (Fig. 4.21). This can be seen as follows:

If E is along y-direction then[6]

$$\nabla \times E = \nabla \times [\hat{j}E_0 \sin(kx - \omega t)]$$

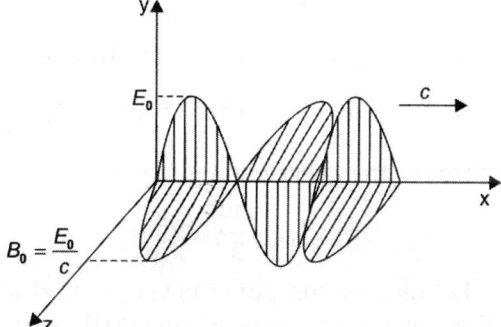

Fig. 4.21

6. We can take one dimensional E and B and $E = E_0 e^{i(kx - \omega t)}$ and $B = B_0 e^{i(kx - \omega t)}$, where E_0 and B_0 are the complex amplitudes of electric and magnetic fields—physical fields are the real parts of E and B.

$$= \left(\hat{i} \frac{\partial}{\partial x} + \hat{j} \frac{\partial}{\partial y} + \hat{k} \frac{\partial}{\partial z} \right) \times [jE_0 \sin (kx - \omega t)]$$

or $\qquad \boldsymbol{\nabla} \times \boldsymbol{E} = \hat{k} E_0 k \cos (dx - \omega t)$

and $\qquad \dfrac{\partial \boldsymbol{B}}{\partial t} = - B_0 \omega \cos (kx - \omega t)$

But $\qquad \boldsymbol{\nabla} \times \boldsymbol{E} = \dfrac{\partial \boldsymbol{B}}{\partial t}$

$\therefore \ B_0 \omega \cos(kx - \omega t) = \hat{k} E_0 k \cos(kx - \omega t)$

or $\qquad B_0 = \dfrac{E_0 k}{\omega} \hat{k} = \dfrac{E_0}{c} \hat{k}$

It signifies that the electric field oscillates in XY-plane and magnetic field in YZ-plane. The wave as a whole is said to be polarised along the Y-direction because by convention, we use the direction of E to specify the polarisation of electromagnetic wave.

Energy and Momentum of an Electromagnetic Wave

We know that the energy density (energy per unit volume) associated with electric field E and magnetic field B is given by

$$U_e = \frac{1}{2} \varepsilon_0 E^2$$

and $\qquad U_b = \dfrac{1}{2} \dfrac{B^2}{\mu_0} = \dfrac{1}{2} \dfrac{E^2}{\mu_0 c^2} = \dfrac{1}{2} \varepsilon_0 E \qquad \left[\because c = \dfrac{1}{\sqrt{\mu_0 \varepsilon_0}} \right]$

which is the electric energy density of an electromagnetic wave and is equal to the magnetic energy density. The total energy density then is

$$U_t = U_e + U_b = \varepsilon_0 E^2 \tag{4.59}$$

As the wave propagates, it carries this energy along with it. The energy flux density (energy crossing per unit area per unit time) transported by the fields is given by

$$I = U_t c = \varepsilon_0 c E^2$$

The average intensity of the electromagnetic waves is

$$I_{av} = c \varepsilon_0 (E^2)_{av}$$

In the case of a harmonic electromagnetic wave

$$(E^2)_{av} = E_0^2 [\sin^2 (kx - \omega t)]_{av} = \frac{1}{2} E_0^2$$

Hence the average intensity is

$$I_{av} = \frac{1}{2} \varepsilon_0 c E_0^2 \tag{4.60}$$

Let us now compute the vector product $E \times B$ for a plane electromagnetic wave. The direction $E \times B$ is perpendicular to the wave front and is therefore, pointing in the direction of propagation of the wave. Its magnitude is

$$|\boldsymbol{E} \times \boldsymbol{B}| = EB = \frac{E^2}{c}$$

Now, we define the vector S called *Poynting vector* as

$$S = c^2\varepsilon_0 E \times B = \frac{1}{\mu_0} E \times B \qquad (4.61)$$

has the magnitude equal to I the energy flux density.

Electromagnetic fields not only carry energy, they also carry momentum. It is an established fact that energy and momentum form a four-vector (a requirement of the principle of relativity). As such, we then expect that an electromagnetic wave in addition to its energy, carries a certain momentum. Since electromagnetic radiations propagate with velocity c, we may use the relation between energy and momentum given by

$$\text{Momentum } P = \frac{\text{Energy} \times v}{c^2} = \frac{\text{Energy}}{c} = \varepsilon_0 |E \times B| \qquad (\because v = c)$$

The dimensions of $\varepsilon_0 |E \times B|$ are $kg \cdot m^{-2} s^{-1}$, i.e., those of momentum per unit volume. Since momentum has the direction as such, we can write in vector form as

$$P = \varepsilon_0 (E \times B) = \frac{1}{c^2} S \qquad (4.62)$$

In the case of light, wavelength is so small ($\sim 10^{-7}$ m) and the period so brief ($\sim 10^{-15}$ s) that any microscopic measurement will encounter many cycles. Typically, therefore, we are not interested in the fluctuating sine-squared terms in the energy and momentum densities. All we want is its average value. Hence, the average value of $\sin^2$ over a complete cycle is $\frac{1}{2}$. Hence

$$\langle U_t \rangle = \frac{1}{2}\varepsilon_0 E_0^2 \qquad (4.62a)$$

$$\langle S \rangle = \frac{1}{2}c\varepsilon_0 E_0^2 \hat{i} = c\langle U_t \rangle \hat{i} \qquad (4.62b)$$

$$\langle P \rangle = \frac{1}{2}\frac{1}{c}\varepsilon_0 E_0^2 \hat{i} = \frac{1}{c}\langle U_t \rangle \hat{i} \qquad (4.62c)$$

The average power per unit area transported by an electro-magnetic wave is called *intensity*

$$I = \langle S \rangle \qquad (4.62d)$$

If an electromagnetic wave has momentum, it has also angular momentum. The angular momentum per unit volume is:

$$L = r \times P = \varepsilon_0 r \times (E \times B) \qquad (4.62e)$$

This is what could be called as *orbital angular momentum of radiation* by similarity with the angular momentum of an orbiting particle. In addition, electromagnetic radiations possess an *intrinsic angular momentum* or *spin* similar to the spin of fundamental particle. For linearly polarised wave, the average value of the component of the spin along the direction of propagation is zero. For *circularly polarised plane wave*, it can be shown that the spin has a component along the direction of propagation equal to $\pm\dfrac{U_t}{\omega}$ depending on whether the polarisation is clockwise or counter clockwise. As such when a charged particle absorbs or emits electromagnetic radiation not only its energy and momentum change but also its angular momentum changes accordingly, a result that has been verified experimentally both directly and indirectly.

Radiation pressure: Since electromagnetic waves carry momentum, they give rise to a certain pressure when they are reflected or absorbed at the surface of a body. The basic principle is same as in case of pressure exerted by a gas on the walls of a container as explained on the basis of kinetic theory of gases.

Suppose, a plane electromagnetic wave falls perpendicularly on a perfectly absorbing surface. The incident momentum per unit volume is P and the amount of momentum in the radiation fall per unit time, on the surface area A of the absorber is obtained by multiplying P by the volume cA. Hence, momentum is PcA. If radiation is completely absorbed per unit time by the surface A, then the force on area A is

$$E = PcA$$

and the radiation pressure

$$P_{rad} = \frac{Force}{A} = \frac{PcA}{A} = Pc = U_t = \varepsilon_0 E^2 \tag{4.63}$$

Thus for normal incidence, the radiation pressure on a perfect absorber is equal to the energy density in the wave.

On the other hand, if the surface is a perfect reflector, the radiation after reflection has a momentum equal in magnitude but opposite in direction to the incident radiation. The change in momentum per unit volume is thus $2P$ and the radiation pressure is accordingly

$$P_{rad} = 2cP = 2U_t = 2\varepsilon_0 E^2 \tag{4.64}$$

These results can be generalised to the case of oblique incidence. In this case, the change in momentum of radiation per unit volume at the perfectly reflecting surface $2P \cos\theta$ and the corresponding radiation pressure is

$$P_{rad} = 2cP \cos\theta = 2U_t \cos\theta.$$

If the radiation propagates in all directions, we must integrate over all directions. Average is found as in case of pressure exerted by the gas on the wall on the basis of kinetic theory of gases and factor $\frac{1}{3}$ comes. Hence

$$P_{rad} = \frac{2}{3}cP = \frac{2}{3}U_t = \frac{2}{3}\varepsilon_0 E^2 \tag{4.64a}$$

when the surface is perfect absorber, the change in momentum normal to the surface is reduced to one-half the previously determined value, i.e.

$$P_{rad} = \frac{1}{3}\varepsilon_0 E^2 \tag{4.64b}$$

Characteristic Impedance

We know that propagation of wave through material medium is controlled by the inertial and elastic properties of the medium. The propagation of electromagnetic waves thus in a medium is also due to the inertial and elastic properties of the medium. Every medium (including vacuum or free space) has inductive property described by what we call the magnetic permeability μ (and for free space μ_0) of the medium. In the SI system of units μ (or μ_0) is expressed in henry per metre. This provides the magnetic inertia of the medium. The elasticity of the medium is provided by the capacitive property called the *electric permittivity* ε (and for free space ε_0) stores the electric field energy. This electromagnetic energy propagates in the medium in the form of electromagnetic waves and the medium offers the resistance as it is offered to other mechanical waves by the medium.

Impedance in the case of an electric circuit is a familiar concept and is the ratio of voltage to the current. In the case of electromagnetic waves, the ratio of electric field (proportional to the voltage) and the magnetising field $H\left(=\dfrac{B}{\mu}\right)$ (proportional to the current denotes the impedance of the dielectric medium. Now, we know that

$$\frac{E}{B} = \frac{E_0}{B_0} = c \quad \text{the velocity of light}$$

and $\qquad B = \mu_0 H.$

Hence characteristic impedance of the free space of electromagnetic waves is

$$Z = \frac{E}{H} = \frac{E_0}{\dfrac{B_0}{\mu_0}} = \mu_0 \frac{E_0}{B_0} = \mu_0 c = \mu_0 \frac{1}{\sqrt{\mu_0 \varepsilon_0}} = \sqrt{\frac{\mu_0}{\varepsilon_0}} \qquad (4.65)$$

and for material medium

$$Z = \sqrt{\frac{\mu}{\varepsilon}} \qquad (4.65a)$$

For free space $\mu_0 = 4\pi \times 10^{-7}$ Henry/m and $\varepsilon_0 = 8.85 \times 10^{-12}$ F/m.

$$\therefore \qquad Z = \sqrt{\frac{4\pi \times 10^{-7}}{8.85 \times 10^{-12}}} = 377\,\Omega$$

4.13 PROPAGATION OF ELECTROMAGNETIC WAVES IN DISPERSIVE MEDIUM

So far, we have considered only the propagation of electromagnetic waves in vacuum. Experiments reveal that the velocity of propagation of an electromagnetic wave through matter is different from its velocity of propagation in vacuum. It is due to the fact that in vacuum, we have assumed no charges and currents but when the electromagnetic waves propagate through matter even if there are no free charges and currents, it induces certain charges and currents in the substance as a result of the polarisation and magnetisation of matter. If the medium is homogeneous and isotropic, it can be proved that the net effect of the *polarisation* and *magnetisation* of the medium by the electromagnetic waves is to replace the constant ε_0 and μ_0 in the Maxwell's equations by electric permittivity ε and magnetic permeability μ, the characteristics of the material. The velocity of the wave in medium then becomes

$$c_m = \frac{1}{\sqrt{\mu \varepsilon}}$$

The ratio between the velocity of electromagnetic waves in vacuum (c) and in matter (c_m) is called *absolute index of refraction* of the substance (n). It is a useful concept for describing the properties of materials in relation to electromagnetic waves. Thus,

$$n = \frac{c}{c_m} = \frac{\sqrt{\mu \varepsilon}}{\sqrt{\mu_0 \varepsilon_0}} = \sqrt{\mu_r \varepsilon_r} \qquad (4.66)$$

where $\varepsilon_r = \dfrac{\varepsilon}{\varepsilon_0}$ and $\mu_r = \dfrac{\mu}{\mu_0}$ are the relative electric permittivity and relative magnetic permeability of the medium respectively. In general μ_r differs very little from 1, for the majority of substances that transmit electromagnetic waves and we can write as a satisfactory approximation

$$n = \sqrt{\varepsilon_r} \qquad (4.66a)$$

This relation as a matter of fact affords a simple experimental method for determining the relative permittivity of the substance if the index of refraction is obtained independently. The consistency of the values of ε_r obtained by this method with those from other kinds of measurements gives a satifsactory foundation to the theory.

To calculate theoretically ε_r, we assume the electrons in a non-conductor to be attached to specific molecules by binding forces whose detailed structure may be extremely complicated and would in any event, requires quantum mechanics for a proper treatment. But we shall picture the electron as attached to the end of spring of force constant k. Hence, binding force (F_b)

$$F_b = -ky = -m\omega_0^2 y$$

where y is displacement from equilibrium, m electron's mass and ω_0 is the natural frequency $\sqrt{\dfrac{k}{m}}$.

When the electron oscillates, there will be some damping force (F_d) which, we assume to be proportional to the velocity, i.e.

$$F_d = -m\beta \frac{dy}{dt}$$

Here the cause of damping does not concern us. The damping must be opposite in direction to the velocity and making it proportional to the velocity is the easiest way to accomplish it.

In presence of an electromagnetic wave of frequency ω, the electron is subjected to a driving force (F_0)

$$F_0 = eE = eE_0 \cos \omega t$$

where e is the charge of the electron and E_0 is amplitude of the wave at the point x, where the electron is situated. Putting all these values and using Newton's second law, we have

$$m\frac{d^2 y}{dt^2} = -m\beta \frac{dy}{dt} - m\omega_0^2 y + eE_0 \cos \omega t$$

or $\quad \dfrac{d^2 y}{dt^2} + \beta \dfrac{dy}{dt} + \omega_0^2 y = \dfrac{eE_0}{m} \cos \omega t \qquad (4.67)$

Equation (4.67) is the equation for the forced oscillator and in the steady state, the system (electron) oscillates with the driving frequency as already discussed in Chapter 2. Hence solution can be written as

$$Y = y_0 \cos(\omega t - \phi) \qquad (4.68)$$

where $\qquad y_0 = \dfrac{E_0 e/m}{\sqrt{(\omega_0^2 - \omega^2)^2 + \beta^2 \omega^2}}$

and $\qquad \tan \phi = \dfrac{\beta \omega}{\omega_0^2 - \omega^2} \qquad (4.69)$

In absence of damping $\beta = 0$

$$y_0 = \frac{E_0 e/m}{\omega_0^2 - \omega^2} \quad \text{and} \quad \phi = 0 \tag{4.70}$$

Neglecting damping, when the electromagnetic radiation of frequency ω propagates through the medium, the electron begins to oscillate and an instantaneous dipole moment associated with the motion of electron is given by

$$P(t) = ey(t) = \frac{e^2/m}{\omega_0^2 - \omega^2} E_0 \cos \omega t \tag{4.71}$$

In general, differently situated electrons within a given molecule exhibit different natural frequencies. Let us say that there are f_i electrons with frequency ω_i and there are N molecules per unit volume then polarisation P is given by

$$P = \frac{Ne^2}{m} \left[\sum \frac{f_i}{\omega_i^2 - \omega^2} \right] E$$

But relative permeability ε_r and electric susceptibility (χ) are related as $\varepsilon_r = 1 + \chi$ with $\chi = \dfrac{P}{\varepsilon_0 E}$. Thus, we have

$$n^2 = \varepsilon_r = 1 + \chi = 1 + \frac{P}{\varepsilon_0 E} = 1 + \frac{Ne^2}{m\varepsilon_0} \left[\sum \frac{f_i}{\omega_i^2 - \omega^2} \right] \tag{4.72}$$

Therefore, the index of refraction depends on the wave frequency and hence also on the wavelength, the variation is shown in Fig. 4.22. Here, $\omega_1, \omega_2, \dots$ are the characteristic frequencies of the emission spectrum of the substance. Consequently, the phase velocity $v = c/n$ of electromagnetic waves in matter also depends on the frequency of radiation. Such medii are called *dispersive medii*. As such, the electromagnetic waves suffer dispersion while propagating through such medii, i.e., a

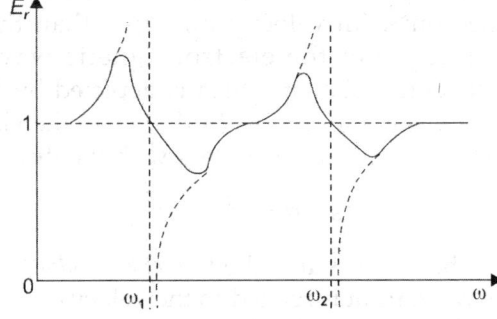

Fig. 4.22

pulse containing several frequencies will be distorted because each component will travel with a different velocity. The group velocity v_g is given by

$$v_g = \frac{d\omega}{dk} = \frac{d}{dk}(vk) = v + k\frac{dv}{dk} = v + k\left(\frac{dv}{d\omega} \frac{d\omega}{dk} \right)$$

or

$$v_g = v + kv_g \frac{dv}{d\omega} = v + kv_g \frac{d(c/n)}{d\omega} = v - \frac{kv_g c}{n^2} \frac{dn}{d\omega}$$

$$= v - \frac{v_g \omega}{n} \frac{dn}{d\omega} \qquad \frac{c}{n} = v \quad \text{and} \quad vk = \omega$$

$$\therefore \qquad v_g = \frac{v}{1 + \dfrac{\omega}{n}\dfrac{dn}{d\omega}} = \frac{\dfrac{c}{n}}{1 + \dfrac{\omega}{n}\dfrac{dn}{d\omega}} = \frac{c}{n + \omega\dfrac{dn}{d\omega}} \tag{4.73}$$

when $\dfrac{dn}{d\omega}$ is positive, the group velocity is less than the phase velocity. Such a situation is called *normal dispersion*. But if $\dfrac{dn}{d\omega}$ is negative, then group velocity is larger than the phase velocity and *anomalous dispersion* result. The possibility exists in this case that the group velocity is larger than c (the velo-

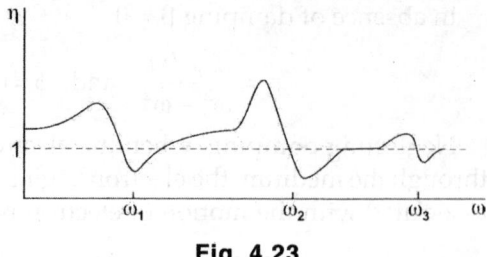

Fig. 4.23

city of light in vacuum) and that an electromagnetic pulse can be transmitted at a velocity greater than c. This is apparently in contradiction to the results derived from the Lorentz transformation and the principle of relativity.

A careful analysis of the transmission of an electromagnetic signal made by Brillouin, Sommerfield and others revealed that it is impossible to transmit a signal with a velocity larger than c. Figure 4.24 shows the variation of phase velocity (v), the group velocity (v_g) and signal velocity v_s near characteristic frequency ω_1. The signal velocity practically coincides with the group velocity except near the characteristic frrequency and is never larger than c even in the region of anomalous dispersion.

When n is larger than 1, so that v is smaller than c, there is the possibility that a charged particle, emitting electromagnetic waves moves in the medium with velocity v_g, larger than the phase velocity v of the electromagnetic waves. The situation is similar which is depicted by Mach or

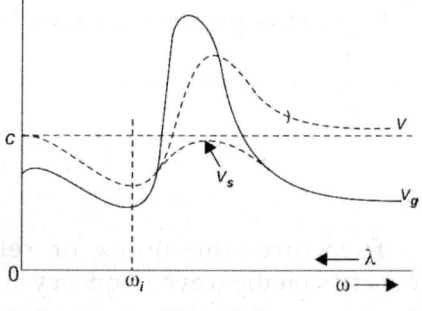

Fig. 4.24

shock waves in fluids. In that case, the electromagnetic waves propagate along conical surfaces making an angle α with the direction of propagation by

$$\sin \alpha = \frac{v_g}{v}$$

These waves are called *Cerenkov radiation*. Because the effective direction of propagation of wave front is related to the velocity of the charged particle, it may be used to measure it. Devices used for this purpose are called *Cerenkov detectors* which are widely used in experiments with fundamental particles because they provide direct information about the velocity of the particle.

We have said that electromagnetic waves appear to propagate in matter with a phase velocity different from their propagation velocity in vacuum. But that difference seems to stem from the fact that permittivity and permeability of matter are different from those of vacuum. This difference is, in turn a consequence of the electric and magnetic polarisation of matter under the action of the incoming electromagnetic wave. Thus, when an electromagnetic wave falls on matter, it induces oscillations in the charged particles of the atoms or molecules, which then emit secondary or scattered waves. These scattered waves are superposed on the original wave giving resultant wave. The phase of secondary wave is in general, different from that of the original wave, since the force oscillator is not always in phase with the driving force. This phase difference affects the resultant wave in such a way that this wave appears to have a phase velocity different from that of a wave in vacuum. This result is particularly satisfying since from the atomic point of view, all charges, both free and bound are equivalent and the electromagnetic waves they emit, must all propagate with the velocity c. It is the wave resulting from the superposition of the individual waves, which have different phases, that, as a consequence of the phase difference appear to have a different velocity of propagation.

4.14 WAVES IN TWO DIMENSIONS

In order to describe a plane wave in two dimensions, the exponential motion is the most convenient. In exponential notations, the progressive wave propagating along positive x-direction is given by

$$\psi(x, t) = A \exp\{ik(vt - x)\}$$

where $k = \dfrac{2\pi}{\lambda}$ is the wave number.

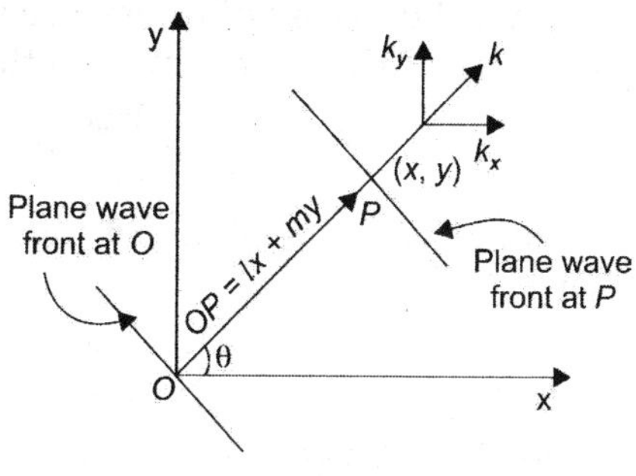

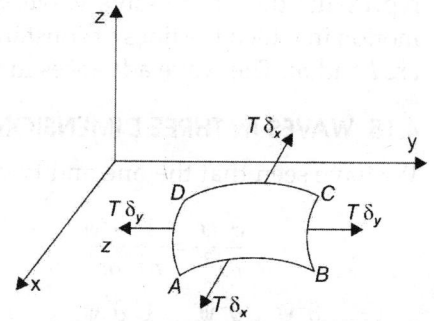

Fig. 4.25

We can now write down the expression for a two-dimensional plane wave travelling in a direction making an angle θ with the X-axis in the plane XY (Fig. 4.25). Let (x, y) be the co-ordinates of the point P, the origin is at 0. Now, the direction cosines of the direction of propagation are $l = \cos \theta$ and $m = \sin \theta$.

Hence, $\qquad OP = lx + my$

Hence the wave propagating along the direction op can be represented as follows:

$$\psi(x, t) = A \exp\{ik(vt - op)\}$$
$$= A \exp\{ik(vt - lx - my)\}$$
$$= A \exp\{ik(kvt - klx - kmy)\}$$

But $kl = k \cos \theta = k_x x$ component of wave vector k and $km = k \sin \theta = k_y$, y component of wave vector k.

$\therefore \qquad \psi(x, y, t) = A \exp\{i(\omega t - k \cdot r)\}$ $\qquad\qquad$ (4.74)

where $k \cdot r = k_x x + k_y y$ and $kv = \omega$. With r as the position vector of point (x, y), k is called the *propagation vector* and gives the direction of propagation of the wave.

Transverse Vibrations of the Stretched Membrane

Now, we shall obtain two dimensional wave equation for stretched membrane which is two-dimensional analogue of the stretched string by using the similar arguments. Let us consider a plane rectangular membrane of negligible thickness in the X–Y plane of a rectangular coordinate system (Fig. 4.26). We shall consider transverse vibration of the membrane, i.e. vibrations in the Z-direction perpendicular to both x and y axis. The displacement $\psi(x, y, t)$ is a function of x, y and t.

Suppose σ is the mass per unit area of the membrane and T is the uniform tension per unit length. This is called *surface tension of the membrane*, i.e. this is the force per unit length of an imaginary line in the surface of the membrane acting tangentially to the membrane and trying to tear apart the membrane along this line.

Fig. 4.26

Figure 4.26 shows a small rectangular element ABCD of sides δx and δy of the membrane in the XY plane. Let $\psi(x, y, t)$ be the displacement of the element at time t in the Z-direction. The force $T\delta x$ and $T\delta y$ are acting on the sides of the element during vibration. The components of these forces in the direction normal to the XY plane constitute the restoring force tending to bring the element back to its equilibrium position.

As we did in case of transverse vibrations in stretched string here also the resultant of force on AD and BC resolved in the direction normal to XY plane

$$= T\delta y \frac{\partial^2 \psi}{\partial x^2} \, \delta x$$

Similarly, the resultant of forces on AB and DC, resolved in direction normal to xy plane

$$= T\delta y \frac{\partial^2 \psi}{\partial y^2} \, \delta y$$

Thus, the net transverse force on the element is

$$F = T\delta x \delta y \left(\frac{\partial^2 \psi}{\partial x^2} + \frac{\partial^2 \psi}{\partial t^2} \right)$$

This is the restoring force which must be equal to Newton's force for dynamic equilibrium. Mass of the membrane element is $\sigma \delta x \delta y$. Hence Newton's force

$$= \sigma \delta x \delta y \frac{\partial^2 \psi}{\partial t^2}$$

Hence the equation of motion of the element is

$$\sigma \delta x \delta y \frac{\partial^2 \psi}{\partial t^2} = T\delta x \delta y \left(\frac{\partial^2 \psi}{\partial x^2} + \frac{\partial^2 \psi}{\partial y^2} \right)$$

or

$$\frac{\partial^2 \psi}{\partial t^2} = \frac{T}{\sigma} \left(\frac{\partial^2 \psi}{\partial x^2} + \frac{\partial^2 \psi}{\partial y^2} \right)$$

$$= v^2 \left(\frac{\partial^2 \psi}{\partial x^2} + \frac{\partial^2 \psi}{\partial y^2} \right) \tag{4.75}$$

where $v = \sqrt{\dfrac{T}{\sigma}}$ is the wave velocity. Equation (4.75) is the equation of two-dimensional wave motion. In case if $\dfrac{\partial^2 \psi}{\partial y^2} = 0$ or $\dfrac{\partial^2 \psi}{\partial x^2} = 0$, the equation reduces to one dimensional wave equation. The solution of the differential Eq. (4.75) is given by Eq. (4.74), which represents the progressive wave in two dimensions. Equation (4.74) represents wave motion in which the lines of constant phase are normal to the line whose direction cosines are l and m. The wave advances in the direction (l, m), with velocity v.

4.15 WAVES IN THREE DIMENSIONS

We have seen that the one and two-dimensional wave equations respectively are

$$\frac{\partial^2 \psi}{\partial x^2} = \frac{1}{v^2} \frac{\partial^2 \psi}{\partial t^2}$$

and

$$\frac{\partial^2 \psi}{\partial x^2} + \frac{\partial^2 \psi}{\partial y^2} = \frac{1}{v^2} \frac{\partial^2 \psi}{\partial t^2}$$

Similarly, the three-dimensional wave equation is written as

$$\frac{\partial^2 \psi}{\partial x^2} + \frac{\partial^2 \psi}{\partial y^2} + \frac{\partial^2 \psi}{\partial z^2} = \frac{1}{v^2}\frac{\partial^2 \psi}{\partial t^2}$$

or $$\nabla^2 \psi = \frac{1}{v^2}\frac{\partial^2 \psi}{\partial t^2} \tag{4.76}$$

This equation represents the wave like behaviour $\psi(x, y, z, t)$ in a three-dimensional medium. For an isotropic medium, i.e., a medium in which the wave velocity is the same in every direction, the quantity ψ is a scalar. For non-isotropic media ψ becomes a vector. Sound and electromagnetic waves are three-dimensional.

Solution of Three-Dimensional Wave Equation

The three-dimensional wave equation can be solved using the method of separation of variables. Let

$$\psi(x, y, z, t) = X(x)\, Y(y)\, Z(z)\, T(t)$$

where $X(x)$, $Y(y)$, $Z(z)$, and $T(t)$ are the functions of x, y, z and t only. Differentiating partially and substituting in Eq. (4.76), then dividing with $XYZT$, we get

$$\frac{1}{X}\frac{\partial^2 X(x)}{\partial x^2} + \frac{1}{Y}\frac{\partial^2 Y(y)}{\partial y^2} + \frac{1}{Z}\frac{\partial^2 Z(z)}{\partial z^2} = \frac{1}{v^2}\frac{1}{T}\frac{\partial^2 T(t)}{\partial t^2} \tag{4.77}$$

Each term in this equation is independent of other. This is possible only when each term is independently constant. Putting

$$\frac{1}{X}\frac{\partial^2 X(x)}{\partial x^2} = -K_x^2 x \qquad \frac{1}{Y}\frac{\partial^2 Y(y)}{\partial y^2} = -k_y^2 y \quad \text{and} \quad \frac{1}{Z}\frac{\partial^2 Z(z)}{\partial z^2} = -K_z^2 z$$

We have

$$\frac{\partial^2 T(t)}{\partial t^2} = -\left(k_x^2 + k_y^2 + k_z^2\right)v^2 T = -k^2 v^2 T \tag{4.78}$$

where $k_x^2 + k_y^2 + k_z^2$ are constants and $k_x^2 + k_y^2 + k_z^2 = K^2$

The solution of equation $\dfrac{1}{X}\dfrac{\partial^2 X(x)}{\partial x^2} = -k_x^2$ is

$$X(x) = A_1 e^{\pm ik_x X} \tag{4.79a}$$

Similarly, solution of $Y(y)$, $Z(z)$, and $T(t)$ equations can be written as

$$Y(y) = A_2\, e^{\pm ik_y y} \tag{4.79b}$$

$$Z(z) = A_3\, e^{\pm ik_z z} \tag{4.79c}$$

$$T(t) = A_4\, e^{\pm ikvt} \tag{4.79d}$$

where A_1, A_2, A_3, A_4 are constants. Equation (4.79d) gives that k must be wave vector and k_x, k_y, k_z are the components of the wave vector k along x, y, z directions respectively. Hence the general solution of three-dimensional wave equation (4.76) can be written as

$$\psi(z, y, z, t) = X(x)Y(y)Z(z)T(t)$$

$$= A e^{\pm ik_x X} \cdot e^{\pm ik_y Y} \cdot e^{\pm ik_z Z} \cdot e^{\pm ikvt}$$

where $A = A_1 A_2 A_3 A_4$, a new constant

$$\therefore \quad \psi(z, y, z, t) = A e^{[\pm ikvt - (x_x X + k_y Y + k_z Z)]}$$

$$= A e^{\pm(\omega t - k \cdot r)} \tag{4.80}$$

where $\omega = kv$ is the angular frequency and r the position vector.

The wave equation (4.76) can be put into other co-ordinate systems depending upon the symmetry of the problem to be solved. As an example for waves spreading out from a point source, spherical polar co-ordinates (r, θ, ϕ) are used. In these coordinates $\Delta^2 \psi$ and wave equation is transferred as

$$\nabla^2 \psi = \frac{\partial^2 \psi}{\partial r^2} + \frac{2}{r} \frac{\partial \psi}{\partial r} + \frac{1}{r^2 \sin \theta} \frac{\partial}{\partial \theta} \left(\sin \theta \frac{\partial \psi}{\partial \theta} \right) + \frac{1}{r^2 \sin^2 \theta} \frac{\partial^2 \psi}{\partial \phi^2} = \frac{1}{v^2} \frac{\partial^2 \psi}{\partial t^2} \tag{4.81}$$

In some case ψ may not vary in one or more co-ordinate directions and full form of the equation may not be required. For example, in special case of waves spreading out from a point source in a homogenous isotropic medium, the waves have spherical symmetry and the wave function ψ depends only on r. In such case Eq. (4.81) reduces to

$$\frac{\partial^2 \psi}{\partial r^2} + \frac{2}{r} \frac{\partial \psi}{\partial r} = \frac{1}{v^2} \frac{\partial^2 \psi}{\partial t^2}$$

or

$$\frac{1}{r} \frac{\partial^2 (\psi r)}{\partial r^2} = \frac{1}{v^2} \frac{\partial^2 \psi}{\partial r^2}$$

or

$$\frac{\partial^2 (\psi r)}{\partial r^2} = \frac{1}{v^2} \frac{\partial^2 (\psi r)}{\partial t^2}$$

This is identical to one-dimensional wave equation and its general solution can be written as

$$r\psi = f_1(r - vt) + f_2(r + vt)$$

For harmonic waves, we have

$$r\psi = A e^{i(\omega t \pm k \cdot r)} \tag{4.83}$$

or

$$\psi = \frac{A}{r} e^{i(\omega t \pm k \cdot r)} \tag{4.84}$$

since k and r have same direction.

This shows that the amplitude of the wave varies as $1/r$ at points far removed from the source. The intensity of the wave is proportional to the square of the amplitude. Thus in a spherical wave, intensity varies inversely as the square of the distance from the source. It is known as the *inverse square law for the intensity of spherical waves*. This is consistent with the consideration of energy in a spreading spherical wave whose surface area increases as the square of radius.

Example 4.14: Calculate the group velocity for very high frequency electromagnetic radiations such as X-rays.

Solution: If ω is much larger than the characteristic frequency ω_i then n^2 (here n is refractive index) is given by

$$n^2 = 1 + \frac{Ne^2}{m\varepsilon_0} \sum_i \frac{f_i}{\omega_i^2 - \omega^2} = 1 - \frac{Ne^2}{m\varepsilon_0 \omega^2} \sum f_i = 1 - \frac{Ne^2}{m\varepsilon_0 \omega^2}$$

Here we have neglected ω_i^2 in comparison to ω^2 and $\Sigma_i f_i = 1$

$$\therefore \qquad n = \left[1 - \frac{Ne^2}{m\varepsilon_0\omega^2}\right]^{1/2} = 1 - \frac{Ne^2}{2m\varepsilon_0\omega^2}$$

Here we have used the binomial theorem and neglected higher powers of $\dfrac{Ne^2}{m\varepsilon_0\omega^2}$.

We observe that refractive index is less than 1 giving phase velocity $V = \dfrac{c}{n}$ greater than c. Now

$$\frac{dn}{d\omega} = \frac{Ne^2}{\varepsilon_0 m\omega^3}$$

and substituting in expression for group velocity V_g, we have

$$V_g = \frac{c}{n + \omega\dfrac{dn}{d\omega}}$$

But

$$n + \omega\frac{dn}{d\omega} = 1 - \frac{Ne^2}{2m\varepsilon_0\omega^2} + \frac{Ne^2}{\varepsilon_0 m\omega^2} = 1 + \frac{Ne^2}{2m\varepsilon_0\omega^2}$$

$$\therefore \qquad V_g = \frac{c}{1 + \dfrac{Ne^2}{2m\varepsilon_0\omega^2}}$$

Therefore, although phase velocity v is larger than c because n is less than one, the group velocity V_g is less than c. The phase velocity can be written as

$$V = \frac{c}{n} = c\,\frac{1}{\left\{1 - \dfrac{Ne^2}{2m\varepsilon_0\omega^2}\right\}} = c\left\{1 - \frac{Ne^2}{2m\varepsilon_0\omega^2}\right\}^{-1}$$

or

$$V = c\left\{1 + \frac{Ne^2}{2m\varepsilon_0\omega^2}\right\} \qquad \text{using binomial theorem}$$

$$\therefore \qquad V\cdot V_g = c^2$$

This relation although not of great validity is satisfied to a very good approximation over wide range of frequencies.

Example 4.15: The intensity of sunlight at normal incidence on the earth's surface is 1.35 kw/m². Calculate the amplitudes of the electric and magnetic fields of the electromagnetic wave at earth's surface. Given $\mu_0 = 4\pi \times 10^{-7}$ H·m^{-1} and $\varepsilon_0 = (36\pi \times 10^9)^{-1}$ F/m.

Solution: The electromagnetic waves emitted by the sun travel in free space of permeability μ_0 and permittivity ε_0. The average value of poynting vector measures the intensity of the wave. The magnitude of poynting vector is given by

$$|S| = \text{Velocity} \times \text{Energy density}$$

$$= c\left\{\frac{1}{2}\varepsilon_0 E^2 + \frac{1}{2\mu_0}B^2\right\} = c\varepsilon_0 E^2 = \frac{cB^2}{\mu_0}$$

Since E and B are harmonic functions as such intensity I is given by

$$I = \langle S \rangle = \frac{1}{2} c \varepsilon_0 E_0^2 = \frac{1}{2} c \frac{B_0^2}{\mu_0}$$

where E_0 and B_0 are the amplitudes of E and B respectively

$$\therefore \quad E_0 = \sqrt{\frac{2I}{c\varepsilon_0}} = \sqrt{\frac{2 \times 1.35 \times 10^3}{3 \times 10^8 \times (36\pi \times 10^9)^{-1}}}$$

$$= \sqrt{\frac{2 \times 1.35 \times 10^3 \times (36\pi \times 10^9)}{3 \times 10^8}} = 1009 \text{ V/m}$$

and

$$B_0 = \frac{E_0}{c} = \frac{1009}{3 \times 10^8} = 3.363 \times 10^{-6} \text{ web/m}^2$$

and

$$H_0 = \frac{B_0}{\mu_0} = \frac{3.363 \times 10^{-6}}{4\pi \times 10^{-7}} = 2.7 \text{ Amp/m}$$

Example 4.16: A particle of charge e and mass m gives rise to an electric field $E = \dfrac{e}{4\pi\varepsilon_0 r^2}$. Calculate the electrostatic energy over the entire space and equating it to the value mc^2, show that the classical radius of electron is 2.8×10^{-15} m.

Solution: The charge distribution equal to an electronic change is ascribed a classical radius r_0 so that the total energy of the charge at the radius becomes equal to the rest mass energy of the electron mc^2.

Consider the electrostatic and magnetic field due to an electron at rest. Assume the electron is in a sphere of radius r_0 and for simplicity assume further that the charge $-e$ of the electron resides entirely on its surface. Thus, the field at any point R distance from the centre of the sphere is

$$E = -\frac{e}{4\pi\varepsilon_0} \frac{R}{R^3} \text{ for } R \geq r_0$$

$$= 0 \text{ for } R < r_0$$

and

$$B = 0$$

Total energy of the field in free space is given by

$$U = \frac{1}{2} \iiint \left\{ \varepsilon_0 E^2 + \frac{1}{\mu_0} B^2 \right\} dv$$

where dv is volume element. In spherical polar co-ordinates R, θ, ϕ the volume element[7]:

$$dv = dR R d\theta R \sin\theta d\phi = R^2 \sin\theta \cdot d\theta \cdot d\phi \cdot dR.$$

$$\therefore \quad U = \frac{1}{2} \int_{R=r_0}^{\infty} \int_{\theta=0}^{\pi} \int_{\phi=0}^{2\pi} \left\{ \frac{\varepsilon_0 e^2}{16\pi^2 \varepsilon_0^2} \cdot \frac{1}{R^4} \right\} R^2 \sin\theta \cdot d\theta \cdot d\phi \cdot dR$$

$$(\because B = 0)$$

$$= \frac{e^2}{32\pi^2 \varepsilon_0} \int_{R=r_0}^{\infty} \int_{\theta=0}^{\pi} \int_{\phi=0}^{2\pi} \frac{\sin\theta}{R^2} dR \cdot d\theta \cdot d\phi$$

$$= \frac{e^2}{32\pi^2 \varepsilon_0} \cdot \frac{2}{r_0} \cdot 2\pi = \frac{e^2}{8\pi\varepsilon_0 r_0}.$$

7. $\int_{R=r_0}^{\infty} \dfrac{dR}{R^2} = \dfrac{1}{r_0}$, $\int_0^{2\pi} d\phi = 2\pi$, $\int_0^{\pi} \sin\theta \cdot d\theta = 2$.

Equation $U = mc^2$, we have

$$\frac{e^2}{8\pi\varepsilon_0 r_0} = mc^2$$

$$r_0 = \frac{e^2}{8\pi\varepsilon_0 me^2} = \frac{(1.6\times10^{-19})^2}{8\pi\times8.85\times10^{-12}\times9.1\times10^{-31}\times9\times10^{16}}$$

$$= 1.4\times10^{-15}\text{ m}$$

The value of $2r_0$ is called the *classical radius of the electron*.

Although the actual radius of the electron, whatever it may be, may not be related to this.

$\therefore$ Classical radius of electron $2r_0 = 2.8\times10^{-15}$ m.

REVIEW QUESTIONS AND PROBLEMS

1. Prove that the velocity of a longitudinal wave in a rod depends upon the Young's modulus and density of the material.
2. (a) State and explain Fourier's theorem. Explain its limitations.
   (b) Analyze the following rectangular wave where T is the time period.

   $$Y(t) = a \quad \text{for} \quad 0 < t < \frac{T}{2}$$

   $$Y(t) = -a \quad \text{for} \quad \frac{T}{2} < t < T$$

3. Discuss the propagation of electromagnetic waves in dispersive media.
4. Derive expression for velocity of transverse wave in a uniformly stretched string.
5. Obtain expression for velocity of longitudinal waves in a solid rod.
6. State Fourier theorem. Under what conditions it holds? How are Fourier coefficients evaluated? Obtain Fourier series for a saw tooth wave.
7. Prove that electromagnetic waves are transverse in nature.
8. Obtain the expression for velocity of electromagnetic wave in a medium and on its basis:
   (i) Find expression for refractive index of the medium.
   (ii) Show that in vacuum electromagnetic waves travel with velocity of light.
9. Define radiation resistance and show that for a free space its value is 377 Ω.
10. Show that the velocity of vector E or B propagating in a medium is $\dfrac{1}{\sqrt{\mu\varepsilon}}$.
11. What is Poynting vector? What does it represent?
12. Write the wave equation in three dimensions and obtain its solution for spherical waves in an isotropic medium.
13. What is meant by Fourier analysis? How is it achieved? Obtain the first three terms of the Fourier series for the following output voltage of a full wave rectifier:

    $$E = E_0 \sin \omega t \quad 0 \le \omega t \le \pi$$
    $$= -E_0 \sin \omega t \quad \pi \le \omega t \le \pi$$

14. What are electromagnetic waves? Prove that in a plane electromagnetic wave, the electric field vector E and magnetic field vector B oscillate in perpendicular planes and in the same phase.

15. Prove that the ratio of velocities of longitudinal and transverse waves in a wire is $\dfrac{1}{\sqrt{\alpha}}$, where α is the strain in wire.

16. Prove that pressure due to sound waves in a gaseous medium is $P = -k\dfrac{\partial y}{\partial x}$, where symbols have their usual meaning.

17. Prove that for electromagnetic waves, the impedance of free space is $377\ \Omega$.

18. Explain the phenomenon of dispersion on the basis of electromagnetic wave principle. What is anomalous dispersion?

19. Write the Maxwell's equations for electromagnetic waves. Derive the expression for energy density of these waves.

20. The output voltage from a half wave rectifier is given by

$$e = e_0 \sin \omega t \qquad 0 \le \omega t \le \pi$$
$$= 0 \qquad \pi \le \omega t \le 2\pi$$

Analyse this output using the Fourier analysis.

21. Obtain solution of differential equation $\dfrac{d^2 y}{dt^2} = v^2 \dfrac{d^2 y}{dx^2}$ and explain the equation for a system for which boundary conditions $y = 0$ for $x = 0$ and $x = 2l$ is obeyed.

22. Prove that the average kinetic energy for a plane progressive wave is equal to its average potential energy. Also derive the expression for the energy density of such a wave.

23. Obtain an expression for the velocity of the sound waves in gases and discuss the Laplace's correction.

24. With the help of Maxwell's equations derive the expression for the velocity of electromagnetic wave in homogeneous isotropic non-conducting medium. Then prove that in free space, this velocity is equal to 3×10^8 m/s.

25. What is Poynting vector? What does it represent? Show that the magnitude of the Poynting vector of an electromagnetic wave is equal to the product of energy density of the electromagnetic field and the velocity of the electromagnetic waves.

26. It is observed that a pulse requires 0.1 sec to travel from one end to the other of a long string, the tension in the string is provided by massing the string over a pulley to a weight which has 100 times the mass of string.
    (a) What is the length of the string?
    (b) What is the equation of the third normal mode?

$$\left[Ans.\ (a)\ 10\ m \quad (b)\ Y = A \sin \dfrac{3\pi x}{L} \cos(30\pi t) \right]$$

27. A long uniform string of linear density 0.1 kg/m is stretched with a force of 50 N. One end of the string ($x = 0$) is oscillated transversely (sinusoidally) with amplitude of 0.02 m and a period of 0.1 sec, so that travelling waves in the positive X-direction are set-up.
    (a) What is the velocity of the waves?
    (b) What is their wavelength?

(c) If at the driving end ($x = 0$), the displacement (y) at $t = 0$ is 0.01 m with $\dfrac{dy}{dt}$ negative, what is the equation of the travelling waves.

[*Ans.* (a) 22.4 m/s, (b) 2.24 m, (c) $y(x, t) = 0.02 \sin\{2.80x - 62.8t + 0.52\}$]

28. A closed loop of uniform string is rotated rapidly at some constant angular velocity ω. The mass of the string is m and the radius is r. A tension T is set-up circumferentially in the string as a result of its rotation.

    By considering the instantaneous centrepetal acceleration of a small segment of the string, show that the tension must be equal to $\dfrac{m\omega^2 R}{2\pi}$.

29. Deduce an expression for the energy of a string vibrating in P loops in terms of its mass, maximum amplitude of vibration and its frequency of vibration.

    Show also that this energy can be expressed as $E = \dfrac{1}{2}mv^2$, where m is the total mass of the string and v is the maximum particle velocity at the antinode.

30. A function is defined by $f(\theta_0) = a \sin \theta$, $0 < \theta < \pi$ and $f(\theta) = 0$, $\pi < \theta < 2\pi$. Determine the Fourier expansion for $f(\theta)$.

31. Two strings of linear density μ_1 and μ_2 are knotted together at $x = 0$ and stretched to a tension F. A wave $Y = A \sin k_1(x - u_1 t)$ in the string of density μ_1, reaches the junction between the 2 strings at which it is partly transmitted into the string of density μ_2 and partly reflected. Call these waves $B \sin k_2(x - u_2 t)$ and $C \sin k_1(x + u_1 t)$ respectively.

    (a) Assuming $k_2 u_2 = k_1 u_1 = \omega$ and that the displacement of the knot arising from the incident and reflected waves is the same as that arising from the transmitted wave show that $A + C = B$.

    (b) It is further assumed that both strings near the knot have the same slope (why, reason it out), i.e. at $x = 0$, $\left(\dfrac{dy}{dx}\right)_1 = \left(\dfrac{dy}{dx}\right)_2$ show that $C = a\dfrac{k_2 - k_1}{k_2 + k_1} = A\dfrac{U_1 - U_2}{U_1 + U_2}$.

    Under what condition is C negative.

32. Show that the displacement of fluid particle inside a three dimensional rectangular box, having rigid boundaries at $x = 0$ and $x = a$, $y = 0$ and $y = b$, and $z = 0$, and $z = c$ under the influence of waves with propagation velocity v can be expressed as

$$Y = A\sin\left(\frac{n_x \pi x}{a}\right)\sin\left(\frac{n_y \pi y}{b}\right)\sin\left(\frac{n_z \pi z}{c}\right)\sin \omega t$$

where $\omega^2 = V^2\pi^2\left[\dfrac{n_x^2}{a^2} + \dfrac{n_y^2}{b^2} + \dfrac{n_z^2}{c^2}\right]$

and n_x, n_y, n_z are integers greater than or equal to 1.

SHORT ANSWER QUESTIONS

1. Lasers are light source which give almost perfectly parallel beam of light. If a 2 kW beam is concentrated by a lens into cross-sectional area of 10^{-6} cm^2, what is the value of the amplitude of electric field and the magnitude of poynting vector?

Ans. $I = \dfrac{1}{2}c\varepsilon_0 E_0^2$ or $E_0 = \sqrt{\dfrac{2I}{c\varepsilon_0}} = 1.23 \times 10^8$ v/m

$|S| = c\varepsilon_0 E_0^2 = 2I = 4 \times 10^{13}$ W/m^2.

2. Sun radiates energy at a rate of 3.8×10^{26} W. The radius of the sun is 7×10^8 m. Calculate the magnitude of Poynting vector at the surface of sun. (b) The intensity of solar radiations incident on earth which is 1.5×10^{11} m away from the sun.

Ans. S = rate of energy radiated per unit surface area

$\therefore \quad S = \dfrac{3.8 \times 10^{26}}{4\pi(7 \times 10^8)^2} = 6.17 \times 10^7$ W/m^2

Intensity of solar radiation on earth

$= \dfrac{3.8 \times 10^{26}}{4\pi(1.5 \times 10^{11})^2} = 1.34 \times 10^3$ W/m^2

3. Assume that all the energy from 1000 W lamp is radiated uniformly, calculate the values of the electric and magnetic fields of the radiation at a distance of 2 m from the lamp.

Ans. Energy flux per unit area $= E_0 H_0 = \dfrac{1000}{4\pi(2)^2} = \dfrac{1000}{16\pi}$

Characteristic impedance $Z = \dfrac{E_0}{H_0} = \sqrt{\dfrac{\mu_0}{\varepsilon_0}}$.

$\therefore \quad E_0^2 = \dfrac{1000}{16\pi}\sqrt{\dfrac{\mu_0}{\varepsilon_0}} = \dfrac{1000}{16\pi} \times 120\pi = 7500$ or $E_0 = 86.6$ v/m

and $H_0 = \dfrac{1000}{16\pi E_0} = 0.23$ A/m

4. Compare the velocities of sound in hydrogen (H$_2$) and carbon dioxide (CO$_2$). The ratio (γ) for H$_2$ and CO$_2$ is 1.4 and 1.3 respectively.

Ans. $v = \sqrt{\dfrac{\gamma RT}{M}}$

$\therefore \quad \dfrac{v_H}{v_{CO}} = \sqrt{\dfrac{\gamma_H}{M_H} \cdot \dfrac{M_{CO}}{\gamma_{CO_2}}} = \sqrt{\dfrac{1.4 \times 44}{1.3 \times 2}} = 4.9$

5. The intensity of sound in a normal conversation at home is about 3×10^{-6} W/m^2 and frequency of normal human voice is about 1000 Hz. Find the amplitude of waves, given density of air at NTP = 1.3 kg/m^3 and velocity of sound 330 m/s.

Ans. $I = 2\pi^2 \rho v^2 A^2 v$ or $A = \dfrac{1}{\pi v}\sqrt{\dfrac{1}{2\rho v}}$

or $A = \dfrac{1}{\pi \times 10^3}\sqrt{\dfrac{3 \times 10^{-6}}{2 \times 1.3 \times 330}} = 1.88 \times 10^{-8}$ m

6. Sunlight strikes the earth with an intensity of 1400 W/m². How many watts of power must an electrical lamp radiate in order to produce at 1 m brightness of sunlight?

   Ans. $P = I \times 4\pi r^2 = 1.4 \times 10^3 \times 4\pi(1)^2 = 17.6$ kW

7. Transverse waves are generated in two uniform steel wires A and B of diameter 1×10^{-3} and 0.5×10^{-3} m respectively by attaching their free end to a vibrating source of frequency 500 Hz. If the two wires are stretched with same tension, the ratio of the wavelength is

   Ans. $V = \sqrt{\dfrac{T}{\mu}} = \sqrt{\dfrac{T}{\pi r^2 d}}$

   and $\lambda = \dfrac{V}{n} = \dfrac{1}{n}\sqrt{\dfrac{T}{\pi r^2 d}}$

   $\therefore \quad \dfrac{\lambda_1}{\lambda_2} = \dfrac{r_2}{r_1} = \dfrac{0.5 \times 10^{-3}}{1 \times 10^{-3}} = 0.5$

8. The velocity of transverse waves in steel wire is 400 m/s and density 7.6×10^3 kg/m³. Then what will be the stress in the wire?

   Ans. $v = \sqrt{\dfrac{T}{\pi r^2 d}} \qquad \dfrac{T}{\pi r^2} = $ Stress

   $\therefore \quad$ Stress $= v^2 d = (400)^2 \times 7.6 \times 10^3 = 1.21 \times 10^9$ N/m²

9. A steel wire 1 m long is stretched by 0.38 mm and then fixed between two supports. Calcualte the frequency of the fundamental mode for transverse vibrations. $(Y = 2 \times 10^{11}$ N/m², $d = 7600$ kg/m³)

   Ans. $n = \dfrac{1}{2l}\sqrt{\dfrac{T}{\pi r^2 d}} = \dfrac{1}{2l}\sqrt{\dfrac{Y \times \text{strain}}{d}}$

   $= \dfrac{1}{2 \times 1}\sqrt{\dfrac{2 \times 10^{11} \times 38 \times 10^{-5}}{7.6 \times 10^3}} = 50$ Hz

10. The wave produced in a rod is given by

    $$y = 2 \times 10^{-5} \sin(15x) \sin(4.8 \times 10^4 t)$$

    Calculate (i) Max particle velocity, (ii) Maximum tensile stress at $x = 3.5$ cm, $y = 8 \times 10^{11}$ N/m².

    Ans. Particle velocity $= \dfrac{dy}{dt} = 2 \times 10^{-5} \times 4.8 \times 10^4 \sin(15x)\cos(4.8 \times 10^4 t)$

    Maximum velocity at $x = 3.5$ cm is

    $V_{max} = 9.6 \times 10^{-1} \sin(15 \times 3.5 \times 10^{-2})$

    $15 \times 3.5 \times 10^{-2} = 52.5 \times 10^{-2}$ rad $= 30°$.

    $\therefore \quad V_{max} = 9.6 \times 10^{-1} \times \dfrac{1}{2} = 0.48$ m/s

    Tensile stress $= Y\dfrac{\partial y}{\partial x} = Y \times 2 \times 10^{-5} \times 15 \times 10^{-2} \cos(15x)\sin(4.8 \times 10^4 t)$

Max. tensile stress at $x = 3.5$ cm i.e.

$$(Stress)_{max} = 8 \times 10^{11} \times 30 \times 10^{-7} \cos 30°$$

$$= 8 \times 30 \times 10^4 \frac{\sqrt{3}}{2}$$

$$= 2.08 \times 10^6 \text{ N} \cdot \text{m}^{-2}$$

11. Solar radiation is received on Earth at the rate of 1.4 k·Wm^{-2}. Assuming the waves to be plane. Calculate the rms values of electric and magnetic field intensities in the waves.

Ans. $E_0 = \sqrt{\dfrac{2I}{c\varepsilon_0}} = 1009 \text{ V/m}$ $\therefore E_{rms} = \dfrac{E_0}{\sqrt{2}} = 713.4 \text{ V/m}$

and $B_0 = \dfrac{E_0}{c} = 3.363 \times 10^{-6} T$ $\therefore \dfrac{B_0}{\sqrt{2}} = 2.38 \times 10^{-6} T$

12. A wave propagating on string has equation $y(x, t) = 0.03 \sin(3x - 2t)$ where y and x are in metres and t in seconds. The ratio of maximum particle velocity and wave velocity is

Ans. Particle velocity $v_1 = \dfrac{dy}{dt} = -0.6\cos(3x - 2t)$

$\therefore$ Max. $v_1 = 0.06$

Wave velocity $v = \dfrac{k}{\omega} = \dfrac{2}{3}$ $\therefore \dfrac{V_1}{v} = \dfrac{0.06 \times 3}{2} = 0.09$

13. A longitudinal wave of frequency 1000 Hz is produced in air whose amplitude is 10^{-3} cm. Find pressure amplitude if the density of air is 1.29 gm/litre and velocity of sound in air is 330 m/s.

Ans. $Y = a\sin\omega\left(t - \dfrac{x}{v}\right)$ and pressure $P = -K\dfrac{dy}{dx}$

or $P = \dfrac{Ka\omega}{v}\cos\omega\left(t - \dfrac{x}{v}\right)$ or pressure amplitude $P_0 = \dfrac{Ka\omega}{v}$

$$P_0 = adv\omega = 10^{-5} \times 1.29 \times 330 \times 2\pi \times 1000 = 26.7 \text{ N/m}^2$$

14. A rope of mass M and length L is freely suspended from roof. If a transverse pulse is generated at the lower end of the rope, then calculate the time taken by the pulse to reach other end.

Ans. Tension at point distance x from free end $= \dfrac{Mg}{L}x$

Hence velocity of pulse at this point $= \sqrt{gx}$

$$v = \dfrac{dx}{dt} = \sqrt{gx} \quad \text{or} \quad \dfrac{dx}{\sqrt{gx}} = dt \quad \therefore \quad t = \int_0^L \dfrac{dx}{\sqrt{gx}} = 2\sqrt{\dfrac{L}{g}}$$

15. Amplitude of a wave propagating in positive x-direction has amplitude expressed as $Y = \dfrac{1}{1 + x^2}$ at time $t = 0$ and $Y = \dfrac{1}{1 + (x + 1)^2}$ at time $t = 2$ sec. Find the wave velocity in m/s.

Ans. $Y = \dfrac{1}{1+x_1^2} = \dfrac{1}{1+(x_2+1)^2}$ or $x_2 + 1 = x_1$ or $x_1 - x_2 = 1\,\text{m}.$

$\therefore \qquad v = \dfrac{1}{2} = 0.5\,\text{m/s}$

MULTIPLE CHOICE QUESTIONS

1. If bulk modulus of water be $0.20 \times 10^{10}\,\text{N/m}^2$, the velocity of sound in water is (m/s).
   (a) $\sqrt{2} \times 10^3$ (b) $\sqrt{2} \times 10^3$ (c) $\sqrt{3} \times 10^3$ (d) $\dfrac{10^3}{\sqrt{2}}$

2. An iron pipe of length 10 m is struck at one end. A man with his ear at the other end hears two sounds. The time interval between two sounds is:
   Given $Y = 2 \times 10^{11}\,\text{N/m}^2$, $d = 7.2 \times 10^3\,\text{kgm/m}^3$ and for air $\gamma = 1.4$ and $\rho = 1.3\,\text{kg/m}^3.$
   (a) 0.028 s   (b) 0.28 s   (c) 2.8 s   (d) 28 s

3. The speed of sound in air is 340 m/s. The frequency of the second harmonic for a pipe 85 cm long open at both ends is (in Hz):
   (a) 200   (b) 400   (c) 600   (d) 100

4. A sting is stretched between fixed supports separated by 75.0 cm. It is observed to have resonant frequencies of 315 and 420 Hz and there is no other resonant frequencies between these two. The wave speed in the string is (in m/s):
   (a) 87.5   (b) 137.5   (c) 157.5   (d) 187.5

5. The length of a sonometer wire is 0.75 m and density $9 \times 10^3\,\text{kg/m}^3$. It can bear a stress of $8.1 \times 10^8\,\text{N/m}^2$. The fundamental frequency that can be produced without breaking is (in Hz):
   (a) 100   (b) 150   (c) 200   (d) 250

6. The fundamental frequency of a closed organ pipe is 150 Hz. The frequency of its second harmonic is given by:
   (a) 75 Hz   (b) 150 Hz   (c) 300 Hz   (d) None of the above

7. The wires of same material length and of radii r and $2r$ respectively are welded together end to end. This combination is used as a sonometer wire and is kept under tension T. What would be the ratio of the number of loops formed in the wires such that the joint is a node when it is set into vibration:
   (a) 1:2   (b) 2:1   (c) 2:3   (d) 3:2

8. An air column is a pipe which is closed at one end, is in resonance with a vibrating tuning fork of frequency 264 Hz. If $v = 330\,\text{m/s}$, the length of the column in cm is:
   (a) 31.25   (b) 62.50   (c) 93.75   (d) 125

9. The linear density of a vibrating string is $1.3 \times 10^{-4}\,\text{kg/m}$. A transverse wave is propagating on it and is described by the equation $y = 0.021 \sin(x + 30t)$ where x and y are in metres and t in sec. The tension in the string is:
   (a) 0.12 N   (b) 0.48 N   (c) 1.2 N   (d) 4.80 N

10. A hollow metallic tube of length L and closed at one end produces resonance with a tuning fork of frequency f. The entire tube is heated carefully so that length changes by l, if the change in velocity v of the sound is u, the resonance will now be produced by tuning fork of frequency:
    (a) $\dfrac{(v+u)}{4(L+l)}$ (b) $\dfrac{(V-u)}{4(L-l)}$ (c) $\dfrac{(V+u)}{4(L-l)}$ (d) $\dfrac{(V-u)}{4(L+l)}$

11. A cylindrical tube open at both ends has fundamental frequency f in air. The tube is dipped vertically in water so that half of it is in water. The fundamental frequency of the air column now is:

(a) $\dfrac{f}{2}$ (b) $\dfrac{3f}{4}$ (c) f (d) $2f$

12. Organ pipe P_1, closed at one end vibrating in its first harmonic and other pipe P_2 open at both ends vibrating in its third harmonic are in resonance with a given tuning fork. The ratio of the length of P_1 to that of P_2 is:

(a) $\dfrac{1}{2}$ (b) $\dfrac{1}{3}$ (c) $\dfrac{1}{5}$ (d) $\dfrac{1}{6}$

13. A man standing between two cliffs, claps his hands and starts hearing a series of echos at an intervals of 2 sec. Since the speed of sound in air is 340 m/s, the distance between the cliffs must be:

(a) 340 m   (b) 680 m   (c) 1020 m   (d) 170 m

14. A 1 watt source of sound produces spherical waves in isotropic medium. The intensity of the wave at a distance of 5 m from source is:

(a) $\pi \times 10^{-2}$ (b) $\dfrac{10^{-2}}{\pi}$ (c) $\dfrac{10^{-3}}{\pi}$ (d) 4×10^{-13}

15. $Y = Y_0 \sin 2\pi\left(nt - \dfrac{x}{\lambda}\right)$ represents a transverse progressive wave. The maximum particle velocity will be 4 times the wave velocity if:

(a) $\lambda = \dfrac{\pi Y_0}{4}$ (b) $\lambda = \dfrac{\pi Y_0}{2}$ (c) $\lambda = \pi Y_0$ (d) $\lambda = 2\pi Y_0$

ANSWERS

1. (a)   2. (a)   3. (b)   4. (c)   5. (c)   6. (d)   7. (a)   8. (a)   9. (a)   10. (a)   11. (c)
12. (d)   13. (c)   14. (b)   15. (b)

5

Superposition of Wave

5.1 INTRODUCTION

Physical systems susceptible to vibrational effects are generally influenced simultaneously by two or more harmonic vibrations. The human ear drum, a microphone diaphragm and radiowave systems are examples of such systems. These vibrations may be of the same or different frequencies; amplitudes and phase relationships. In this chapter, we will examine their resultant effect on the vibrational motion of the system. In doing so, we will assume that the response of the system is linear and that the resultant displacement is simply the superposition of the individual displacements. The superposed vibrations may be either along the same direction, i.e., the X-axis or along other directions. Vibrations along the oblique directions can be projected along the mutually orthogonal directions and could then be treated as linear combinations. The two special cases considered here are the superposition of the two vibrations acting (i) along the same direction, and (ii) along the two mutually perpendicular directions. These two cases cover all types of motion.

Out of these two, the second case, i.e., two vibrations acting along mutually perpendicular directions on superposition give the phenomenon known as *Lissajous figures. These are the closed curves which depend on the ratio of frequencies of two superposing waves, their amplitude ratio and phase difference.* These we have discussed in chapter 1. As such, we will discuss here only the first case, i.e. superposition of two vibrations acting along the same direction.

5.2 PRINCIPLE OF SUPERPOSITION AND LINEARITY

The principle of superposition states that *"when two (or more) waves propagate simultaneously in the elastic medium, the resultant displacement of any particle at any instant is equal to the vector sum of the displacements of that particle, corresponding to the separate waves at that instances".*

The principle of superposition means that of many waves, each one moves independently as if the other were not present at all; and that their individual shapes and other characteristics are not changed due to the presence of another.

To illustrate the principle of superposition, suppose y_1 is the displacement of a point x at time t, due to a wave described by a function $y_1(x, t)$. It must be a solution of the general differentiation equation of the wave motion $\dfrac{\partial^2 Y}{\partial x^2} = \dfrac{1}{v^2} \dfrac{\partial^2 Y}{\partial t^2}$, that is

$$\frac{\partial^2}{\partial x^2} Y_1(x,t) = \frac{1}{v^2} \frac{\partial^2 Y_1}{\partial t^2}(x,t) \qquad (5.1)$$

Similarly, let y_2 be the displacement of the same point x at the same time t due to another wave described by $Y_2(x, t)$. This must also be a solution of the differential equation. That is

$$\frac{\partial^2}{\partial x^2} Y_2(x,t) = \frac{1}{v^2} \frac{\partial^2 Y_2}{\partial t^2}(x,t) \qquad (5.2)$$

Adding these equations, we have

$$\frac{\partial^2}{\partial x^2} \{Y_1(x,t) + Y_2(x,t)\} = \frac{1}{v^2} \frac{\partial^2}{\partial t^2} \{Y_1(x,t) + Y_2(x,t)\}$$

or

$$\frac{\partial^2 Y}{\partial x^2}(x,t) = \frac{1}{v^2} \frac{\partial^2 Y}{\partial t^2}(x,t) \qquad (5.3)$$

where

$$Y(x,\ t) = Y_1(x,\ t) + Y_2(x,\ t) \qquad (5.4)$$

Equation (5.3) shows that the sum of the two wave functions described by Eq. (5.4) also satisfies the general differential equation as each separate wave functions does. Hence $Y(x, t)$ is a proper function to describe the displacement of the point x at time t. Equation (5.4) shows that the displacement Y of the point x at time t is equal to the sum of the displacements y_1 and y_2 of that point due to the separate waves at the same time. Thus, the principle of superposition is a consequence of the *form* of the differential equation of the wave motion.

The word form means the differential equation of the wave motion is linear and homogenous. A differential equation is said to be linear if it contains the terms that depend only on the first powers of the variable and its derivatives. Further, an equation is said to be homogeneous if it contains no terms independent of the variable Y.

Linear homogeneous differential equations have a very interest-ing and important property. "The sum of any two solutions is itself a solution". This is also the statement of superposition principle. Thus, the superposition principle holds only for linear differential equations. It does not apply if the equations are not linear because the sum of two solutions of a non-linear equation is not itself a solution of the equation.

Thus, the principle of superposition hold for those waves only, whose equation of motion are linear, i.e. obey Hooke's law. Thus, it does not hold for *shock waves* created by explosions and *water waves*.

5.3 PHENOMENON ARISING FROM SUPERPOSITION OF WAVES

Superposition of waves gives rise to various phenomenon like interference, diffraction, beats and stationary waves.

Interference and Diffraction

When two harmonic wave trains of the same frequency and having constant phase relation, travel simultaneously in same or nearly same direction through the medium, the resultant wave intensity at any point is different from the sum of the intensities due to separate wave trains. This phenomenon is called *interference*. It arises because at each point of observation, the phase difference between the superposing waves depends upon the different paths travelled by the individual waves and the resultant amplitude (and hence the intensity) may be greater or less than that of any single wave. Thus, there is a redistribution of intensity in space. This redistribution of energy is known as "*interference pattern*". This interference may be called as *spacial interference* or *interference in space*.

When an advancing wave is obstructed by some barrier (obstacle), then each unobstructed point on the original wave front acts as a new source and the disturbance beyond the obstacle is the superposition of all the waves spreading out from these new secondary sources. As a result, there is a characteristic intensity distribution in the region beyond the barrier (obstacle). This distribution is known as *"diffraction pattern"*.

Beats

The principle of superposition results in another type of interference which may be called *"interference in time* or *temporal interference"*. It occur when two wave trains of slight different frequencies travel through the same region or nearly the same direction, then the amplitude of the resultant wave at any point is not constant but varies with time harmonically. This causes the variation in the intensity with time. This variation of intensity at any point with time is called *beats*.

Stationary (or Standing) Waves

When two harmonic wave trains of the same frequency travel along the same line in opposite directions, their superposition gives a pattern having alternately points of zero displacement (nodal points) and points of maximum displacement (anti-nodal points). This is known as *stationary wave pattern* having no energy transmission.

5.4 INTERFERENCE

When two waves of same frequency having constant phase relation, travelling in the same direction arrive at a point simultaneously, the resultant intensity of sound at that point is different from the sum of intensities due to each wave separately. This redistribution in the intensity as a result of superposition of the waves is called *interference*. If the waves arrive at the point in same phase, i.e., the crest coinciding with crest and trough with trough as shown in Fig. 5.1(a), they reinforce each other and the resultant wave has amplitude which is the sum of the amplitudes of the individual waves and the resultant intensity is maximum. *The interference in this case is said to be constructive.* This situation will always occur when the phase difference between the waves is

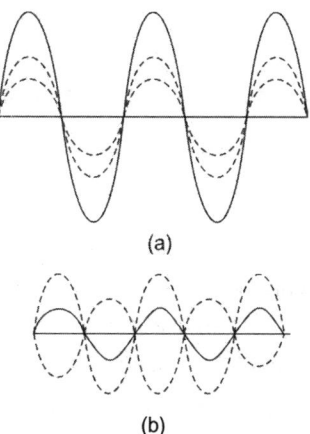

(a)

(b)

Fig. 5.1

$$\phi = 0, 2\pi, 4\pi, \ldots = 2n\pi$$

or path difference

$$\Delta = 0, \lambda, 2\lambda, \ldots = n\lambda$$

Figure 5.1(b) represents the two wave trains arriving at the point in opposite phase such that the crest of one coincides with the trough of the other and vice versa; the resultant amplitude is equal to the difference of amplitudes of the two waves and the intensity of sound is minimum. Interference in this case is called *destructive interference*. This situation will always occur when the phase difference between the waves is

$$\phi = \pi, 3\pi, 5\pi, \ldots = (2n - 1)\pi$$

or path difference

$$\Delta = \frac{\lambda}{2}, \frac{3\lambda}{2}, \frac{5\lambda}{2}, \ldots = (2n - 1)\frac{\lambda}{2}$$

Analytical Treatment

When two wave trains propagate through a medium simultaneously in the same direction, each particle of the medium performs a motion which is the resultant of the two motions due to each wave train separately.

Let us consider two simple harmonic waves of the same frequency travelling in the same direction. Let a_1 and a_2 be the amplitudes of the waves and ϕ the phase difference between them at any point in the medium. If the displacement due to these waves at that point be y_1 and y_2 then

$$y_1 = a_1 \sin \frac{2\pi}{\lambda}(vt - x) \tag{5.5}$$

and

$$y_2 = a_2 \sin \left\{ \frac{2\pi}{\lambda}(vt - x) + \phi \right\} \tag{5.6}$$

where v is the wave velocity in the medium amd λ the wavelength. Since the two displacements y_1 and y_2 are along the same line, the resultant displacement is given by the principle of superposition as their algebraic sum. Hence the resultant displacement

$$y = y_1 + y_2 = a_1 \sin \frac{2\pi}{\lambda}(vt - x) + a_2 \sin \left\{ \frac{2\pi}{\lambda}(vt - x) + \phi \right\}$$

$$= a_1 \sin \frac{2\pi}{\lambda}(vt - x) + a_2 \sin \frac{2\pi}{\lambda}(vt - x)\cos\phi + a_2 \cos \frac{2\pi}{\lambda}(vt - x)\sin\phi$$

$$= (a_1 + a_2 \cos\phi)\sin \frac{2\pi}{\lambda}(vt - x) + a_2 \sin\phi \cos \frac{2\pi}{\lambda}(vt - x)$$

As a_1, a_2 and ϕ are constants and we can replace them by other constants. As such, we can put

$$a_1 + a_2 \cos\phi = a \cos\theta \tag{5.7}$$

and

$$a_2 \sin\phi = a \sin\theta \tag{5.8}$$

Here a and θ are new constants. Thus

$$y = a\cos\theta \cdot \sin \frac{2\pi}{\lambda}(vt - x) + a\sin\theta \cdot \cos \frac{2\pi}{\lambda}(vt - x)$$

$$= a \left\{ \sin \frac{2\pi}{\lambda}(vt - x) + \theta \right\} \tag{5.9}$$

Equation (5.9) is similar to the Eqs (5.5) and (5.6). Hence, the resultant wave is also simple harmonic and of the same period but having new amplitude a and a new phase constant θ. The values of a and θ are known in terms of initial wave constants and can be easily determined. Squaring and adding Eqs (5.7 and 5.8), we have

$$a^2 \cos^2\theta + a^2 \sin^2\theta = (a_1 + a_2 \cos\phi)^2 + a_2^2 \sin^2\phi$$

or

$$a^2 = a_1^2 + 2a_1 a_2 \cos\phi + a_2^2 (\cos^2\phi + \sin^2\phi)$$

$$= a_1^2 + a_2^2 + 2a_1 a_2 \cos\phi$$

or

$$a = \sqrt{a_1^2 + a_2^2 + 2a_1 a_2 \cos\phi} \tag{5.10}$$

Dividing Eq. (5.8) by (5.7), we have

$$\tan \theta = \frac{a_2 \sin \phi}{a_1 + a_2 \cos \phi} \qquad (5.11)$$

Thus, we have the resultant wave train of the same frequency but with a different phase and amplitude.

The resultant intensity I being proportional to the square of the amplitude, is given by

$$I = a^2 = a_1^2 + a_2^2 + 2a_1a_2 \cos \phi \qquad (5.12)$$

Thus, the resultant intensity at any point depends upon the phase difference ϕ between the two waves at that point.

Maximum Intensity

The intensity I will be maximum at points for which $\cos \phi = +1$ that is phase difference

$$\phi = 0, 2\pi, 4\pi, ...$$
$$= 2n\pi \quad \text{where} \quad n = 0, 1, 2, ...$$

or path difference $\Delta = n\lambda$

$[\because$ path difference $= \dfrac{\lambda}{2\pi} \times$ phase difference$]$

Hence

$$I_{max} = (a_1^2 + a_2^2 + 2a_1a_2) = (a_1 + a_2)^2$$

Thus, at points where the two interfering waves meet in the same phase ($\phi = 0, 2\pi$, 4π, ...), the resultant intensity $(a_1 + a_2)^2$ is greater than the sum of intensities $(a_1^2 + a_2^2)$ of the two waves and the waves are said to have *interfere constructively*.

Minimum Intensity

The intensity I will be minimum at points for which $\cos \phi = -1$, that is

$$\text{phase difference} \quad \phi = \pi, 3\pi, 5\pi, ...$$
$$= (2n-1)\pi \quad \text{where } n = 1, 2, 3, ...$$

and path difference $\Delta = (2n-1)\dfrac{\lambda}{2}$

and $$I_{min} = (a_1^2 + a_2^2 - 2a_1a_2) = (a_1 - a_2)^2$$

Thus, at points where the two interfering waves meet in opposite phase ($\phi = \pi, 3\pi$, 5π, ...), the resultant intensity $(a_1 - a_2)^2$ is less than the sum of the intensities $(a_1^2 + a_2^2)$ of the two waves and the waves are said to have *interfere destructively*.

At intermediate points, the intensity is found to be between its maximum and minimum value. Thus, if we move over the space of overlapping of the two waves, we shall pass alternately through the regions of intensity maximum and minimum. The pattern formed by these regions is called the *interference pattern*.

Conditions for Interference

From the above discussion, we come to the conclusion that following conditions must be fulfilled to obtain interference between two waves.

(1) The two waves must have the same frequency (wavelength).

(2) The two waves should propagate along the same direction.

(3) For constructive interference, the phase difference between the waves should be even multiple of π, i.e.

$$\phi = 2n\pi \quad \text{where } n = 0, 1, 2, ...$$

or path difference should be even multiple of $\dfrac{\lambda}{2}$, i.e.

$$\Delta = 2n \cdot \frac{\lambda}{2} = n\lambda \quad \text{where } n = 0, 1, 2, ...$$

(4) For destructive interference, the phase difference between the waves should be odd multiple of π, i.e.

$$\phi = (2n - 1)\pi \quad \text{where } n = 1, 2, ...$$

or path difference should be odd multiple of $\dfrac{\lambda}{2}$, i.e.

$$\Delta = (2n - 1)\frac{\lambda}{2} \quad \text{where } n = 1, 2, 3, ...$$

(5) If the amplitudes differ appreciably, then there will be sufficient intensity even at points where these waves meet in opposite phase (i.e. cause destructive interference). Hence, the interference will not be clear because the contrast between maximum intensity and minimum intensity will be poor. For better contrast, the amplitudes of the waves should be exactly equal. In that case (i.e., $a_1 = a_2 = A$) intensity at any point where phase difference is ϕ, will be given by

$$I = A^2 + A^2 + 2A^2 \cos\phi$$

$$= 4A^2 \cos^2 \frac{\phi}{2}$$

The variation of intensity with phase is shown in Fig. 5.2. I_{max}, i.e. for constructive interference $\phi = 0, 2\pi, 4\pi = 2n\pi$

$$I_{max} = 4A^2 = 4I$$

where $I = A^2$, the intensity of the individual wave. For I_{min}, i.e., for destructive interference $\phi = \pi, 3\pi, 5\pi, ... = (2n - 1)\pi$

$$I_{min} = 0$$

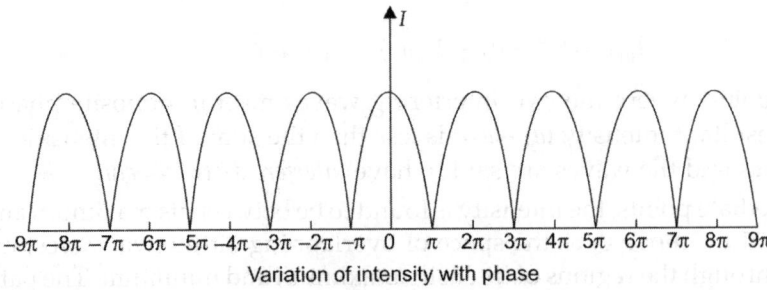

Variation of intensity with phase

Fig. 5.2

(6) If the phase difference between the two waves at a point varies with time, then the intensity of sound at that point will also vary with time and the interference will not be sustained. Hence, *the phase difference between the interfering waves should remain constant.*

The phase difference between two waves reaching at a point depends on two factors:
(1) The different paths traversed by the two waves,
(2) The phase difference between two sources.

The phase difference due to different paths traversed by the two waves do not change with time once the experimental set-up is done. Hence the phase difference between two sources should not change with time. Such sources, having *same frequency and have a phase difference between them which does not change with time are called* **coherent sources**. Two coherent sources located at different points, the waves emitted from them maintain a constant phase difference at a given point in space in the region of overlap. Hence, a point that is one of maximum intensity at a given time will always be a point of maximum intensity and likewise a point of minimum intensity will always be a point of minimum intensity. In other words, a stationary interference pattern will be observed. This coherency between the two sources is an essential condition to obtain stationary (sustained) interference pattern.

Interference and Conservation of Energy

In interference phenomenon, the energy of the medium due to the passage of the two waves remains conserved. It is simply redistributed over the interference pattern. This is proved mathematically as follows:

The resultant intensity at a point in the interference produced by two waves of amplitudes a_1 and a_2 can be written as

$$I = a_1^2 + a_2^2 + 2a_1a_2 \cos\phi$$

where ϕ is the phase difference between the waves at that point. The average intensity between the range $\phi = 0$ to $\phi = 2\pi$ is:

$$\langle I \rangle = \frac{\int_0^{2\pi} I d\phi}{\int_0^{2\pi} d\phi} = \frac{1}{2\pi}\int_0^{2\pi}(a_1^2 + a_2^2 + 2a_1a_2 \cos\phi)\,d\phi$$

$$= a_1^2 + a_2^2$$

because $\int_0^{2\pi} \cos\phi = 0$

Thus, the average intensity is equal to the sum of the separate intensities. That is, whatever the energy apparently disappears at the minima is actually present at maxima. Thus, there is no loss of energy in the interference phenomenon.

Interference Between Waves from Independent Sources

In practice, we do not have point sources. A practical source of light has a huge number of atoms which emit discontinuously wave trains of light at short intervals ($\sim 10^{-8}$ sec). These intervals are different for different atoms. Therefore, the phase of the light wave train emitted by the innumerable atoms in the source changes with time in a random way. Hence the waves from two independent light sources will meet at a point with a phase difference changing with time. Therefore, if at a certain instant, the two waves interfere at that point constructively, producing maximum intensity, a short time ($\sim 10^{-8}$ sec), later they will interfere at that same point destructively producing minimum intensity. Thus, the two waves will produce a time averaged uniform intensity throughout, i.e. no interference pattern would be observed.

In the case of sound waves, the sources producing them are sometimes (but rarely) single and they can be matched in frequency more or less exactly. Therefore, interference between waves from separate sources, like two tuning forks is observable. However, waves from two violins cannot give a sustained interference effect because a violin is not a single source.

In practice, interference effects are obtained by means of two coherent sources obtained by special means. Two vertical rods touching the source of water and driven by a common device produce an interference pattern on the water surface. Two parallel slits receiving light from a common narrow illuminated slit produce an interference pattern of light on a screen. Sound waves emitted by two loudspeakers driven by a common audio-oscillator produce an interference pattern of sound. In all these cases, any change in the phase of the common driver is transmitted simultaneously to the two sources, so that at any point a constant phase difference is maintained between the two waves.

Quincke's Interference Tube

In a more direct method of demonstrating interference between two trains of sound waves, we make use of the branched tube shown in Fig. 5.3 known as *Quincke's tube*. A tube A divides into two branches S and T, which reunite at L. The part T of the tube is made to slide telescopically over the other part so that the length of the part ATL can be varied at will. A tuning fork F of high pitch[1] say

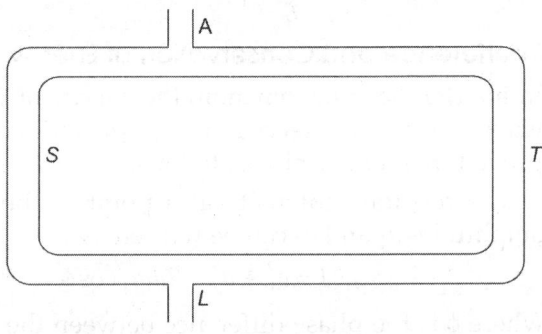

Fig. 5.3

about 1000 Hz is placed over the end A and a rubber tube attached to a side tube at L, is used to convey to the ear the sound which arrives at L via the paths ASL and ATL. If the sliding tube is adjusted so that the two paths are of the same length, the waves take the same time to travel by either path from A to L and therefore, the two sets always arrive at L in the same phase. They will therefore, reinforce each other at the ears of listener and produce a loud sound. If now T is drawn out so that the path ATL is made longer than the path ASL by $\dfrac{\lambda}{2}$, the compression of one path arrive at L at the same time as a rarefaction of the other, there will be complete neutralisation of the two wave trains and no sound will be heard at L. Drawing T still further out the sound is again heard and intensity goes on increasing and becomes maximum, when the path difference amounts to one wavelength. No sound will again be heard when the path difference $ATL - ASL = \dfrac{3\lambda}{2}$ or in general $(2n-1)\dfrac{\lambda}{2}$, where n is integer (i.e. $n = 1, 2, 3, ...$). Maximum sound will be heard when $ATL - ASL = 2n\dfrac{\lambda}{2} = n\lambda$.

The experiment besides illustrating interference provides a good laboratory method of measuring the velocity of sound in gases. Adjust the apparatus so that the two paths ASL and ATL are equal. Next draw the sliding tube out till the sound is reduced to a minimum. When this occurs, the sliding tube will have been moved through a distance

---

1. A Galton's whistle may be used as a source of sound at A and detector which may be sensitive flame, is arranged at the opposite opening at L.

$\frac{\lambda}{4}$, which can be measured by means of graduations on the tube. A displacement of the tube through a distance $\frac{\lambda}{2}$ produces a path difference of λ and hence the reinforcement. Proceeding in this way, a series of readings can be taken and a mean value of $\frac{\lambda}{2}$ and hence λ is obtained. If n be the frequency of the fork $v = n\lambda$.

One can ask what has become of the energy of the sound waves in this case. The answer is that, it has been reflected from L back to the and A. As the two wave trains are in opposite phase at L, there are no variations of pressure at this point. We have already discussed in case of normal modes of open organ pipes that it is on account of this condition of no variation of pressure at the open end that reflection takes place with a reversal of phase. Here L is acting similar to the open end and reflects the sound energy back to A.

Acoustic Filter

This is an acoustical device used to filter out tones of certain frequencies from a complex sound. Its simple form consists of a uniform pipe AB (Fig. 5.4) to which a side tube CD closed at the lower end D is attached. The length of the tube CD is $\lambda/4$, where λ is the wavelength of the component to be removed.

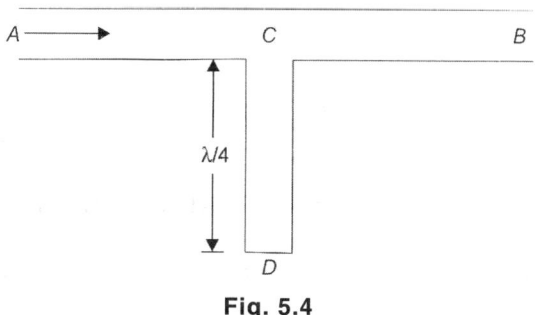

Fig. 5.4

The filter depends for its action upon the phenomenon of interference. The complex sound entering through A is divided at C into two parts, one going along CB and the other entering the tube CD. When the second part after reflection at D returns to C, it is $\lambda/2$ behind the part of travelling along AB. Therefore, the two parts will interfere producing a minimum of that component whose wavelength is λ. Hence, the component will be weakened. By arranging a number of such side tubes of different appropriate lengths, a number of components can be filtered out.

Example 5.1: In a large room a person receives direct sound waves from a source 120 m away from him. He also receives waves from the same source which reach him being reflected from the 25 m high ceiling, a point half way between them. For which wavelengths will these two sound waves (i) interfere constructively, (ii) interfere destructively.

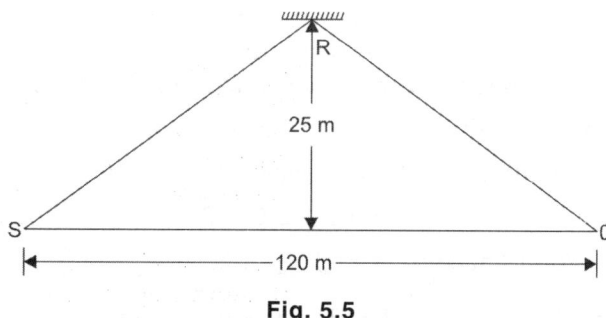

Fig. 5.5

Solution: As shown in Fig. 5.5, the path $SR = \sqrt{(60)^2 + (25)^2} = 65$ m. Hence path $SRO = 130$ m and path $SO = 120$ m. Path difference $= 130 - 120 = 10$ m. In addition to this path difference, the wave on reflection at R suffers a phase change of π or path difference $\lambda/2$. Hence net path difference

$$\Delta = 10 + \frac{\lambda}{2}$$

(i) For constructive interference $\Delta = n\lambda$.

$$\therefore \qquad 10 + \frac{\lambda}{2} = n\lambda$$

$$\text{or} \qquad (2n-1)\frac{\lambda}{2} = 10$$

$$\text{or} \qquad \lambda = \frac{20}{2n-1}$$

$\therefore$ Wavelengths which interfere constructively are $\lambda_1 = 20$ m, $\lambda_2 = \frac{20}{3}$ m, $\lambda_3 = \frac{20}{5} = 4$ m etc.

(ii) For destructive interference $\Delta = (2n-1)\frac{\lambda}{2}$

$$\therefore \qquad 10 + \frac{\lambda}{2} = (2n-1)\frac{\lambda}{2}$$

$$\text{or} \qquad (2n-2)\frac{\lambda}{2} = 10$$

$$\text{or} \qquad \lambda = \frac{10}{m} \quad \text{where } m = n-1$$

$\therefore$ Wavelengths which interfere constructively are $\lambda_1 = 10$ m, $\lambda_2 = 5$ m, $\lambda_3 = \frac{10}{3}$ m, etc.

Example 5.2: Two speakers connected to the same source of fixed frequency are placed 2.0 m apart in a box. A sensitive microphone placed at a distance of 4.0 m from their mid-point along the perpendicular bisector shows maximum response. The box is slowly rotated till the speakers are in a line with the microphone. The distance between the mid-point of the speakers and the microphone remains unchanged. Exactly 5 maximum responses are observed in the microphone in doing this. Calculate the wavelength of sound wave.

Solution: The initial and final positions of the speakers L_1 and L_2 and microphone O are shown in Fig. 5.6(a) and (b) respectively. Initially, the path difference from L_1 to O and L_2 to O is zero,

i.e. $\qquad L_1O - L_2O = 0$

Finally the path difference

$$\Delta = LO - L_1O = 2 \text{ m}$$

For n^{th} maxima $\Delta = n\lambda$

$$\therefore \qquad 2 = 5\lambda \quad \text{or} \quad \lambda = \frac{2}{5} = 0.4 \text{ m}$$

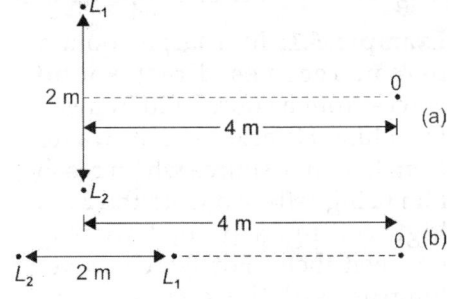

Fig. 5.6

Example 5.3: In Fig. 5.7, S is a sound source and C an observer at a horizontal distance d. The direct wave from S and the wave reflected from point A at a horizontal level at altitude H are in same phase. When the layer rises a distance h and the wave is reflected from point B, no signal is detected at C. Given that the incident and reflected rays make the same angle with the reflecting layer. Find an expression for the wavelength λ of the wave in terms of D, H and h.

Solution: In first case (i.e. reflection from A) the path difference between two rays reaching C is

$$\Delta_1 = (SA + AC) - SC = 2(SA) - SC$$

$$= 2\sqrt{H^2 + \frac{d^2}{4}} - d = \left\{\sqrt{4H^2 + d^2} - d\right\}$$

in second case (i.e., the reflection from B), the path difference between rays reaching C is

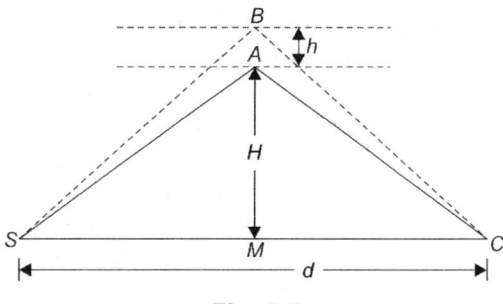

Fig. 5.7

$$\Delta_2 = 2SB - SC = \left\{2\sqrt{H + h)^2 + \frac{d^2}{4}} - d\right\}$$

$$= \left\{\sqrt{4(H + h)^2 + d^2} - d\right\}$$

Hence the difference between these two path differences is

$$\Delta_2 - \Delta_1 = \left\{\sqrt{4(H + h)^2 + d^2} - \sqrt{H^2 + d^2}\right\}$$

Since no signal is received at O, hence this path difference is $\frac{\lambda}{2}$.

∴

$$\frac{\lambda}{2} = \sqrt{4(H + h)^2 + d^2} - \sqrt{4(H^2) + d^2}$$

or

$$\lambda = 2\left\{\sqrt{4(H + h)^2 + d^2} - \sqrt{4H^2 + d^2}\right\}$$

Here phase change due to reflection is not considered as it cancels out.

5.5 BEATS

When two wave trains of nearly same frequency are travelling along the same line in the same direction, the intensity of the resultant sound at any point rises and falls alternately with time. This waxing and waning of intensity at any point with time is called *beats*. It arises as a result of the interference between the sound wave trains emitted from the two sources.

If two wave trains of nearly the same frequency are travelling along the same line in the same direction, then phase condition at any fixed point is continuously changing and the resultant displacement alternately waxes and wanes in amplitude on account of the wave trains getting in and out of phase at a regular intervals. If at a given point, they are in same phase, the compression or rarefaction of one falls on the compression or rarefaction of the other, the result is that there is maximum disturbance. Now, as the frequency of one is shorter than that of the other, they gradually get out of phase. The wave train with higher frequency will be ahead in phase than other until after a certain number of vibrations depending upon the difference of their frequencies, it has fallen ahead by half a wavelength so that the compression of one arrives at the same point at the same time as the rarefaction of the other producing a minimum intensity. Again some time later, the higher frequency wave train has gained one entire vibration and is again in phase with the other and two will reinforce again as in the beginning. Thus, a single note is heard whose intensity rises and falls with time. These periodic alterations of sound between a maximum and minimum intensity, we call them *beats*.

The phenomenon of beats can be demonstrated graphically which is shown in Fig. 5.8. The displacements produced by two wave trains at a point with time are plotted in Fig. 5.8(a), where continuous and dotted curve represent two wave trains with slightly different frequencies. The resultant displacement at that point is the sum of the individual displacements and is shown in Fig. 5.8(b). At time t_1, the two displacements are in phase and add to each other. But on account of difference in frequencies, they gradually get more and more out of phase until at time t_2, they practically cancel each other. Again they get more and more in

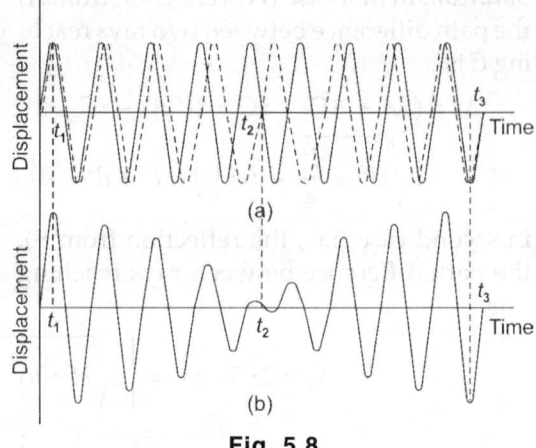

Fig. 5.8

step until once again at time t_3 they reinforce each other. The resultant curve clearly shows the variations in amplitude and hence in intensity.

From the graph, we see that during the interval t_1 to t_3, one wave train gains one vibration over the other and one beat is produced. If the frequencies of the wave trains are n_1 and n_2, then one gains $(n_1 - n_2)$ vibrations over the other in one second. Each time, one gains one complete vibration over the other, the two are in same phase and reinforce each other; there are $(n_1 - n_2)$ maximas per second and equal number of minimas per second, i.e. the frequency of the beats is equal to the difference in the frequencies of the two notes. These beats play a very important part in acoustical determinations and in the theory of music.

5.6 MATHEMATICAL TREATMENT OF BEATS

Let n_1 and n_2 be the frequencies of the two wave trains such that $(n_1 \sim n_2)$ is small. For simplicity, let us suppose that at a particular point through which they pass the two wave trains are in the same phase when $t = 0$ and their respective amplitudes are each equal to a. The displacements of the point due to the two wave trains are represented by

$$y_1 = a \sin 2\pi n_1 t \tag{5.13}$$

and $$y_2 = a \sin 2\pi n_2 t \tag{5.13a}$$

Using principle of superposition, the resultant displacement is given by

$$y = y_1 + y_2 = a \sin 2\pi n_1 t + a \sin 2\pi n_2 t$$

$$= a\{\sin 2\pi n_1 t + \sin 2\pi n_2 t\}$$

$$= 2a \sin \frac{2\pi(n_1 + n_2)}{2} t \cos \frac{2\pi(n_1 - n_2)t}{2}$$

$$= 2a \cos \frac{2\pi(n_1 - n_2)t}{2} \cdot \sin \frac{2\pi(n_1 + n_2)t}{2} \tag{5.14}$$

Therefore, the resultant wave has frequency $= \dfrac{(n_1 + n_2)}{2}$ and amplitude

$$A = 2a \cos(2\pi)\frac{(n_1 - n_2)t}{2}$$

which is function of time, i.e., the amplitude at the site of superposition varies with time, the amplitude A has maximum value $2a$, when $\cos 2\pi \dfrac{(n_1 - n_2)t}{2} = \pm 1$ and a minimum value zero, when $\cos 2\pi \dfrac{(n_1 - n_2)t}{2} = 0$. This rise and fall in resulting amplitude with time causes waxing and waning of the intensity which is referred to as beats. We can now calculate beat frequency.

For maximum value of amplitude and hence intensity:

$$\cos \pi(n_1 - n_2)t = \pm 1$$

or $\qquad \pi(n_1 - n_2)t = m\pi \qquad$ where $m = 0, 1, 2, \ldots,$ etc.

or $\qquad\qquad t = \dfrac{m}{n_1 - n_2}$

Hence maximum intensity occur at times

$$t_0 = 0,\ t_1 = \dfrac{1}{n_1 - n_2},\ t_2 = \dfrac{2}{n_1 - n_2},\ t_3 = \dfrac{3}{n_1 - n_2}, \ldots,\ \text{etc.}$$

Hence time interval between two consecutive maximas is

$$t_1 - t_0 = t_2 - t_1 = t_3 - t_2 = \ldots = \dfrac{1}{n_1 - n_2}$$

Hence the frequency of maximum is $= (n_1 - n_2)$ per sec. For minimum value of amplitude and hence intensity

$$\cos \pi(n_1 - n_2)t = 0$$

or $\qquad \pi(n_1 - n_2)t = (2p - 1)\dfrac{\pi}{2}$

where $\qquad\qquad P = 1, 2, 3, \ldots,$ etc.

or $\qquad\qquad t = \dfrac{2p - 1}{2(n_1 - n_2)}$

Hence minimum intensity $(= 0)$ occur at times

$$t_1 = \dfrac{1}{2(n_1 - n_2)},\ t_2 = \dfrac{3}{2(n_1 - n_2)},\ t_3 = \dfrac{5}{2(n_1 - n_2)} \text{ and so on.}$$

So, that the minimas are regularly timed in between the maximas and the interval between two consecutive minimas is also $\dfrac{1}{n_1 - n_2}$ and their frequency is also $n_1 - n_2/\text{sec}$.

Since the interval between successive maxima and that between successive minima is $\dfrac{1}{n_1 - n_2}$. Hence the beat frequency $= n_1 - n_2/\text{sec}$.

Thus, the resultant of two simple harmonic waves of nearly equal frequencies may be regarded approximately a simple harmonic wave of almost the same frequency $\left(n = \dfrac{n_1 + n_2}{2}\right)$, but differing in phase from the component waves and whose amplitude is alternately the sum and difference of the component amplitudes, and the variation having a frequency equal to the difference of the component frequencies.

5.7 APPLICATIONS OF BEATS

There are a number of applications of the phenomenon of beats.

(i) Determination of Frequency of Tuning Fork

The phenomenon of beats can be used to determine the frequency of a tuning fork. The tuning fork whose frequency is to be determined say A and another tuning fork B whose frequency is known are sounded together. The number of beats produced in a measured interval of time is counted. If n be the frequency of B and p be the number of beats produced per second, then the frequency of A must be

$$n_a = n \pm p$$

Now, one prong of unknown fork A is loaded with a little wax and the two forks are sounded again. If the number of beats per second is decreased, then the frequency of the fork $n_a = n + p$ but if number of beats per second is increased, then the frequency of A must be

$$n_a = n - p$$

The frequency of the fork A is lowered when it is loaded with wax. Thus, if its frequency was already greater than B, then on loading, the difference between the frequencies of A and B must decrease and hence the number of beats per second must also decrease. On the other hand, if the frequency of A was smaller than that of B, then on loading A, the difference between the frequencies of A and B must increase. Hence the number of beats per second must also increase.

(ii) Tuning of Musical Instrument

The musical instrument is sounded together with another with which it is to be tuned. If the frequencies are slightly unequal, beats are heard. Now, the frequency of the given instrument is altered until the beats disappear. This is called *tuning*.

(iii) Detection of Harmful Gases in Mines

The apparatus used to detect harmful gases in mines has two similar small pipes, one filled by pure air and the other by mine air. These pipes are blown together. If the mine air is pure, the pipes emit sound of the same frequency and no beats are heard. When fire-damp (harmful gases) are present, the density of the mine air decreases and the velocity of sound in it is changed. Consequently, the frequency of the sound produced by the mine air pipe is slightly changed (increased) and the beats are heard. The miners have tables with them to estimate the percentage of firedamp from the number of beats heard. This gives warning before hand to the miners before the mine air is dangerous enough to cause an explosion.

(iv) In Radio Reception

Beats are used in the heterodyne reception of radio waves. The incoming high frequency signals are mixed with the waves of slightly different frequency generated at the reception station. These two combined together produce pulsating frequency lying in the audible range.

5.8 BEATS AND AMPLITUDE MODULATION

When two waves of equal amplitude a but of frequencies n_1 and n_2 differing by a small amount superpose, they give rise to a wave system of frequency $(n_1 + n_2)/2$ but with amplitude varying with time, periodically between $2a$ and 0. The maximum and minimum amplitudes occur at a frequency $n_1 \sim n_2$. This is the phenomenon of beats in which the amplitude is said to be modulated. In this example, the amplitude varies between $2a$ and zero. This is called *complete or 100% modulation*.

Amplitude modulation is very common in radio transmission. The audio frequency sound waves (speech or music) converted into electrical signal at the broadcasting station cannot be transmitted as such to distant places because of two reasons: (i) their intensity decreases rapidly with distance, (ii) an enormously large transmission antenna will be required. Hence in practice, a radio frequency wave (produced by oscillator) is used to carry the desired audio frequency signal. The process of superposing the audio frequency signal on the carrier radio frequency wave is called *modulation*. In this process, the amplitude of the carrier waves varies in accordance with the audio frequency signal which is to be transmitted.

Fig. 5.9 shows (a) the radio frequency carrier wave, (b) the audio frequency modulation signal, and (c) the resultant modulated wave.

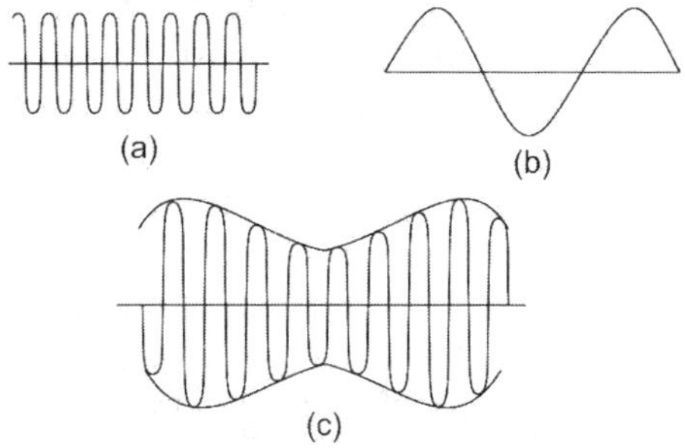

Let the carrier wave be represented by equation

$$E_c = E_{oc} \cos \omega_c t$$

and modulating signal as

$$E_m = E_{om} \cos \omega_m t$$

Fig. 5.9

where $\omega_m \ll \omega_c$. Since the amplitude of the modulated carrier waves varies in accordance with the modulating signal, it should be of the form

$$E = E_{oc} + kE_{om}$$

where k is a proportionality factor. Thus, the modulated wave may be represented as

$$E = e \cos \omega_c t = [E_{oc} + kE_{om} \cos \omega_m t]\cos \omega_c t$$

or $\qquad E = E_{oc} \cos \omega_c t + mE_{oc} \cos \omega_m t \cos \omega_c t$

where $\qquad m = \dfrac{kE_{om}}{E_{oc}}$

$\therefore \qquad E = E_{oc} \cos \omega_c t + \dfrac{1}{2} mE_{oc} \left[\cos(\omega_c + \omega_m)t + \cos(\omega_c - \omega_m)t\right]$

here m is called *modulation index*. It determines the degree of modulation. It is equal to 1 for 100% modulation, where the amplitude of modulated wave will vary between $2E_{oc}$ and zero. The last expanded expression indicates the frequency spectrum of the modulated wave. It consists of three components each having a constant amplitude and frequency. The component frequencies are ω_c, $(\omega_c + \omega_m)$ and $(\omega_c - \omega_m)$. The wave of frequency ω_c and amplitude E_c is just the carrier wave, while the other two waves of frequencies $(\omega_c + \omega_m)$ and $(\omega_c - \omega_m)$ are of same amplitude $\dfrac{1}{2} mE_{oc}$ are known as *upper side band and lower side band* respectively. The generation of side bands, however, lead to the crowding of ratio frequencies and interference between stations.

5.9 DEMONSTRATION OF BEATS

The phenomenon of beats can be demonstrated in several ways which are discussed below:

(i) Take two tuning forks of the same frequency mounted on resonance boxes and place them on the table with the open ends of the resonance boxes facing each other. Set them vibrating simultaneously. You will not observe any rise and fall in the intensity of sound. Attach a small quantity of wax to the prongs of one of them, so as to lower its frequency slightly. On sounding the forks together again, beats will be heard. By varying the amount of wax or its position on the prong the number of beats per second can be changed.

(ii) Beats can be heard between a vibrating fork held with its stem pressing against the monochord board and a vibrating string of the monochord when their frequencies are nearly equal. Very often, we employ a wooden disc of 3″ in diameter to the centre of which a wooden rod 8″ or so is attached to hear the beats. The end of the rod is kept in contact with the top of the sounding box and the disc is placed near the ear. Beats can be heard very distinctly if there is a slight difference in the frequencies. The method is employed for tuning the vibrating string to the frequency of the fork by altering the position of the movable wedge.

(iii) The beat curve may be demonstrated optically by an arrangement shown in Fig. 5.10. Two large electrically driven turning forks P and Q of equal frequencies are mounted vertically parallel to each other. Small mirror strips m_1 and m_2 are attached to their prongs facing each other near the tips. A narrow beam of light from a powerful arc lamp A passes through a condensing lens L and is successively reflected from mirrors m_1 and m_2. It is then made to fall on a

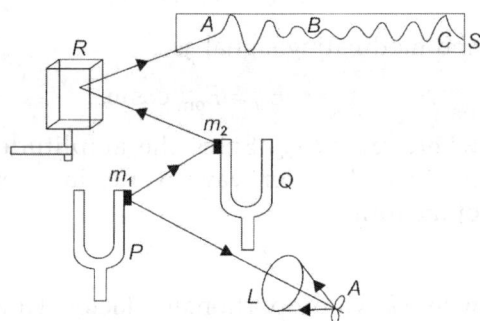

Fig. 5.10

revolving mirror R and is finally brought to focus on a screen S giving a bright spot of light. When the fork P vibrates the spot of light is spread out into a band. If Q also vibrate simultaneously, the band will expand or shrink depending on the relative phases of the forks. Now, one of them is loaded with a small piece of wax on a prong, which slightly lowers its frequency. Then the forks are made to vibrate again simultaneously. When the beats will be heard and on revolving the mirror R, the light beam traces out a beat curve on the screen indicating the rise and fall in the intensity of sound. This curve will be similar as shown in Fig. 5.10.

5.10 DISTINCTION BETWEEN STATIONARY INTERFERENCE AND BEATS

1. Interference occurs by the superposition of two wave trains travelling in the same or opposite direction whereas beats are produced by the superposition of two wave trains travelling only in the same direction.

2. For interference, the frequencies of the two wave trains should be exactly equal, whereas for beats, the frequencies of two wave trains should be slightly different.

3. The interference pattern consists of alternate fixed positions of maximum and minimum disturbance and is permanent, i.e. throughout the time, there are fixed

places of maximum and minimum effects. The beats pattern consisting of periodic maximum and minimum sounds, travels outwards with the speed of sound and is heard at all places, i.e. throughout the space, there are regularly occurring times of maximum and minimum effects.

4. For interference, there is a constant phase difference between the waves meeting at a point at all the times, i.e. at some points the wave trains are in the same phase to produce a maximum and at others in opposite phase to produce a minimum, whereas phase difference at a particular point in beats continually changes between 0 and π to produce a maximum or a minimum effect.

5. In interference, the amplitude of the resultant vibration varies from point to point but remains fixed for a given point, whereas in case of beats the amplitude of the resultant vibrations at a given point rises and falls periodically as a consequence sound of waxing and waning intensity passes through the ear with time.

5.11 COMBINATION TONES

When two pure tones of frequencies p and q are sounded together and sustained, a series of new tones is produced in addition to the primary tones. The new tones are called *combination tones*. The strongest of them is one having a frequency $(p - q)$ and is called the *first order difference tone*. This may further combine with one of the primary tones to produce a *second order difference tone* of frequency $(p - 2q)$, a *third order difference tone* of frequency $(p - 3q)$ and so on. There is also a *first order summation tone* having frequency $(p + q)$ which is much weaker than the first difference tone. It also combines with one of the primary tones to produce a *second order summation tone* of frequency $(p + 2q)$ or $(2p + q)$ and so on. Besides these, there are tones of frequencies $2p, 3p, ...$ and $2q, 3q, ...,$ known as *self combination tones*. Thus, theoretically an infinite number of combination tones are possible.

The first difference tone may be noticed in a policeman's whistle. The whistle consists of two short pipes side by side, each alone giving a feeble high-pitch tone slightly different in frequency with the other. When the whistle is blown, the pipes emit a penetrating low-pitched tone which gives the characteristic quality of the whistle. This is the first difference tone. If, while the whistle is being blown, a finger is placed over one of the pipes, the low pitched difference tone disappears and a feeble high pitched tone is heard.

Subjective and Objective Combination Tones

Young and Koning have maintained that the combination tones are subjective, i.e., they are formed in the ear and have no existence in air. But Helmholtz pointed out that at least in some cases, the tones are objective, i.e., formed in air outside the ear. This occurs in instruments like double siren where the same mass of air is violently agitated by two tones simultaneously. Helmholtz detected the objective tones by means of resonators tuned to the expected tone.

Rucker and Edser performed an experiment directly proving the existence of the combination tones in air, the ear not being used at all. Sound from a double siren (worked by the same wind chest) was picked up by a horn and directed on to a fork of frequency 64 Hz, one prong of which carried one of the mirrors of a Michelson's interferometer (Fig. 5.11).

When the fork was at rest, steady interference fringes were observed in the telescope. As the speed of the double siren was adjusted so as to give either a difference or summation tone of frequency 64 (the same as the natural frequency of the fork), the fringes were at once disturbed. This proved the objective existence of a combination tone of frequency 64 in air which excited resonant vibrations in the fork and hence disturbed the fringes.

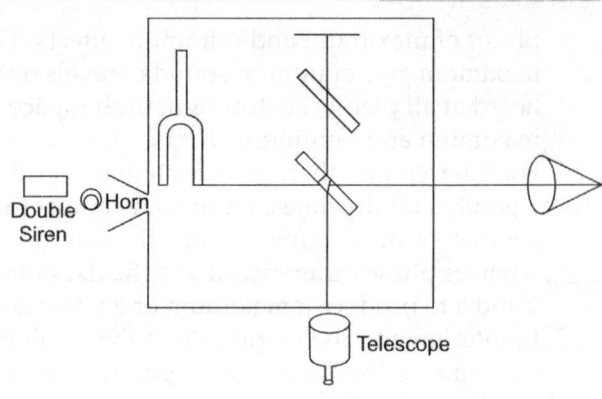

Fig. 5.11

Origin of Combination Tones

The question as to how the combination tones arise has been a controversial one.

(i) Koning's Beat Tone Theory

The first difference tone which was the first to be noticed has a frequency equal to the difference in frequencies of the two primary tones. This led Young, Koning and others to conclude that combination tones and beats are identical. When the beats follow in rapid succession, they merge into a tone. But this theory has been rejected on the following grounds:

    (a) According to this theory, the combination tones would be subjective only, which is actually not the case.

    (b) It gives no explanation of the summation tones.

    (c) Combination tones usually occur with load primary tones while beats can be heard with quite feeble tones also.

(ii) Helmholtz's Theory

The fact that the combination tones are usually produced by loud primary tones as suggested by Helmholtz that the vibrating medium, whether a membrane, eardrum or a limited air cavity, when subjected to large vibrations, the restoring forces called into play are no longer proportional to the displacement but vary in a more complicated manner. He added a new term "proportional to the square of the displacement" and put the equation of motion (neglecting damping forces) as

$$m\frac{d^2y}{dt^2} + sy + s'y^2 = F_1 \sin pt + F_2 \sin(qt + \phi)$$

where $F_1 \sin pt$ and $F_2 \sin(qt + \phi)$ represent the primaries. The solution of the equation contains terms involving $\sin 2pt$, $\sin 2qt$, $\sin(p-q)t$, $\sin(p+q)t$, showing that the resultant sound consists of tones of frequencies $2p$, $2q$, $(p-q)$, $(p+q)$ in addition to p and q. The amplitudes of these tones are found to be proportional to the product F_1F_2 and so requires large values of the primary amplitudes F_1F_2 and so requires large values of the primary amplitudes F_1 and F_2 in order to become appreciable. Thus, the production of various combination tones is explained.

This theory is open to the following objections:

    (a) Combination tones are too loud to be explained by adding $s'y^2$ term which is negligible to first approximation.

    (b) In practice, the summation tone is much weaker than the difference tone. Helmholtz's theory fails to explain this.

    (c) Sometimes comparatively feeble primary tones give rise to combination tones, a fact not covered by the above theory.

(iii) Waetzmann's General Asymmetry Theory

Waetzmann held the view that the vibrating systems producing combination tones have general asymmetry which is responsible for these tones. He realised such a system by loading a membrane with a central mass on one side only. The amplitude of free vibrations of this membrane were asymmetrical about the zero position (Fig. 5.12(a)). He next applied to the membrane double

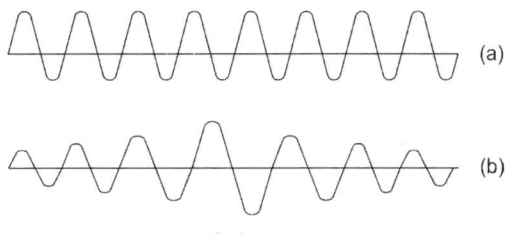

Fig. 5.12

forcing by two simple primary tones and recorded the resultant forced vibrations. A curve like one shown in Fig. 5.12(b) was obtained. On analysing the curve by Fourier's theorem, he got the original primary tones of frequencies p and q, the *difference tones* of frequency $(p - q)$ an amplitude greater than that of either primary, a weak tone of frequency $(2p - q)$ and occasionally the summation tone of frequency $(p + q)$. This loaded membrane may be taken as eardrum which is loaded asymmetrically on one side by bones and hence gives subjective tones. Other systems also show similar proportions so that we may expect to find objective tones due to the same cause.

The theory thus explains the production of combination tones both as regards amplitude as well as frequency.

Use of Combination Tones in Tuning to an Octave (Frequency Ratio 2:1)

When two tuning forks nearly *an octave* apart are sounded together, beats are heard. If for example, the frequency of one fork is 100 and that of the other is 198, then 2 beats per second are produced. Such beats are produced by combination tones. The forks produce a first order diffrerence tone of frequency $(198 - 100) = 98$. This tone gives 2 beats per second with the tone of lower fork. The two forks can be tuned to an exact octave by making use of these beats. If the frequency of one of the fork is adjusted until the beats disappear, the ratio of the frequencies becomes 2:1.

Distinction Between Beats and Combination Tones

When two tones of slightly different frequencies say p and q are sounded together, the intensity of the resultant sound alternately rises and falls $(p \sim q)$ times per second. This is an example of beats. As the difference between the frequencies increases, beats can no more be distinguished separately and give rise to discord.

On the other hand, if two tones of frequencies p and q which may be sufficiently different, are sounded together loudly and sustained a new series tones of frequencies $(p - q)$, $(p - 2q)$, $(p - 3q)$, ... $(p + q)$, $(p + 2q)$, $(2p + q)$, ... $2p$, $3p$, ... $2q$, $3q$ ... are produced. These are the different types of combination tones. The combination tone of frequency $(p - q)$, same as the frequency of beats is the strongest and was first to be noticed. Therefore, Young, Koning and others held the view that beats and combination tones are identical. They argued that the beats occurring in rapid succession merge into a tone. This argument, was however rejected. Following are the points of distinction between the beats and combination tones:

(i) Combination tones usually occur with loud tones whereas beats occur equally plainly with faint ones.

(ii) Combination tones can be heard with two primary tones separated by a large frequency interval, but beats are clearly heard when the interval is small.

(iii) Beats have only one frequency $(p - q)$ while combination tones are of various types having different frequencies.

(iv) Combination tones along with primary tone may give rise to beats.

(v) In the origin of beats, the principle of superposition holds good but it fails in the origin of combination tones.

Example 5.4: Two radio stations broadcast their programmes at the same amplitudes A and at slightly different frequencies ω_1 and ω_2 respectively, where $\omega_2 - \omega_1 = 10^3$ Hz. A detector receives the signals from the two stations simultaneously. It can only detect signals of intensity $\geq 2A^2$.

(i) Find the time interval between successive maximas of the intensity of the signal received by the detector.

(ii) Find the time for which detector remains idle in each cycle of the intensity of the signal.

Solution: (i) Let the signal waves are given by

$$y_1 = A \sin 2\pi\omega_1 t \quad \text{and} \quad y_2 = A \sin 2\pi\omega_2 t$$

The resultant disturbance is given by

$$y = y_1 + y_2 = A \sin 2\pi\omega_1 t + A \sin 2\pi\omega_2 t$$

$$= 2A \sin 2\pi \frac{(\omega_1 + \omega_2)}{2} t \cos 2\pi \frac{\omega_2 - \omega_1}{2} t$$

$$= 2A \sin 2\pi \frac{(\omega_2 - \omega_1)}{2} t \sin 2\pi \frac{\omega_1 + \omega_2}{2} t$$

Thus, the resultant disturbance has the amplitude

$$= 2A \cos 2\pi \frac{(\omega_2 - \omega_1)}{2} t$$

For maxima

$$\cos \pi (\omega_2 - \omega_1) t = \pm 1$$

or $\quad \pi (\omega_2 - \omega_1) t = m\pi \quad$ where $\quad m = 0, 1, 2, 3, \ldots$

or $\quad t = \dfrac{m}{\omega_2 - \omega_1} = 0, \dfrac{1}{\omega_2 - \omega_1}, \dfrac{2}{\omega_2 - \omega_1}, \ldots$

Hence the time interval between successive maxima

$$\frac{1}{\omega_2 - \omega_1} = \frac{1}{10^3} = 10^{-3} \text{ sec}$$

(ii) The resultant intensity is given by

$$I = A_1^2 + A_2^2 + 2A_1 A_2 \cos \delta$$

$$= 4A^2 \cos^2 \frac{\delta}{2}$$

where $\delta = 0$, intensity is maximum. $I_{max} = 4A^2$

$\delta = \dfrac{\pi}{2}$, intensity $I = 2A^2$

$\delta = \pi$, intensity $I = 0$

$\delta = \dfrac{3\pi}{2}$, intensity $I = 2A^2$

$\delta = 2\pi$, intensity $I = 4A^2$.

The detector remains idle from $\delta = \dfrac{\pi}{2}$ to $\dfrac{3\pi}{2}$ or for each half cycle. Hence required time

$$t = \frac{T}{2} = \frac{10^{-3}}{2} = 5 \times 10^{-4}\,\text{s}$$

Example 5.5: Three sources of sound of frequencies 400, 401 and 402 of equal intensities are sounded together. Calculate the number of beats per second.

Solution: Let a be the amplitude and $(n - 1)$, n and $(n + 1)$, the frequencies of the three waves. The equation for the resultant displacement is

$$y = a \sin 2\pi(n - 1)t + a \sin 2\pi nt + a \sin 2\pi(n + 1)t$$

adding the first and third term, we have

$$y = a \sin 2\pi nt + 2a \sin 2\pi nt \cos 2\pi t$$
$$= a\,(1 + 2 \cos 2\pi t) \sin 2\pi nt$$

This shows that the resultant amplitude is $a(1 + \cos 2\pi t)$. It is maximum when

$$\cos 2\pi t = +1$$

or $\qquad\qquad 2\pi t = 2m\pi$ where $m = 0, 1, 2, 3, \ldots$

or $\qquad\qquad t = m = 0, 1, 2, \ldots,$ s

Clearly time interval between successive maxima is 1 sec, i.e., the frequency of maximas is 1 Hz.

The intensity is minimum when $1 + 2 \cos 2\pi t = 0$

or $\qquad\qquad 2\cos 2\pi t = -1$

or $\qquad\qquad \cos 2\pi t = -\dfrac{1}{2}$

or $\qquad\qquad 2\pi t = 2m\pi + \dfrac{2\pi}{3}$ where $m = 0, 1, 2, 3, \ldots$

or $\qquad\qquad t = \left(m + \dfrac{1}{3}\right) = \dfrac{1}{3}, \dfrac{4}{3}, \dfrac{7}{3}, \ldots$

The time interval between successive minimum is 1 s, i.e. frequency of minimas is 1 Hz.

$$\text{Frequency of heats} = \text{Frequency of maxima}$$
$$= \text{Frequency of minima}$$
$$= 1 \text{ per second}$$

Example 5.6: A metal wire of diameter 1 mm is held on two knife edges at a distance of 50 cm. The tension in the wire is 100 N. The wire vibrating with its fundamental frequency and a vibrating tuning fork together produce 5 beats/s. The tension in the wire is then reduced to 81 N. When the two are excited, beats are heard at the same rate. Calculate (i) frequency of the fork, and (ii) density of the material of the wire.

Solution: Let the frequency of the fork be n. When tension in the string is reduced, the number of beats remain unchanged. This means initially the frequency of wire is higher than that of the fork. As number of beats per second is 5. Hence initially the frequency of wire is $(n + 5)$.

$$\therefore \qquad n + 5 = \frac{1}{2l} \sqrt{\frac{T_1}{\mu}} \qquad \text{where } T_1 = 100 \text{ N}$$

When tension is reduced to $T_2 = 81$ N, the frequency of the wire $n_2 = n - 5$.

$$\therefore \qquad n - 5 = \frac{1}{2l} \sqrt{\frac{T_2}{\mu}}$$

$$\therefore \qquad \frac{n + 5}{n - 5} = \sqrt{\frac{100}{81}} = \frac{10}{9}$$

or $\qquad 9n + 45 = 10n - 50$

or $\qquad n = 95$ Hz

(ii) $n_2 = n - 5 = 90$ Hz when tension is 81 N.

$$\therefore \qquad N_2 = \frac{1}{2l} \sqrt{\frac{T_2}{\mu}} = \frac{1}{2l} \sqrt{\frac{T_2}{\pi r^2 d}}$$

$$\therefore \qquad 90 = \frac{1}{2 \times 0.5} \sqrt{\frac{81}{\pi d (5 \times 10^{-4})^2}}$$

$$\therefore \qquad d = \frac{81}{\pi \times (90)^2 \, (5 \times 10^{-4})^2} = 12.7 \times 10^3 \text{ kg/m}^3$$

5.12 STATIONARY WAVES

When two identical waves, either transverse or longitudinal travel through a medium along the same line in opposite directions, they superpose to produce a new type of waves, which appear stationary in space. These waves are called *stationary waves* or *standing waves*. For example, when a wave is sent along a string or along the air-column of a pipe, it is reflected at the other end and superpose upon the incident wave to produce stationary waves. When such waves are formed, then certain particles of the medium remain permanently at rest, while some other particles undergo maximum displacement compared to the others. The former are called the *nodes* and the latter *antinodes*.

Conditions for Stationary Waves

Obviously for the formation of stationary waves, the medium must not be infinite in length, i.e., it should have a boundary. In such a medium any travelling wave will, by reflection at the boundary produce an identical wave travelling in the opposite direction so that stationary waves will be formed. Hence, the presence of a bounded medium is the essential condition for the formation stationary waves.

5.13 REFLECTION AND TRANSMISSION OF TRANSVERSE WAVES AT A BOUNDARY BETWEEN TWO STRINGS

Let us suppose that two strings 1 and 2 of different linear densities are joined at a point to form a composite string. Let us assume that both strings are stretched with the same tension T. The characteristic impedances of the strings are $z_1 = \mu_1 v_1$ and $z_2 = \mu_2 v_2$, where

μ_1 and μ_2 are the linear densities of the strings and $v_1 \left(= \sqrt{\dfrac{T}{\mu_1}} \right)$ and $v_2 \left(= \sqrt{\dfrac{T}{\mu_2}} \right)$ are the wave velocities in strings 1 and 2 respectively.

Let us suppose a wave (called *incident wave*) is travelling in the positive X-direction on string 1. The particle displacements of strings 1 are given by

$$y_i(x, t) = A_i \sin(\omega t - k_1 x)$$

where $k_1 = \dfrac{2\pi}{\lambda_1} = \dfrac{2\pi v}{v_1}$ and A_i is the ampli-
tude of the incident wave. When this wave reaches the boundary (which we shall take at $x = 0$) separating the two strings, it is partly transmitted and partly reflected at the boundary. The reflected wave travels on string 1 is the negative X-direction and transmitted wave travels on string 2 in the positive X-direction (Fig. 5.13). The particle displacement due to these waves are therefore given by

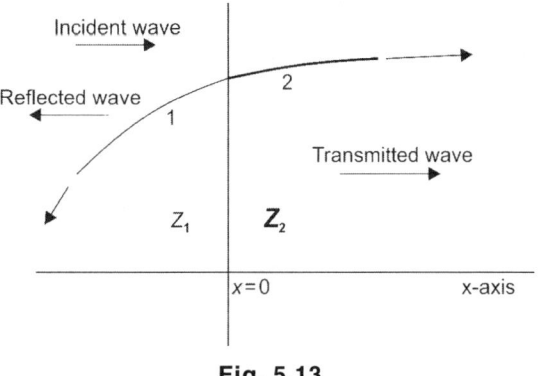

Fig. 5.13

$$y_r(x, t) = A_r \sin(\omega t + k_1 x)$$

and $\qquad y_t(x, t) = A_t \sin(\omega t - k_2 x)$

where $k_2 = \dfrac{2\pi}{\lambda_2} = \dfrac{2\pi v}{v_2}$ and A_r and A_t are the amplitudes of the reflected and transmitted waves respectively.

Boundary Conditions

The point $x = 0$ at the boundary undergoes oscillations under the combined influence of the incident and reflected waves in string 1. It then acts as a source of transmitted waves travelling in string 2. The boundary conditions to be satisfied at $x = 0$ are:

(i) The displacement is the same immediately to the left and to the right of the boundary at $x = 0$, i.e., $y(x, t)$ is continuous across the boundary at $x = 0$. Therefore, velocity $\dfrac{\partial y}{\partial t}(x, t)$ is also continuous.

(ii) The restoring force or transverse component of tension $\left[-T \dfrac{\partial y}{\partial x}(x, t) \right]$ is continuous across the boundary at $x = 0$. To understand this, let us imagine that there is an infinitesimal element of mass at $x = 0$. If the gradient $\dfrac{\partial y}{\partial z}$ is not continuous, this would give rise to a net transverse force acting on this infinitesimal small mass of the string at $x = 0$. Consequently, this element will have an infinite acceleration which is not permitted.

Recalling that $y(x, t)$ in the string 1 is due to the superposition of the displacements due to the incident and reflected waves and (x, t) in the string 2 is due only to the transmitted wave and using the boundary conditions, we have

$$y_i(x, t) + y_r(x, t) = y_t(x, t)$$

or $\quad A_i \sin(\omega t - k_1 x) + A_r \sin(\omega t + k_1 x) = A_t \sin(\omega t - k_2 x)$

and setting $x = 0$

$$A_i \sin \omega t + A_r \sin \omega t = A_t \sin \omega t$$

or $\qquad\qquad A_i + A_r = A_t$ $\qquad\qquad\qquad\qquad$ (5.15)

Boundary condition (ii) at $x = 0$ requires that

$$-T\frac{\partial y_i}{\partial x} - T\frac{\partial y_r}{\partial x} = -T\frac{\partial y_t}{\partial x}$$

or $A_i k_i T \cos(\omega t - k_1 x) - A_r k_1 T \cos(\omega t - k_1 x) = A_t k_2 T \cos(\omega t - k_2 x)$

and setting $x = 0$, we have

$$Tk_1(A_i - A_r) = k_2 T A_t$$

$$k_1 T = 2\pi v \frac{T}{v_1} = 2\pi v \mu_1 v_1$$

$$= 2\pi v z_1 \quad (T = \mu_1 v_1^2, z = \mu_1 v_1)$$

and $\qquad\qquad k_2 T = 2\pi v z_2$

Then, we have

$$z_1(A_i - A_r) = z_2 A_i \qquad\qquad\qquad\qquad (5.16)$$

Equations (5.15) and (5.16) yield

$$\frac{A_r}{A_i} = r_{12} = \frac{z_1 - z_2}{z_1 + z_2} \qquad\qquad\qquad\qquad (5.17)$$

and $\qquad\qquad \dfrac{A_t}{A_i} = t_{12} = \dfrac{2z_1}{z_1 + z_2} \qquad\qquad\qquad\qquad (5.18)$

where r_{12} and t_{12} are the reflection and transmission amplitude co-efficients, when a wave travels from string 1 to string 2. The ratio $\dfrac{A_r}{A_i}$ is the fraction of the incident amplitude reflected at the boundary and the ratio $\dfrac{A_t}{A_i}$ is the fraction of the incident amplitude transmitted across the boundary. These fractions depend only on the impedances and are independent of the angular frequency ω of the incident wave.

If the string is rigidly fixed at $x = 0$ say by attaching this end to a wall, then the medium 2 is infinitely massive which means $z_2 = \infty$. In this case $\dfrac{A_t}{A_i} = 0$ giving $A_t = 0$ indicating that there is no transmitted wave and $\dfrac{A_r}{A_i} = -1$ or $A_r = -A_i$ which means that the incident wave is completely reflected with a reversal in amplitude. The reversal of amplitude means a phase change of π on reflection. It is clear from Eq. (5.17) that if $z_2 > z_1$, the ratio A_r / A_i will be negative indicating a phase change of π on reflection. Thus, we conclude that if a wave travelling in a medium of lower impedance meets the boundary of a medium of higher impedance (i.e., a medium in which the wave velocity is smaller), the wave reflected at the boundary undergoes a phase change of π.

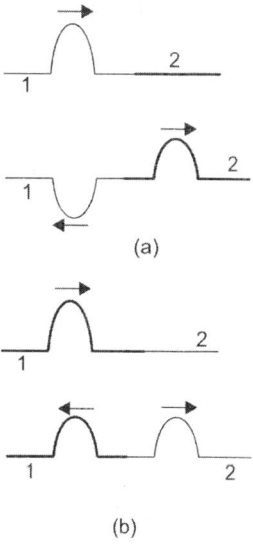

(a)

On the other hand, if $z_2 < z_1$, the ratio $\dfrac{A_r}{A_i}$ is positive. Thus, if a wave is reflected at the boundary of a lower impedance (i.e., a medium in which the wave velocity is higher), the reflected wave does not undergo any phase change.

Furthermore Eq. (5.18) shows that the ratio $\dfrac{A_t}{A_i}$ always remains positive independent of whether z_2 is less than or more than z_1. This shows that the transmitted wave does not undergo any phase change. This is illustrated in Fig. 5.14. Figure 5.14(a) depicts reflection and transmission of pulse when string 2 has a

(b)

Fig. 5.14

higher impedance. Note a phase change of π in reflected pulse whereas Fig. 5.14(b) depicts reflection and transmission of pulse, when the second string has a lower impedance. Note that the reflected pulse does not undergo any phase change in this case. The transmitted pulse also does not undergo any phase change.

If the wave travels from string 2 to string 1, the amplitude reflection and transmission co-efficients are given by

$$r_{21} = \frac{z_2 - z_1}{z_2 + z_1} \quad \text{and} \quad t_{21} = \frac{2z_2}{z_1 + z_2}$$

Reflection and Transmission of Energy at the Boundary

Waves are a very useful mechanism for the transport of energy in a medium. It is interesting to consider what happens to the energy in a wave when it meets a boundary between two media of different impedances.

As the wave propagates along the string, each part of the string is thrown into harmonic oscillations with the passage of time. We have already seen in Chapter 4 that the rate at which energy is carried per unit length along the string is given by

$$P = \frac{1}{2}\mu v \omega^2 A^2 = \frac{1}{2} z \omega^2 A^2$$

Here we will now calculate the rates at which the energy is incident, reflected and transmitted at the boundary at $x = 0$. The rate of incident energy is the rate at which the energy is carried by the incident wave which is given by

$$P_i = \frac{1}{2} z_1 \omega^2 A_i^2$$

Similarly, the rates of the reflected and transmitted energies respectively are

$$P_r = \frac{1}{2} z_1 \omega^2 A_r^2$$

and

$$P_t = \frac{1}{2} z_2 \omega^2 A_r^2$$

Using Eqs (5.17 and 5.18), we have

$$P_r = \frac{1}{2}z_1\left(\frac{z_1 - z_2}{z_1 + z_2}\right)^2 A_i^2\omega^2$$

and

$$P_t = \frac{1}{2}z_2\frac{4z_1^2}{(z_1 + z_2)^2}A_i^2\omega^2 = \frac{2z_2 z_1^2}{(z_1 + z_2)^2}A_i\omega^2$$

Notice that

$$P_r + P_t = \frac{1}{2}z_1\omega^2 A_i^2\left(\frac{z_1 - z_2}{z_1 + z_2}\right)^2 + \frac{1}{2}\frac{4z_2 z_1^2}{(z_1 + z_2)^2}A_i\omega^2$$

$$= \frac{1}{2}z_1\omega^2 A_i^2 = P_i$$

In other words, the rate at which energy arrives at the boundary with the incident wave is equal to the rate at which energy leaves the boundary with the reflected and transmitted waves. This is in consistent with the conservation of energy at the junction of two media. Thus energy is conserved. All the energy arriving at the boundary with the incident wave leaves the boundary with the reflected and transmitted waves. This is expected since we have assumed that there is no absorption of energy at the boundary.

The reflected and transmitted energy co-efficients are given by

$$\frac{\text{Reflected energy}}{\text{Incident energy}} = \frac{P_r}{P_i} = \left(\frac{z_1 - z_2}{z_1 - z_2}\right)^2$$

$$\frac{\text{Transmitted energy}}{\text{Incident energy}} = \frac{P_t}{P_i} = \frac{4z_1 z_2}{(z_1 + z_2)^2}$$

∴ Reflection co-efficient + Transmission co-efficient = 1. Also, if $z_1 = z_2$, no energy is reflected and the impedances are said to be matched.

5.14 STANDING WAVES ON A STRING OF FIXED LENGTH

Consider a string of length L stretched with a tension T along the X-axis with its end $A(x = 0)$ and $B(x = L)$ rigidly fixed (Fig. 5.15). A transverse wave is produced by creating a harmonic disturbance near to end A. The wave travels along the string in the positive x-direction and gets reflected at the

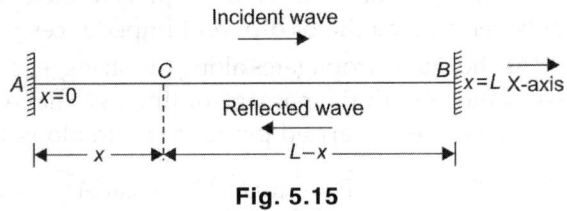

Fig. 5.15

fixed end B giving rise to a reflected wave which travels in the negative x-direction. The superposition between the two oppositely travelling waves give rise to standing wave on the string. Let us consider the simplest case of a monochromatic incident wave of single frequency ω and amplitude a_i. The displacement at time t of particle say C, located at x due to the incident wave travelling in the positive x-direction is given by

$$y_i(x, t) = A_i \sin(\omega t - kx) = A_i \sin\{k(vt - x)\} \tag{5.19}$$

where $k = \dfrac{2\pi}{\lambda}$ and λ being the wavelength of the wave and $v = \dfrac{\omega}{k}$ is the wave velocity.

We will now find the displacement of this particle C due to the wave reflected from fixed end B. The incident wave created at A is reflected from end B. Hence

$$y_r(x, t) = A_r \sin\{k(vt - x')\}$$

where A_r is the amplitude of the reflected wave and x' is the distance travelled by the wave from A to B and then to C, i.e., $x' = L + L - x = (2L - x)$. Hence, displacement at time of particle C at x due to reflected wave is given by

$$y_r(x, t) = A_r \sin\{k[vt - (2L - x)]\}$$
$$= A_r \sin\{k(vt + x - 2L)\} \tag{5.20}$$

Here the argument of sin function involves $(vt + x)$ which represents a wave in the negative X-direction, i.e., the direction of proportion of the wave reflected at B.

Using the principle of superposition, the resultant displacement at time 't' at any point X is given by

$$y(x, t) = y_i(x, t) + y_r(x, t)$$
$$= A_i \sin\{k(vt - x)\} + A_r \sin\{k(vt + x - 2L)\} \tag{5.21}$$

Since end $x = 0$ and $x = L$ are rigidly fixed, the boundary conditions are (i) $y = 0$ at $x = 0$, (ii) $y = 0$ at $x = L$, at all times, using condition (ii) Eq. (5.21) becomes

$$0 = A_i \sin\{k(vt - L)\} + A_r \sin\{k(vt - L)\}$$
$$= (A_i + A_r) \sin\{k(vt - L)\}$$

which is satisfied for all the values of t if

$$A_i + A_r = 0 \quad \text{or} \quad A_i = -A_r = A \text{ (say)}$$

The reversal of amplitude of the reflected wave implies a phase change of π on reflection at the rigid end $x = L$. Setting $A_r = -A_i = -A$ in Eq. (5.21), we have

$$y(x, t) = A[\sin\{k(vt - x)\} - \sin\{k(vt + x - 2L)\}]$$
$$= 2A \sin\{k(L - x)\} \cos\{k(vt - L)\} \tag{5.22}$$

The correctness of Eqs (5.21 and 5.22) can be easily checked by substituting $x = L$ which gives $y = 0$. This is true as the end $x = L$ is rigidly fixed and has no resultant motion.

Now, the first boundary condition, i.e., $y = 0$ at $x = 0$ gives

$$\sin kL = 0 \quad \text{i.e., } kL = 0, \pi, 2\pi, \dots$$

or

$$k_n L = n\pi$$

or

$$k_n = \frac{n\pi}{L} \tag{5.23}$$

where n is an integer having values 0, 1, 2, 3, ...

Hence the allowed frequencies are $\quad (\because \omega = kv)$

$$\omega_n = \frac{n\pi v}{L}$$

or

$$v_n = \frac{\omega_n}{2\pi} = \frac{nv}{2L} = \frac{n}{2L}\sqrt{\frac{T}{\mu}} \tag{5.24}$$

Using Eq. (5.23) in (5.22), we have

$$y_n(x, t) = 2A\sin\left\{\frac{n\pi}{L}(L - x)\right\}\cos\left\{\frac{n\pi}{L}(vt - L)\right\}$$

$$= -2A\sin\left(\frac{n\pi x}{L}\right)\cos\left(\frac{n\pi vt}{L}\right) \tag{5.25}$$

or $\qquad\qquad y_n = -2A\sin(k_n x)\cos(\omega_n t) \tag{5.25a}$

Equation (5.25a) is the equation of a standing or stationary wave. It is obvious that this equation does not represent a travelling wave. Since it no longer has the characteristic form involving $(\omega t - kx)$ or $(\omega t + kx)$ in the argument of sine or cosine function.

Modes of Vibrations

We observe that the frequencies given by Eq. (5.25) are the frequencies of the normal modes of a string fixed at both ends which we have already discussed in chapter 4. The mode with $n = 0$ is physically not possible since $v = 0$ implies that there is no vibration at all.

The mode with $n = 1$ is called the *fundamental mode*. The frequency and particle displacements in this mode are given by Eqs (5.24 and 5.25).

$$v_1 = \frac{1}{2L}\sqrt{\frac{T}{\mu}}$$

and $\qquad y(x, t) = -2A\sin\left(\frac{\pi x}{L}\right)\cos\left(\frac{\pi vt}{L}\right)$

Notice that points $x = 0$ and $x = L$ are permanently at rest. These points on the string are called *nodal points* or *nodes*. But the point $x = L/2$ has a maximum displacement. These points are called *anti-nodes*. Figure 5.16(a) shows particle displacement in the fundamental mode.

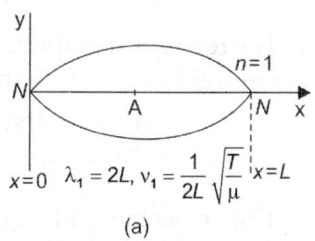

(a)

The mode with $n = 2$ is called the *second harmonic*. The frequency and particle displacements in this mode are given by

$$v_2 = \frac{1}{L}\sqrt{\frac{T}{\mu}} = 2v_1$$

$$y_2(x, t) = -2A\sin\left(\frac{2\pi x}{L}\right)\cos\left(\frac{2\pi vt}{L}\right)$$

We notice that, in addition to nodes at $n = 0$ and $x = L$,

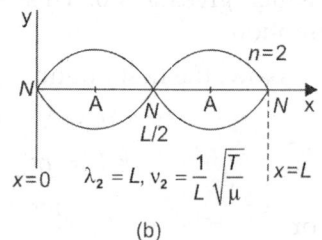

(b)

there is an other node at $x = \dfrac{L}{2}$. The antidotes are at $\dfrac{L}{4}, \dfrac{3L}{4}$. This mode is shown in Fig. 5.16(b). Figure 5.16(c) shows the third harmonic. We observe that in the nth harmonic, there are $(n - 1)$ positions (between the fixed ends) equally spaced along the string where the displacement is always zero. These points marked N are the nodes in standing wave pattern.

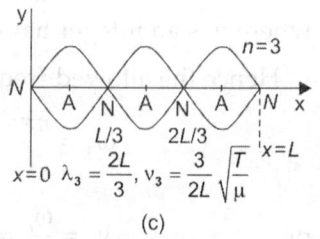

(c)

Fig. 5.16

Characteristic Features of the Standing Wave

Standing waves are not waves in the true sense of the terms, they can be at most called *harmonic vibrations of modes* with the following characteristic features:

(1) In a travelling wave, the amplitude of particle oscillations is the same at all positions of the medium. On the other hand, in a stationary (standing) wave, the amplitudes of oscillations are different at different places. At nodes, the amplitude is zero and at antinodes, the amplitude is maximum, being equal to the sum of the amplitude of the constituent waves. At the intermediate points, the amplitude varies between these two limits. The distance between two consecutive nodes or antinodes is half the wavelength ($\lambda/2$) of the constituent waves, whereas the distance between consecutive nodes and antinode is $1/4$ the wavelength ($\lambda/4$) of the constituent waves.

(2) In a travelling wave the particle velocity is given by

$$\frac{dy}{dt} = Akv\{k(vt - x)\}$$

The velocity varies from place to place and from time to time. The maximum value is $\pm Av$ or $\pm A\omega$ and minimum is 0. Values of velocity are the same for all the particles but are reached at different times. On the other hand, in standing wave the particle velocity is given by

$$\frac{dy}{dt} = 2kAv\sin(kx)\sin(kvt)$$

The velocity varies from place to place and from time to time. The maximum and minimum values of particle velocity are different for different positions but are reached at the same time. The velocity when the particle crosses mean position is maximum for particles at anti-nodes and zero at nodes.

(3) In a travelling wave, the phase of particle oscillations in $\omega\left(t - \dfrac{x}{v}\right)$ which depends on both x and t. Therefore, different particles acquire their maximum displacement at different intervals of time. This interval is determined by the wave velocity. In a standing wave, the phase of particle oscillation is ωt. Which is independent of position, i.e., at a given time, the particles between two consecutive nodal points are in same phase. Therefore, all points acquire their maximum displacement at one time and zero displacement also at one time. The points on either side of a nodal point are in opposite phase.

(4) The most striking feature of a standing wave is that energy is not transported along the medium. The reason is that the energy cannot conceivably flow past nodes of the medium which are permanently at rest. There is nothing like a wave front in standing wave and the phase is same at all points, changing only with time. Hence there is no energy flux. Further in a standing wave twice in each cycle, all particles of the medium have zero displacement simultaneously so that the potential energy is zero and entire energy of the system is then kinetic. A quarter of the period later, all particles of the medium simultaneously attain their maximum displacement and at this instant, the kinetic energy is zero then entire energy of the medium becomes potential. At other instants, the energy is partly potential and partly kinetic. Hence in standing wave, the energy periodically becomes entirely kinetic and entirely potential. In a travelling wave, the entire energy of the medium is half potential and half kinetic.

Standing Wave Ratio (SWR)

Standing waves are formed due to a superposition of waves travelling in opposite directions. If the amplitudes of these travelling waves are equal and opposite (resulting from complete reflection at rigid boundaries, nodes (i.e., points of zero amplitude) will exist in the standing wave pattern. In practice, however, the boundaries are not perfectly rigid resulting in incomplete or partial reflection.

If we have waves which are partially reflected {i.e. $|A_r| < |A_i|$} from a bopundary, they combine with the incoming waves to form a standing wave pattern which is, however, superposed on the travelling waves so that the amplitude at nodes is not zero. In such a case, we define standing wave ratio (SWR) as

$$SWR = \frac{\text{Maximum amplitude}}{\text{Minimum amplitude}}$$

To calculate SWR, let us consider two monochromatic waves of different amplitudes travelling in opposite directions in a medium. The displacement at a point x at time t due to the two waves are given by

$$y_i(x,t) = A_i \sin(\omega t - kx)$$

$$y_r(x,t) = A_t \sin(\omega t + kx)$$

The resultant displacement at a point x due to superposition of the two waves is given by

$$y(x, t) = y_i(x,t) + y_r(x,t) = A_i \sin(\omega t - kx) + A_r \sin(\omega t + kx)$$
$$= (A_i + A_r)\cos kx \sin \omega t - (A_i - A_r)\sin kx \cos \omega t$$

This equation can be written as

$$y(x,t) = A \sin(\omega t - \phi)$$

where the resultant amplitude A and phase ϕ are given by

$$A^2 = (A_i^2 + A_r^2 + 2A_iA_r \cos 2kx)$$

or

$$A = (A_i^2 + A_r^2 + 2A_iA_r \cos 2kx)^{\frac{1}{2}}$$

and

$$\tan \phi = \frac{A_i + A_r}{A_i - A_r} \tan kx.$$

At antinode, amplitude is maximum (given by $\cos 2kx = +1$) and maximum amplitude is given by

$$A_{max} = A_i + A_r$$

and at nodes (given by $\cos 2kx = -1$) amplitude is minimum given by

$$A_{min} = A_i - A_r$$

Hence

$$SWR = \frac{A_{max}}{A_{min}} = \frac{A_i + A_r}{A_i - A_r} = \frac{1 + \dfrac{A_r}{A_i}}{1 - \dfrac{A_r}{A_i}} \qquad (5.26)$$

But $r = \dfrac{A_r}{A_i}$ where r is called *amplitude reflection co-efficient.*

∴

$$SWR = \frac{1+r}{1-r} \qquad \text{or} \qquad r = \frac{SWR - 1}{SWR + 1}$$

where we have assumed that the phase change on reflection is zero or π because in both cases the amplitude reflection co-efficient r will come out the same.

The standing wave ratio is calculated by measuring the maximum and minimum values of amplitude A and the reflection co-efficient is then determined using Eq. (5.27). This is the usual way of determining the reflection co-efficient and hence of the impedance of a boundary. For a perfectly rigid boundary, the reflection co-efficient is unity and impedance infinite.

5.15 REFLECTION AND TRANSMISSION OF LONGITUDINAL WAVES (SOUND WAVES) AT A BOUNDARY BETWEEN TWO MEDIA

When a sound wave meets a boundary separating two media of different acoustic impedance, it is partly reflected and partly transmitted at the boundary. Consider a plane sound wave travelling in a medium 1 of density ρ_1 and incident normally on a plane boundary at $x = 0$ separating medium 1 from another medium 2 of density ρ_2. The acoustic impedance of the two media respectively are

$$z_1 = \rho_1 v_1$$

and
$$z_2 = \rho_2 v_1$$

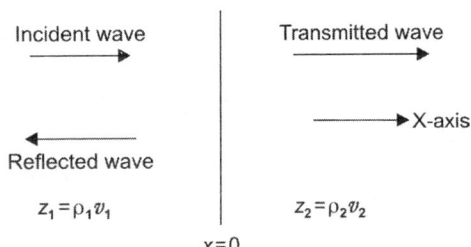

Fig. 5.17

where v_1 and v_2 are the sound speeds in medium 1 and medium 2 respectively (Fig. 5.17). The incident, reflected and transmitted waves respectively are

$$\xi_i(x,t) = A_i \sin(\omega t - k_1 x)$$

$$\xi_r(x,t) = A_r \sin(\omega t + k_1 x)$$

$$\xi_t(x,t) = A_t \sin(\omega t - k_2 x).$$

The boundary conditions are:

1. The particle displacement $\xi(x, t)$ is continuous across the boundary. Hence, the particle velocity $\dfrac{\partial \xi}{\partial t}$ is also continuous.

2. The excess pressure $\left(P = \Delta P = -E \dfrac{\partial \xi}{\partial x} = -\gamma P_0 \dfrac{\partial \xi}{\partial x} \right)$ is continuous across the boundary

(see Eq. (4.31a) of chapter 4). Here $\gamma = \dfrac{C_p}{C_v}$ ratio of the two specific heats (C_p at constant pressure and C_v at constant volume) of the gaseous medium and P_0 is the equilibrium pressure. Thus, the two boundary conditions to be satisfied are (at $x = 0$)

$$\xi_i(x,t) + \xi_r(x,t) = \xi_t(x,t)$$
and $$p_i + p_r = p_t$$

or $$\frac{\partial \xi_i}{\partial x} - \frac{\partial \xi_r}{\partial x} = -\frac{\partial \xi_t}{\partial x}$$

The two boundary conditions give

$$A_i + A_r = A_t \qquad (5.28)$$

and $$k_1(A_r - A_i) = k_2 A_t$$

But

$$k_1 = \frac{\omega}{v_1} = \frac{\omega}{\rho_1 v_1^2} \quad \rho v_1 = \frac{\omega z_1}{\gamma P_0} \quad \left(\because z_1 = \rho_1 v_1, \quad v_1 = \sqrt{\frac{\gamma P_0}{\rho_1}} \right)$$

and

$$k_2 = \frac{\omega}{\gamma P_0} \cdot z_2$$

$$\therefore \qquad z_1(A_r - A_i) = z_2 A_t \tag{5.29}$$

From Eqs (5.28 and 5.29), we have

$$r_{12} = \frac{A_r}{A_i} = \frac{z_1 - z_2}{z_1 + z_2} \tag{5.30}$$

and

$$t_{12} = \frac{A_t}{A_i} = \frac{2z_1}{z_1 + z_2} \tag{5.31}$$

Equation (5.30) gives amplitude reflection co-efficient and Eq. (5.31) gives amplitude transmission co-efficients. It is clear that if $z_1 > z_2$, A_r/A_i is positive indicating that the incident and reflected displacement are in phase. But if $z_1 < z_2$, A_r/A_i is negative showing that the reflected wave undergoes a phase change of π, with respect to the incident wave. Furthermore, it is clear from Eq. (5.31) that A_t/A_i remains positive independent of whether z_1 is less or more than z_2 which means that the transmitted wave does not undergo any phase change.

At a rigid wall where z_2 is infinity, $A_r = -A_i$ showing that the wave is completely reflected.

Reflection and Transmission of Sound Energy

The intensity of a sound wave of amplitude A and frequency v travelling with speed v in a medium of density ρ is given by [chapter 4 Eq. (4.37)]

$$I = 2\pi^2 v^2 A^2 \rho v = 2\pi^2 v^2 A^2 Z$$

where z is the characteristic acoustic impedance offered by the medium. The intensity co-efficients of reflection and transmission are therefore given by

$$\text{Reflection co-efficient} = \frac{I_r}{I_i} = \frac{2\pi^2 v^2 A_i^2 z_1}{2\pi^2 v^2 A_i^2 z_1}$$

$$= \frac{A_r^2}{A_i^2} = \frac{(z_1 - z_2)^2}{(z_1 + z_2)^2} \tag{5.32}$$

$$\text{Transmission co-efficient} = \frac{I_t}{I_i} = \frac{2\pi^2 v^2 A_t z_2}{2\pi^2 v^2 A_i^2 z_1}$$

$$= \frac{z_2 A_t^2}{z_1 A_i^2} = \frac{(4z_1 z_2)}{(z_1 + z_2)^2} \tag{5.33}$$

We observe that $\dfrac{I_r}{I_i} + \dfrac{I_t}{I_i} = 1$ or $I_r + I_t = I_i$ which means that energy is conserved.

Experiments show that there is almost total reflection of sound energy at the air-water interface, whereas in the case of steel-water interface, the reflection co-efficient is about 0.85 or 85%. Thus, only about 15% of sound energy is transmitted at a steel-water interface.

This severely limits the transmission and detection device used in submarines using ultrasonic waves.

5.16 STANDING WAVES IN PIPES

Before discussing standing waves in air columns in pipes, let us determine the reflection and transmission co-efficients for particle pressures and velocities at the boundary. The particle velocities in incident, reflected and transmitted waves at the boundary at $x = 0$ are

$$\left(\frac{\partial \xi_i}{\partial t}\right)_{x=0} = v_i = A_i \omega \cos \omega t$$

$$\left(\frac{\partial \xi_r}{\partial t}\right)_{x=0} = v_r = A_r \omega \cos \omega t$$

$$\left(\frac{\partial \xi_t}{\partial t}\right)_{x=0} = v_t = A_t \omega \cos \omega t$$

$\therefore$ Velocity reflection co-efficient

$$\frac{v_r}{v_i} = \frac{A_r}{A_i} = \left(\frac{z_1 - z_2}{z_1 + z_2}\right) \tag{5.34}$$

and velocity transmission co-efficient

$$\frac{v_t}{v_i} = \frac{A_t}{A_i} = \frac{2z_1}{z_1 + z_2} \tag{5.35}$$

Similarly, pressure co-efficient for reflection

$$\frac{P_r}{P_i} = \frac{v_r}{v_i} = \frac{z_2 - z_1}{z_1 + z_2} \tag{5.36}$$

and pressure transmission co-efficient

$$\frac{P_t}{P_i} = \frac{z_2 z_1}{z_1 v_i} = \frac{2z_1}{z_1 + z_2} \tag{5.37}$$

Thus, we see that if $z_1 > z_2$, the incident and reflected particle velocities are in phase but the incident and reflected acoustic pressures are out of phase by 180° (or π). On the other hand, if $z_1 < z_2$, the acoustic pressures are in phase but the particle velocities are out of phase by 180° (or π).

At rigid wall where z_2 is infinite, we have

$$\frac{v_r}{v_i} = -1 \quad \text{and} \quad \frac{P_r}{P_i} = +1$$

Standing Waves in a Closed Pipe

Consider a closed pipe of length L along the X-axis with its open end at $x = 0$ and closed end at $x = L$. At the closed end, there is rigid boundary i.e., z_2 is infinite. Hence, the particle velocity has a reflection co-efficient –1 at the closed end which means that $A_i = -A_r = +A$ (say). Therefore, the incident wave (travelling in the positive X-direction) and the reflected wave (travelling in the negative x-direction) are given by

$$\xi_i(x,t) = A \sin\{k(vt - x)\}$$

$$\xi_r(x,t) = -A \sin\{k(vt + x - 2L)\} \tag{5.38}$$

The superposition of these oppositely travelling waves give standing waves in pipe, which, as in the case of string are described by

$$\xi(x,t) = 2A\sin\{k(L-x)\}\cos k(vt-L) \tag{5.39}$$

and the excess pressure at x at any time t is given by

$$P = \Delta P = -\gamma P_0 \frac{\partial \xi}{\partial x} = z\gamma P_0 Ak\cos\{k(L-x)\}\{\cos k(vt-L)\} \tag{5.40}$$

The open end at $x = 0$, represents a condition of zero pressure variation during the oscillation. This means that, at the open end, the excess pressure is zero, i.e., the pressure is equal to the equilibrium value P_0. Thus, at $x = 0$, we have $P = 0$ for all time. Putting $x = 0$ and $P = 0$ in Eq. (5.40), we have

$$0 = 2\gamma P_0 Ak\cos kL\cos\{k(v_t - L)\}$$

This is true for all value of t if

$$\cos kL = 0$$

or $$kL = \frac{\pi}{2}, \frac{3\pi}{2}, \frac{5\pi}{2}$$

or $$k_m L = \frac{(2m-1)\pi}{2}$$

or $$k_m = \frac{(2m-1)\pi}{2L} \tag{5.41}$$

or $$\lambda = \frac{4L}{(2m-1)} \tag{5.42}$$

where $m = 1, 2, 3, \dots$

Hence the frequency of the vibration of air column inside the pipe is given by

$$\nu_m = \frac{v}{\lambda_m} = \frac{(2m-1)}{4L}\sqrt{\frac{E}{\rho}} \tag{5.43}$$

where $E = \gamma P_0$ is the modulus of elastically of the gas. The particle displacements in the pipe are obtained by using Eqs (5.42 and 5.38), we have:

$$\xi(x,t) = 2A\sin\left\{\frac{(2m-1)\pi}{2L}(L-x)\right\}\cos\left\{\frac{(2m-1)\pi}{2L}(vt-L)\right\} \tag{5.44}$$

Here, we again observe that the frequencies given by Eq. (5.43) are just the frequencies of the normal modes of a pipe closed at one end (as discussed in chapter 4). We have already shown the first three modes of vibration in chapter 4. In this case, the particle displacements are longitudinal. Hence, a pipe closed at one end has only odd harmonics of frequencies 3, 5, 7, ... times the fundamental frequency (ν_1). All the even harmonics of frequencies $2\nu_1$, $4\nu_1$, etc. are absent. The frequency of the fundamental ($m = 1$) mode is given by

$$\nu_1 = \frac{1}{4L}\sqrt{\frac{E}{\rho}}$$

Standing Wave in an Open Pipe

Here, we will first discuss the mechanism of reflection at an open end of a pipe. Consider a pipe terminated by an opening into a large room. The equilibrium pressure P_0 of the air in the pipe is equal to the pressure of air in the room. At the open end of the pipe, air can freely rush in or out. When a compressional wave reaches the open end, the particles emerge out of the pipe spreading freely in all directions. Consequently, expansion takes place just outside the pipe in all directions tending to bring the pressure of the emerging air equal to the equilibrium pressure P_0. This happens at a certain distance from the opening (called the *end correction*). Right at the opening, the pressure is not exactly equal to P_0. This difference in pressure just inside and outside the pipe gives rise to a force, which causes the expansion of air which spreads out sideways. The spreading of air in the outside region produces rarefaction at the open end and hence to annul this rarefaction air from behind compression rushes forward. This again results in a rarefaction in the back region from where the air has rushed forward and so on. In this way, a wave of rarefaction starts from the open end and travels backward, i.e. in a direction opposite to the wave of compression. Thus, a compression on arrival at the open end of pipe is reflected back into the pipe as a rarefaction. Since the open end represents a place of zero pressure variation (this happens at a certain point just outside the opening), the excess pressure is zero at the opening. This means that the pressure reflection co-efficient is −1 at the open end. It is clear from Eqs (5.34 and 5.36) that the velocity or amplitude reflection co-efficient is +1 at the open end, showing that there is no reversal of an amplitude as a result of reflection at the open end. Thus, a compression is reflected from air open end as a rarefaction and vice versa without any reversal of sign of the amplitude.

We shall now discuss standing waves in a pipe of length L lying along the X-axis with its open ends at $x = 0$ and $x = L$. A sound wave of amplitude A_i and frequency ω ($\omega = kv$) is sent in the positive X-direction. The particle displacements due to these incident waves are given by

$$\xi_i(x, t) = A_i \sin\{k(vt - x)\}$$

The wave reflected at the end $x = L$ is given by

$$\xi_r(x, t) = A_r \sin\{k(vt + x - 2L)\}$$

Since the amplitude reflection co-efficient $\dfrac{A_r}{A_i}$ at the open end is +1, we have $A_r = A_t = A$. Hence, resultant displacement is given by

$$\xi(x, t) = A[\sin\{k(vt - x)\} + \sin\{k(vt + x - 2L)\}]$$
$$= 2A \cos\{k(L - x)\} \sin\{k(vt - L)\}$$

Therefore
$$\frac{\partial \xi}{\partial x} = 2kA \sin\{k(L - x)\} \sin\{k(vt - L)\}$$

The excess pressure is given by

$$P = \gamma P_0 \frac{\partial \xi}{\partial x} = -2\gamma P_0 kA \sin\{k(L - x)\} \sin\{k(vt - L)\}$$

The excess pressure P at the other open end at $x = 0$ must also be zero. Putting $P = 0$ at $x = 0$ gives

$$0 = 2\gamma P_0 kA \sin kL \sin\{k(vt - L)\}$$

which is satisfied for all values of t if

$$\sin kL = 0$$

or
$$k_n L = n\pi \quad (n = 1, 2, 3, ...)$$

The frequency of vibration of air column in the pipe is given by

$$v_n = \frac{n}{2L}\sqrt{\frac{E}{\rho}}$$

and the particle displacements are given by

$$\xi(x,t) = 2A\cos\left(\frac{n\pi x}{L}\right)\sin\left(\frac{n\pi vt}{L}\right)$$

First three modes of vibrations for a pipe open at both ends are discussed in chapter 4. In this case, all the even and odd harmonics are present.

Example 5.7: The vibrations of a string of length 60 cm fixed at both the ends are represented by equation:

$$y = 4\sin\left\{\frac{\pi x}{15}\right\}\cos(96\pi t)$$

where x and y are in centimetres and t in seconds. (a) What is the maximum displacement at $x = 5$ cm? (b) Where are the nodes located along the string? (c) What is the velocity of the particle at $x = 7.5$ cm and $t = 0.25$ s? (d) Write down the equations of the component waves, whose superposition gives the above wave.

Solution: For $x = 5$ cm

$$y = 4\sin\left(\frac{5\pi}{15}\right)\cos(96\pi t)$$

$$= 2\sqrt{3}\cos(96\pi t)$$

So, y will be maximum when $\cos(96\pi t) = +1$

∴ $$y_{max} = 2\sqrt{3}\,\text{cm}$$

(b) At nodes, the amplitude of waves is zero. Hence the position of nodes is given by

$$\sin\left(\frac{\pi x}{15}\right) = 0$$

or $$\frac{\pi x}{15} = n\pi$$

or $$x_n = 15n$$

Hence the position of nodes is given by

$x = 0, 15$ cm, 30 cm, 45 cm, 60 cm   (∵ Length of string is 60 cm.)

(c) As $y = 4\sin\left(\frac{\pi x}{15}\right)\cos(96\pi t)$. Hence particle velocity

$$v = \frac{dy}{dt} = -4(96\pi)\sin\left(\frac{\pi x}{15}\right)\sin(96\pi t)$$

Hence particle velocity at $x = 7.5$ cm and $t = 0.25$ sec is

$$v = -4(96\pi)\sin\left(\frac{\pi \times 7.5}{15}\right)\sin(96\pi \times 0.25)$$

$$= 0$$

Since

$$\sin(96\pi \times 0.25) = \sin(24\pi) = \sin 12 \times 2\pi = \sin 2\pi = 0$$

(d) We know that

$$2\sin A \cos B = \sin(A + B) + \sin(A - B)$$

So

$$y = 4\sin\left(\frac{\pi x}{15}\right)\cos(96\pi t)$$

$$= 2\left[\sin\left(\frac{\pi x}{15} + 96\pi t\right) + \sin\left(\frac{\pi x}{15} - 96\pi t\right)\right]$$

$$= 2\left[\sin\left(96\pi t + \frac{\pi x}{15}\right) - \sin\left(96\pi t - \frac{\pi x}{15}\right)\right]$$

$$= 2\sin\left(96\pi t + \frac{\pi x}{15}\right) - 2\sin\left(96\pi t + \frac{\pi x}{15}\right)$$

Hence the wave is the superposition of two waves

$$y_1 = 2\sin\left\{96\pi t + \frac{\pi x}{15}\right\} \quad \text{and} \quad y_2 = -2\sin\left\{96\pi t - \frac{\pi x}{15}\right\}$$

Example 5.8: A metallic rod of length 1 m is rigidly clamped at its mid-point. Longitudinal stationary waves are set-up in the the rod in such a way that there are two nodes on either side of the mid-point. The amplitude of an anti-node is 2×10^{-6} m. Write the equation of motion at a point 2 cm from the mid-point and those of constituent waves in the rod ($y = 2 \times 10^{11}$ N/m^2 and $\rho = 8 \times 10^3$ kg/m^3).

Solution: In rods like strings, clamped point is a node while free end antidote, so the situation in accordance with given condition is as shown in Fig. 5.18.

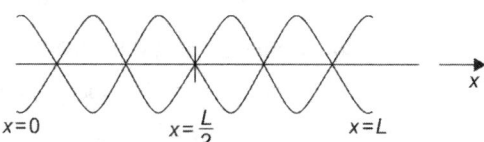

Now, as the distance between two consecutive nodes is $\lambda/2$ while between node and anti-node is $\lambda/4$.

Fig. 5.18

$$\therefore \quad 4\left(\frac{\lambda}{2}\right) + 2\left(\frac{\lambda}{4}\right) = L$$

or $\qquad 2.5\lambda = 1 \quad \text{or} \quad \lambda = \dfrac{1}{2.5} = 0.4 \text{ m}$ \hfill (i)

In rods, the velocity is given by

$$v = \sqrt{\frac{Y}{\rho}} = \sqrt{\frac{2 \times 10^{11}}{8 \times 10^3}} = 5000 \text{ m/s}$$

and frequency $\qquad \nu = \dfrac{v}{\lambda} = \dfrac{5000}{0.4} = 12500 \text{ Hz}$ \hfill (ii)

Now, if incident and reflected waves along the rod are

$$y_1 = A\sin(\omega t - kx) \quad \text{and} \quad y_2 = A\sin(\omega t + kx + \phi)$$

the resultant wave is

$$y = y_1 + y_2 = 2A\cos\left(kx + \frac{\phi}{2}\right)\sin\left(\omega t + \frac{\phi}{2}\right)$$

Now, as free end of the rod is an anti-node, i.e., amplitude is maximum at $x = 0$, so that

$$\cos\left[kx_0 + \frac{\phi}{2}\right] = \text{max} = 1 \quad \text{i.e., } \phi = 0$$

and $\qquad A_{max} = 2A = 2 \times 10^{-6} \quad \text{(given)}$

so $\qquad y = 2 \times 10^{-6} \cos kx \sin \omega t$

$$= 2 \times 10^{-6} \cos\left(\frac{2\pi}{\lambda}x\right)\sin(2\pi vt)$$

Using the values of λ and v, the equation becomes

$$y = 2 \times 10^{-6} \cos\left(\frac{2\pi}{4}x\right)\sin(2\pi \times 12500\,t)$$

Now, for a point 2 cm from the mid-point $x = (0.5 \pm 0.02)$.

Hence, equation is

$$y = 2 \times 10^{-6} \cos 5\pi\,(0.5 \pm 0.02)\sin(25000\,\pi t)$$

This is the required equation.

The equations of the constituent waves are

$$y_1 = 10^{-6}\sin(25000\,\pi t - 5\pi x)$$

and $\qquad y_2 = 10^{-6}\sin(25000\,\pi t + 5\pi x)$

Example 5.9: The following equations represent transverse waves

$$z_1 = A\cos(kx - \omega t), \quad z_2 = A\cos(kx + \omega t),$$

and $\qquad z_3 = A\cos(ky - \omega t).$

Identify the the combination (a) of the waves which will produce standing waves. (b) A wave travelling in the direction making an angle of 45°, with the positive X- and positive Y-axis. In each case, find the positions at which the resultant intensity is always zero.

Solution: (a) Standing waves are produced due to superposition of two waves of equal amplitude and frequency travelling in opposite direction with same speed. Here amplitude, frequency and velocity of all waves are equal but z_1 is travelling along positive X-axis and z_2 along negative X-axis and z_3 along positive Y-axis. Hence superposition of z_1 and z_2 will produce stationary waves, i.e.

$$z = z_1 + z_3 = A\cos(\omega t - kx) + A\cos(\omega t + kx)$$

$$= 2A\cos kx \cos \omega t$$

The amplitude of standing waves is $2A\cos kx$.

Intensity $I \alpha A^2$. Hence intensity is zero means amplitude is zero. Hence the positions where amplitude is zero is given by

$$\cos kx = 0$$

or $\qquad kx = (2n-1)\dfrac{\pi}{2}$

or $\qquad \dfrac{2\pi}{\lambda}x = (2n-1)\dfrac{\pi}{2}$

$\therefore \qquad x = (2n-1)\dfrac{\lambda}{4}$

Hence the intensity is zero at positions

$$x = \dfrac{\lambda}{4}, \dfrac{3\lambda}{4}, \dfrac{5\lambda}{4}, \dots$$

(b) Since the waves z_1 and z_3 are travelling with same speed $v = v\lambda = \dfrac{\omega}{k}$ along positive X- and Y-axes respectively

$$(v)_x = (v)_y \qquad \text{i.e.,} \qquad \dfrac{dx}{dt} = \dfrac{dy}{dt}$$

or $\qquad y = x + c$

So, the superposition of z_1 and z_3 will give rise to a wave propagating along $y = x + c$, i.e., in a direction of 45° inclined to positive x and y-axes. So, the required wave

$$z = z_1 + z_3 = A\cos(\omega t - kx) + A\cos(\omega t - ky)$$

$$= 2A\cos\dfrac{k(x-y)}{2}\cos\left\{\omega t - \dfrac{k(x+y)}{2}\right\}$$

The amplitude of this wave is

$$A_r = 2A\cos k\dfrac{(x-y)}{2}$$

and amplitude will be zero when

$$\cos\dfrac{k(x-y)}{2} = 0$$

or $\qquad \dfrac{k(x-y)}{2} = (2n-1)\dfrac{\pi}{2}$

or $\qquad x - y = (2n-1)\dfrac{\lambda}{4} \times 2$

or $\qquad x - y = (2n-1)\dfrac{\lambda}{2}$

or $\qquad x - y = \dfrac{\lambda}{2}, \dfrac{3\lambda}{2}, \dfrac{5\lambda}{2}, \dots$

Example 5.10: The displacement of the medium in a sound wave is given by the equation $y = A\cos(ax + bt)$, where A, a and b are positive constants. The wave is reflected by an obstacle situated at $x = 0$. The intensity of the reflected wave is 0.64 times that of the incident wave. (a) What is the wavelength and frequency of the incident wave? (b) Write the equation for the reflected wave. (c) In the resultant wave formed after reflection, find the maximum and minimum values of the particle speed in the medium. (d) Express the resultant wave as a superposition of a standing wave and a travelling wave. What are the positions of the antinodes of the standing wave? What is the direction of propagation of the travelling wave?

Solution: (a) Comparing the given equation with standard equation of travelling wave, i.e.

$$y = A \sin(\omega t + kx + \phi)$$

Amplitude = A, angular frequency $\omega = b$, wave vector $k = a$ and phase $\dfrac{\pi}{2}$. Hence frequency

$$v = \frac{\omega}{2\pi} = \frac{b}{2\pi} \text{ and wave vector } k = \frac{2\pi}{\lambda} = a \text{ or wavelength } \lambda = \frac{2\pi}{a}.$$

(b) In case of reflection of displacement wave from hard boundary, there is a phase change of π and as the intensity is 0.64 of the incident wave, hence amplitude of the reflected wave is 0.8 A.

Hence equation of the reflected wave is

$$y_r = 0.8A \cos\{bt - ax + \pi\}$$
$$= -0.8A \cos(bt - ax)$$

(c) In a wave, the velocity of the particle executing SHM and having displacement y is given by

$$v = \omega\sqrt{A^2 - y^2}$$

In standing wave amplitude varies from point to point, so particle velocity will be maximum when amplitude is maximum (i.e. at anti-nodes) and $y = \min = 0$, i.e., particle is passing through its mean position, so that

$$v_{max} = \omega A_{max} = b(A + 0.8A) = 1.8bA *$$

And the particle velocity will be minimum where $y = \max = A$ (whatever it is), i.e., at extreme position and so

$$v_{min} = \omega\sqrt{A^2 - A^2} = 0$$

(d) A stationary wave is the result of superposition of two waves of equal amplitude and frequency travelling in opposite direction at the same speed. Considering the incident wave as superposition of two waves of amplitude 0.8 A and 0.2 A, we get

$$y = 0.8A[\cos(bt + ax) - \cos(bt - ax)] + 0.2A \cos(bt + ax)$$
$$= -1.6A \sin ax \sin bt + 0.2A \cos(bt + ax) = y_{sta} + y_{pro}$$

where $y_{sta} = -1.6A \sin ax \sin bt$
$y_{pro} = 0.2A \cos(bt + ax) **$

Now, as anti-nodes are points of maximum displacement in stationary wave, so anti-nodes are given by

$$\sin ax = \pm 1$$

or $$ax = (2n - 1)\frac{\pi}{2} \quad \text{where } n = 1, 2, 3, ...$$

∴ $$x = \frac{(2n - 1)\pi}{2a}$$

or $$x = \frac{\pi}{2a}, \frac{3\pi}{2a}, \frac{5\pi}{2a}, ...$$

---

* In this problem as amplitude is different at different position so v_{max} is different at different positions being maximum at antinodes where $A = A_{max} = A_1 + A_2$. So, $v_{max} = 1.8 bA$ and min at nodes where $A_{min} = A_1 \sim A_2$ and so $(v_n)_{min} = 0.2 bA$.

** In this problem $A_1 \neq A_2$ so the wave is not perfectly standing, i.e., at nodes amplitude is minimum $A_1 \sim A_2$ but not zero. So, nodes are not permanently at rest and energy will escape through them. This gives SWR = 1.8/0.2 = 9.

As there is positive sign before x in the equation of progressive wave, the wave is travelling along negative X-axis, i.e., in the direction of the incident wave.

Example 5.11: Three strings 1, 2 and 3 are joined together and the composite string is stretched with tension 10 N. The linear densities of string 1 and 3 are 1×10^{-3} kg/m and 4×10^{-3} kg/m respectively. A transverse wave of frequency 100 Hz is produced in string 1. Calculate the linear density and length of the intermediate string 2 so that the wave is completely transmitted through the composite string without any loss due to reflection at the boundaries.

Solution: $T = 10$ N $\mu_1 = 1 \times 10^{-3}$ kg/m and $\mu_3 = 4 \times 10^{-3}$ kg/m. Hence, wave velocities in string 1 and 3 respectively are

$$v_1 = \sqrt{\frac{T}{\mu_1}} = \sqrt{\frac{10}{10^{-3}}} = 100 \text{ m/s}$$

$$v_3 = \sqrt{\frac{T}{\mu_3}} = \sqrt{\frac{10}{4 \times 10^{-3}}} = 50 \text{ m/s}$$

Characteristic impedances of the string is

$$z_1 = \mu_1 v_1, z_2 = \mu_2 v_2 \quad \text{and} \quad z_3 = \mu_3 v_3$$

The condition for impedance matching is

$$z_2^2 = z_1 z_2$$

or $$\mu_2^2 v_2^2 = \mu_1 \mu_2 v_1 v_3$$

or $$\mu_2^2 \cdot \frac{T}{\mu^2} = \mu_1 \mu_3 v_1 v_3 = \mu_1 \mu_3 \sqrt{\frac{T}{\mu_1} \cdot \frac{T}{\mu_3}}$$

$$= \mu_1 \mu_3 \frac{T}{\sqrt{\mu_1 \mu_3}}$$

or $$\mu_2 = \sqrt{\mu_1 \mu_3}$$

$\therefore$ $$\mu_2 = (1 \times 10^{-3} \times 4 \times 10^{-3})^{\frac{1}{2}} = 2 \times 10^{-3} \text{ kg/m}$$

If L is the length of string 2, the second condition for impedance matching is

$$L = \frac{\lambda_2}{2}$$

where λ_2 is the wavelength in string 2. Now, since frequency cannot change on reflection and transmission, we have

$$v_1 = v\lambda_1, \quad v_2 = v\lambda_2$$

$\therefore$ $$\frac{v_1}{v_2} = \frac{\lambda_1}{\lambda_2}$$

But

$$\lambda_1 = \frac{v_1}{v} = \sqrt{\frac{T}{\mu_1}} \cdot \frac{1}{v} = \sqrt{\frac{10}{10^{-3}}} \cdot \frac{1}{10^2} = 1 \text{ m}$$

$\therefore$ $$\frac{v_2}{v_1} = \sqrt{\frac{\mu_1}{\mu_2}} = \frac{1}{\sqrt{2}}$$

$$\therefore \qquad \frac{\lambda_2}{\lambda_1} = \frac{1}{\sqrt{2}} \quad \text{or} \quad \lambda_2 = \frac{\lambda_1}{\sqrt{2}} = \frac{1}{\sqrt{2}} = 0.7 \text{ m}$$

$$\therefore \qquad L = \frac{\lambda_2}{4} = \frac{0.7}{4} = 0.175 \text{ m} = 17.5 \text{ cm}$$

Example 5.12: Show that in waves in a linear bounded medium (stationary waves), the net transmission of energy is zero.

Solution: We know that in stationary waves, the pressure variation P in the medium is given by

$$P = E \cdot \frac{4\pi a}{\lambda} \cos \frac{2\pi x}{\lambda} \cdot \cos \frac{2\pi vt}{\lambda}$$

$$= P_x \cdot \cos \frac{2\pi vt}{\lambda}$$

where $P_x = E \frac{4\pi a}{\lambda} \cdot \cos \frac{2\pi x}{\lambda}$,

E is modulus of elasticity of mediium and is equal to γP_0.

Similarly, particle velocity U_p is given by

$$U_p = \frac{\partial y}{\partial t} = \frac{4av}{\lambda} \cdot \sin \frac{2\pi x}{\lambda} \sin \frac{2\pi vt}{\lambda}$$

$$= U_x \sin \frac{2\pi vt}{\lambda}$$

where v is wave velocity and $U_x = \frac{4av}{\lambda} \sin \frac{2\pi x}{\lambda}$.

Therefore, the energy transmitted (workdone) per unit area is small interval

$$dt = PU_p dt$$

Hence the energy transmitted in period

$$T = \int_0^T PU_p dt$$

$\therefore$ Rate of energy transmission $= \dfrac{1}{T} \displaystyle\int_0^T P_x \cos \frac{(2\pi vt)}{\lambda} U_x \sin \frac{(2\pi vt)}{\lambda} dt$

$$= \frac{P_x U_x}{2T} \int_0^T \sin \frac{(4\pi vt)}{\lambda} dt$$

$$= \frac{P_x U_x}{2T} \left[\frac{-\lambda}{4\pi vt} \cos \frac{4\pi vt}{\lambda} \right]_0^T$$

$$= -\frac{P_x U_x \lambda}{\pi vt} \left[\cos \frac{2\pi vt}{\lambda} - 1 \right] = 0$$

Since $\dfrac{v}{\lambda} = \dfrac{1}{T}$ and $\cos \dfrac{2\pi vt}{\lambda} = \cos 2\pi = 1$

REVIEW QUESTIONS AND PROBLEMS

1. State and explain the principle of superposition of waves and describe various phenomenon arising due to superposition of two waves.
2. Explain the phenomenon of interference in sound waves. Find the resultant of two plane simple harmonic waves of the same period travelling in same direction but differing in phase and amplitude. Hence give the conditions for interference.
3. When two waves interfere, does one wave affect the progress of the other? Is there loss of energy in this progress?
4. Explain why interference fringes of sound waves can be produced by two independent sources but not so in case of optical interference.
5. Explain why interference is not ordinarily observed between sound waves from two violins.
6. What are beats? Explain graphically and mathematically their production and derive an expression for the frequency of beats.
7. Write short notes on:
   (i) Distinction between interference and beats.
   (ii) Applications of beats.
8. Discuss briefly the beats and amplitude modulation.
9. What are combination tones? Give the various theories of their origin and discuss the evidence in support of their existence, objective or otherwise.
10. How can you use the combination tones in tuning two forks to an exact octave?
11. Distinguish between beats and combination tones.
12. What are stationary waves and how are they produced?
13. Distinguish clearly between progressive and stationary waves. Discuss the energy distribution in both types of waves.
14. Explain the phenomenon of stationary waves when two wave trains of exactly similar waves are travelling along the same string in opposite directions. Obtain the expression for the distance between two nodes.
15. Give the theory of the formation of stationary waves. Discuss their formation in a pipe which is closed at one end and in one which is open at both ends.
16. What are nodes and antinodes? How will you demonstrate their existence?
17. Explain the manner in which reflection of a transverse wave occurs, when the wave reaches a fixed point in the string. Show with the help of diagram how stationary waves are formed in string.
18. A transverse wave travels in the positive x-direction in a string of characteristic impedance z_1, and meets the junction of another string of characteristic impedance z_2. The composite string is stretched with tension T. Deduce the expressions for the reflection and transmission co-efficient of amplitude and energy.
19. What is a standing wave? Derive the equation that describes a standing wave on a string of length L fixed rigidly at both ends. State the characteristic features which distinguish a standing wave from a travelling wave.
20. A transverse wave incident on a boundary is reflected with a reduced amplitude. Obtain an expression for this standing wave ratio in terms of the amplitude reflection co-efficient. What is the significance of this ratio?
21. Obtain the amplitude and energy reflection and transmission co-efficients when a plane wave (sound) travelling in a medium of impedance z_1 is incident normally on a plane boundary of medium of impedance z_2.

22. Discuss analytically standing waves in (a) an open pipe, and (b) a closed pipe.

23. Two strings of linear densities μ_1 and μ_2 are joined together and composite string is stretched with certain tension. A transverse wave travelling in the first string is incident on the junction separating the second string. Calculate the fraction of incident amplitude reflected and transmitted at the junction, if $\dfrac{\mu_1}{\mu_2}$ is (a) 9, (b) 0.25, (c) 4, and (d) ∞.

$$\left[Ans.\,(a)\,-1,\,0;\,(b)\,-\frac{1}{3},\frac{2}{3};(c)\,\frac{1}{3},\frac{4}{3};(d)\,1,\,2 \right]$$

24. Calculate the reflection and transmission energy co-efficients in each case in problem 23.

$$\left[Ans.\,(a)\,1,\,0;\,(b)\,-\frac{1}{9},\frac{8}{9};(c)\,\frac{1}{9},\frac{8}{9};(d)\,1,\,0 \right]$$

25. Stationary waves are produced by a superposition of two waves which are

$$y_1 = 0.02\sin\pi(t - 2x) \quad\text{and}\quad y_2 = 0.02\sin\pi(t + 2x)$$

where all the quantities are measures in MKS units.
(a) Determine the amplitude, period, wavelength and velocity of each wave.
   [*Ans.* 0.02 m, 2 s, 1 m, 0.5 m/s]
(b) Prove that the resultant displacement of a particle at x at time t is given by
   $y = 0.04\cos 2\pi x \sin\pi t$.
(c) At what value of x, y is zero for all values of t.

$$\left[Ans.\,x = (2n - 1)\frac{\lambda}{4} \right]$$

(d) What is the difference between the two nearest values of x at which $y = 0$? Is this difference related to the wavelength of the either wave? If so, how?

$$\left[Ans.\,x = \frac{\lambda}{2} = 0.5\text{ m} \right]$$

26. Sound waves produced under water are incident normally on water-air interface. If the speed of sound in water is 1500 m/s and in air 350 m/s, calculate the percentage of incident sound energy transmitted out into air. Density of air 1.3 kg/m³.
   [*Ans.* 0.1% approx.]

27. Impedance matching is achieved between two strings 1 and 3 by inserting between them a string 2 of intermediate impedance. A transverse wave of frequency 100 Hz is generated in string 1. If the speed of waves in the string 1 is 100 ms⁻¹ and in the string 3, the speed is 25 ms⁻¹, what is the speed of the wave in the intermediate string? What is the length of this string?   [*Ans.* v = 50 ms⁻¹ and L = 12.5 cm]

28. A longitudinal wave travelling in a rod encounters a discontinuity where the Young's modulus suddenly doubles, the density remaining the same. Calculate the reflection and transmission amplitude co-efficients.   [*Ans.* –0.17, 0.83]

29. A string vibrates according to the equation $y = 5\sin\dfrac{\pi x}{3}\cos 40\pi t$ where x and y are in centimetres and t in seconds. Find out the amplitude and velocity of the two component waves whose superposition can give rise to this vibration. What is the

distance between adjacent nodes? What is the velocity of a particle of string at

$x = 1.5$ cm at $t = \dfrac{9}{8}$ seconds?

[*Ans.* $a = 2.5$ cm, $\lambda = 6$ cm, $v = 120$ cms^{-1}, $n = 20$ Hz, $d = \dfrac{\lambda}{2} = 3$ cm, $v = 0$]

30. A longitudinal stationary wave in a rod is expressed as
$$y = 0.002 \sin(0.15x) \sin(4.8 \times 10^4 t)$$
Calculate the maximum particle velocity and maximum tensile stress at the point $x = 3.5$ cm. The Young's modulus of the material of the rod is 8.0×10^{10} N/m^2.

[*Ans.* $v_{max} = 48$ m/s, Stress $= 2.1 \times 10^7$ N/m^2]

31. Define a linear bounded medium. Prove that two waves travelling in opposite directions are produced in a linear bounded medium when disturbed. Also prove that such a medium can vibrate with certain discrete frequencies.

32. Show that in a linear bounded medium:
    (i) When both the boundaries are rigid, certain discrete frequencies are possible with all harmonics.
    (ii) When one boundary is rigid and the other is free, only odd harmonics are present.
    (iii) When both the boundaries are free, all harmonics are present.

SHORT ANSWER QUESTIONS

1. A harmonic wave is given by $y = 0.5 \sin 16\pi \left[t - \dfrac{x}{40} \right]$. If for normal incident, its amplitude falls to 60% on reflection from a plane boundary. Write the equations of reflected wave for rigid and soft boundaries.

   Ans. Amplitude of reflected wave is 0.3. For rigid boundary, there is a phase change of π. Hence reflected wave equation $y = -0.3 \sin 16\pi \left[t + \dfrac{x}{40} \right]$ and for non-rigid wall, there is no phase change, hence equation is $y = 0.3 \sin 16\pi \left[t + \dfrac{x}{40} \right]$.

2. Distinguish between beats and interference.

   Ans. In beats at a given position, intensity varies periodically with time with periodicity $\dfrac{1}{f_1 - f_2}$ while in interference at a given time, intensity varies periodically with position with periodicity λ.

3. Which of the following two waves on superposition can give rise to (a) beats, (b) stationary waves, (c) interference.

   $$y_1 = A \cos 2\pi \left[f_1 t + \dfrac{x}{\lambda_1} \right], \quad y_2 = A \cos 2\pi \left[f_1 t + \dfrac{x}{\lambda_1} + \pi \right]$$

   $$y_3 = A \cos 2\pi \left[f_2 t + \dfrac{x}{\lambda_2} \right], \quad y_4 = A \cos 2\pi \left[f_1 t - \dfrac{x}{\lambda_1} \right]$$

   Ans. (a) y_1 and y_3, (b) y_1 and y_4, (c) y_1 and y_2.

4. What factors determine the pitch of a tuning fork?

Ans. For tuning fork $f = \dfrac{d}{L^2}\sqrt{\dfrac{Y}{\rho}}$, d is thickness of prongs in the direction of vibration and L-length of prong. Y is Young's modulus of the material and ρ the density.

5. In case of vibration of two taut strings, strings are identical in all respects with length of one say B, being 4 times that of A. Which frequencies of A will match with that of B if first eight harmonics are considered?

Ans. $n_b = \dfrac{p_b}{8L}\sqrt{\dfrac{T}{\mu}}$ and $n_a = \dfrac{p_a}{2L}\sqrt{\dfrac{T}{\mu}}$ ∴ $p_b = 4p_a$ for $p_a = 1$ and 2, $p_b = 4$ and 8. Hence first and second harmonic of A coincides with 4th and 8th of B.

6. If the fundamental frequency of an organ pipe is 150 Hz, what is the frequency of (a) Second harmonic, if it is closed at one end. (b) Second harmonic, if it is open at both ends. (c) Second overtone, if it is closed at one end. (d) Second overtone, if it is open at both ends?

Ans. (a) In closed pipe even harmonics do not exist. (b) For a pipe open at both ends of length l, the fundamental frequency is $v = \dfrac{v}{2l}$ and second harmonic has frequency $2v = 300$ Hz. (c) Fundamental frequency of closed pipe is $v = \dfrac{v}{4l}$ then its second overtone has frequency $5v = 5 \times 150 = 750$ Hz. (d) Second overtone of open pipe has frequency $v = 3 \times 150 = 450$ Hz.

7. A sound wave of 40 cm wavelength enters the tube as shown in Fig. 5.19. What must be smallest radius r such that minimum will be heard at the detector?

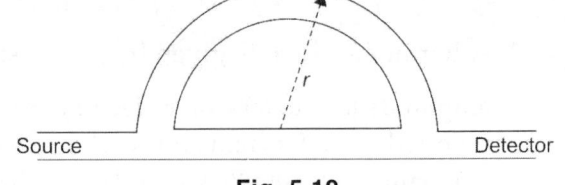

Fig. 5.19

Ans. $\Delta = \pi r - 2r = (2n-1)\dfrac{\lambda}{2}$ for minimum radius $n = 1$

∴ $r = \dfrac{\lambda}{2(\pi-2)} = \dfrac{40}{2(\pi-2)} = 17.5$ cm

8. You are given four tuning forks, the lowest frequency of the fork is 300 Hz. By striking two tuning forks at a time 1, 2, 3, 5, 7 and 8 Hz, beat frequencies are heard. What are the possible frequencies of the other three forks?

Ans. 301, 303, 308 Hz.

9. Two wires of same length but of radii r and $2r$ are welded together end to end. The welded point is midway between two bridges and is under tension T. Find the ratio of the no. of loops formed in the wires such that the welded point is node when stationary waves are produced in the wire.

Ans. $\dfrac{P_1}{r_1} = \dfrac{P_2}{r_2}$ ∴ $\dfrac{P_1}{P_2} = \dfrac{r}{2r} = \dfrac{1}{2}$.

10. The tunnel leading through a hill greatly amplifies tones at 135 Hz and 138 Hz. Find the shortest length of the tuned if velocity of sound in air is 330 m/s.

Ans. The tunnel may be treated as open pipe. Hence

$$\frac{330 \times n}{2l} = 135 \text{ and } \frac{330(n+1)}{2l} = 138 \quad \therefore n = 45$$

and $\qquad l = \dfrac{330 \times 45}{2 \times 135} = 55 \text{ m}$

11. A column of air and a tuning fork produce 4 beats per second when sounded together. Turning fork gives the lower tone. The temperature of air is 15°. When temperature falls to 10°C, the two produce 3 beats per second. Find the frequency of the fork.

Ans. $n_a - n = 4$ and $n_a = \dfrac{v_{15}}{2l}$ $\quad n_a' - n = 3$ and $n_a' = \dfrac{v_{10}}{2l}$

$\therefore \quad n_a - n_a' = 1$ and $\dfrac{n_a}{n_a'} \dfrac{v_{15}}{v_{10}} = \sqrt{\dfrac{288}{283}} = \left(1 + \dfrac{5}{283}\right)^{\frac{1}{2}}$

or $\qquad \dfrac{n_a}{n_a'} = 1 + \dfrac{5}{2 \times 283} \quad \therefore \dfrac{n_1 - n_a'}{n_a'} = \dfrac{5}{2 \times 283}$

or $\qquad n_a' = \dfrac{2 \times 283}{5}, \; n_a' = 113.2 \quad \therefore n = n_a' - 3 = 110.2 \text{ Hz}$

12. The linear density of a vibrating string is 1.3×10^{-4} kg/m. A transverse wave is propagating on the string and is described by equation $y = 0.021 \sin(x + 30t)$ where x and y are in metres and t in seconds. The characteristic impedance offered by the medium to the wave propagation is?

Ans. $z = v\mu, \quad v = \dfrac{\omega}{k} = \dfrac{30}{1} = 30 \text{ m/s}$ and $z = 30 \times 1.3 \times 10^{-4} \text{kgs}^{-1}$,

$z = 39 \times 10^{-4} = 3.9 \times 10^{-3} \text{ kg/s}$

13. A hollow metallic tube of length L and closed at one end produces resonance with a tuning fork of frequency f. The entire tube is heated carefully so that length changes by l. If the change in velocity V of sound is v, calculate the frequency at which new resonance will occur.

Ans. Length becomes $(L + l)$ and velocity $V + v$, hence resonance frequency $= \dfrac{(V + v)}{4(L + l)}$

14. A set of 56 turning forks is arranged in series of increasing frequencies. If each fork gives 4 beats/sec. with the preceding one and the last fork is found to be an octave of the first, find the frequency of the first fork.

Ans. Frequency of the first fork $= n$, the frequency of the 56th fork

$$N = n + 4 \times 55 = n + 220 = 2n \quad \therefore n = 220 \text{ Hz}.$$

15. Transverse waves are generated in two uniform wires A and B of same material by attaching their free ends to a vibrating source of frequency 200 Hz. The cross-sectional area of A is half that of B while the tension on A is twice that on B. Find the ratio of the wavelengths of transverse waves in A and B.

Ans. $\lambda = \dfrac{v}{n} = \dfrac{1}{n}\sqrt{\dfrac{T}{Ad}} \quad \therefore \dfrac{\lambda_a}{\lambda_b} = \sqrt{\dfrac{T_a \cdot A_b}{T_b A_a}} = 2$

MULTIPLE CHOICE QUESTIONS

1. Out of the following equations, the equations which may represent beats, inter-ference and standing waves respectively are:

(i) $2A\cos\left\{\dfrac{\Delta\omega}{2}\right\}\cos(\omega t - kx)$, (ii) $2A\cos kx \sin\omega t$,

(iii) $2A\cos\dfrac{\phi}{2}\sin(\omega t - kx + \theta)$

(a) (i), (ii), (iii)   (b) (i), (iii), (ii)   (c) (iii), (ii), (i)   (d) (ii), (i), (iii)

2. Given below are some functions of x and t which represent the displacement of transverse or longitudinal elastic waves:
(i) $y = 2\cos 3x \sin(10t)$,
(ii) $y = 3\sin(5x - 0.5t) + 4\cos(5x - 0.5t)$,
(iii) $y = 4\cos^3(5x - 0.5t)$

State which of these may represent, harmonic travelling wave, standing wave and complex periodic wave respectively:
(a) (iii), (i), (ii)   (b) (i), (ii), (iii)   (c) (ii), (i), (iii)   (d) (i), (iii), (ii)

3. The maximum possible wavelength of standing waves in 1 m long string, if it is touched in the middle when clamped at both ends (in metres) is:
(a) 2   (b) 4   (c) 0.5   (d) 1

4. The equation $y = A\sin^2(\omega t - kx)$ represents a wave:

(a) with amplitude A and frequency $\dfrac{\omega}{2\pi}$

(b) with amplitude $\dfrac{A}{2}$ and frequency $\dfrac{\omega}{\pi}$

(c) which is a harmonic progressive wave
(d) which is stationary wave

5. In case of superposition of waves at $x = 0$
$$y_1 = 4\sin(1026\pi t) \quad \text{and} \quad y_2 = 2\sin(1014\pi t)$$
The beat frequency observed is:
(a) 4   (b) 2   (c) 6   (d) 8

6. In above question, the ratio of maximum and minimum intensity is:
(a) 2   (b) 4   (c) 3   (d) 9

7. Stationary wave is represented by $y = A\sin(100t)\cos(0.01x)$, where y and A are in millimetres, t in seconds and x in metre. The velocity of the wave is:
(a) 1 m/s   (b) 10^2 m/s   (c) 10^4 m/s   (d) None of the above

8. A wave represented by the equation $y = a\cos(kx - \omega t)$ is superposed with another wave to form stationary wave. Such that the point $x = 0$ is a node. The equation for the other wave is:
(a) $a\sin(kx + \omega t)$ (b) $-a\cos(kx + \omega t)$ (c) $-a\cos(kx - \omega t)$ (d) $-a\sin(kx - \omega t)$

9. Sound waves from a vibrating body reach a point by two paths, where path differs by 8 cm or 24 cm. There is silence at the point. The frequency of the body if the velocity of sound in air is 332 m/s is:
(a) 2075 Hz   (b) 1037.5 Hz   (c) 3112.5 Hz   (d) 4150 Hz

10. A cylindrical tube open at both ends has a fundamental frequency f in air. The tube is dipped vertically in water so that half of it is in water. The fundamental frequency of air column now is:

(a) $\dfrac{f}{2}$ (b) $\dfrac{3f}{4}$ (c) f (d) $3f$

ANSWERS

1. (b)   2. (c)   3. (a)   4. (b)   5. (c)   6. (d)   7. (d)   8. (b)   9. (a)   10. (c)

6

Acoustics

6.1 INTRODUCTION

The branch of physics that deals with the process of generation, reception and propagation of sound is termed *acoustics*. Truly speaking, this branch covers many fields and is closely related to various branches of engineering, e.g. (i) design of acoustical instruments (ii) electroacoustics, viz., the branch relating to the methods of sound production and recording (microphones, amplifiers, loudspeakers, etc.), (iii) architectural acoustics dealing with the design and construction of buildings, operas, music halls, recording rooms in radio and television broadcasting stations (in general architectural acoustics deals with the behaviour of sound waves in a closed space) and (iv) musical acoustics dealing with the design of musical instruments, etc.

The subject has developed to such an extent that it is classified as *acoustical engineering*.

6.2 ARCHITECTURAL ACOUSTICS

In a good auditorium particularly theatres, concert halls, classrooms, etc. sound produced by the speaker or by a source should be heard with sufficient loudness and clarity. It is found that some auditoriums are acoustically good and some bad, i.e. in some auditoriums the sounds produced are distinctly audible while in others the sounds lack in distinctness. In a good auditorium the following conditions should be satisfied:
  i. The sound heard must everywhere be sufficiently loud and no echoes should be present.
 ii. The 'quality' of sound, i.e. speech and music must remain unaltered, i.e. the relative intensity of several components of a complex sound must be maintained.
iii. The successive syllables spoken must be clear and distinct. Each syllable should die away sufficiently quickly to avoid overlapping with the next syllable.
 iv. There should be no undesirable echoes. Reverbation should be quite proper.
  v. There should be neither any concentration of sound nor any zone of silence in any part of the hall.
 vi. There should be no undue noise and no resonance within the hall or building.
vii. There should be no echelon effect.

To achieve this, various factors need attention especially, the one known as *reverbation*.

6.3 REVERBATION

A common defect in halls, large rooms and auditoriums is the undue persistence or prolongation of the sound of the speaker or the singer. This arises due to the successive

304

reflections of the sound from the walls, ceiling, floors, etc. of the hall or room. A short sound made in a hall reaches a listener directly as well as after successive reflections from the walls, ceiling and floor of the hall. The listener, therefore, receives a series of sounds of diminishing intensity (since part of energy is lost at each reflection) and, instead of single sound, he/she hears a roll of sound. This prolongation or persistence of sound is termed *reverbation*. Reverbation can also be defined as the persistence of audible sound after the source has stopped to emit any sound.

Similarly, when a continuous source of sound starts sounding, some of the sound waves reach the listener directly, while the other reach successively later on after reflection and add to the direct waves still reaching there. During this time the energy is partly absorbed by walls and other materials in the room and partly escape through the open windows and ventilators, and after a few seconds a balance is reached between the energy emitted from the source and the energy lost. Then the intensity of sound attains a steady value. If now the source of sound is stopped, the sound does not stop instantaneously but the human ear continues to pick up the successive reflections until they fall below the minimum audibility. We may note that louder the original sound, the longer will this process take. Again this gradual decay of sound is known as *reverbation*.

Figure 6.1a shows the rise and fall of sound. When a succession of different notes are sounded the effect is as shown in Fig. 6.1b.

The time gap between the initial direct note and the reflected note upon the minimum audibility level is called *reverbation time*. In other words, we can say that the time taken for the sound to fall below the minimum audibility, measured from the instant of its generation (in the case of short sounds) or the stopping of the source (in the case of continuous note) is termed *time of reverbation*.

In a hall or auditorium if the reverbation time is too large, then there is overlapping of successive sounds which causes confusion and results in loss of clarity in hearing. On the other hand, if the reverbation time is too small, then the loudness will be inadequate. Obviously, the reverbation time for a hall or auditorium should neither be too large nor too small. Obviously, the reverbation time must have a definite value which may be satisfactory for the speakers and audience. Thus, an adjustment of the appropriate time of reverbation time is essential requirement of good acoustics. The perfect time for reverbation is termed *optimum reverberation time*.

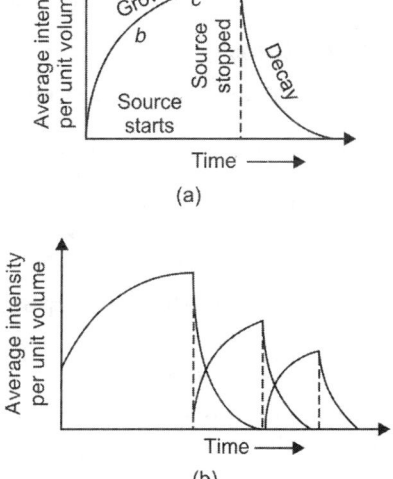

Fig. 6.1. Growth and decay of sound

Prof WC Sabine made the extensive study of the problem of reverbation. He found that there is an optimum period of reverbation depending upon the size of the hall at which best results are obtained. This period is found greater for music than for speech. For example, in a hall of 400,000 cubic feet the maximum intelligibility of speech is obtained for a reverbation period between 1.0 s and 1.5 s, while the music is not appealing between 1.5 s and 2 s. Below this value the intensity of sound is weak and the music appears lifeless, while above this the syllables overlap and become indistinct.

Derivation for Reverbation Time

Sabine developed the reverbation time formula to express the rise and fall of sound in an auditorium. He observed that the reverbation time depends upon the size of the hall and the absorbing power of the objects in it and its walls. His formula for reverbation time is based on the assumption that the sound energy in the room is distributed uniformly all around. While deriving Sabine's formula, following steps have to be considered:

i. To calculate, in terms of energy density E, the rate at which the energy is incident upon the walls and other surfaces of the hall or auditorium and hence the rates at which it is being absorbed.

ii. To calculate the final steady value of energy density E in terms of the rate of emission of power P of the sound emitting source.

iii. To calculate the standard reverbation time T.

iv. Sabine by conducting numerous experiments involving the measurement of reverberation time found that the reverberation time T was:

    a. directly proportional to the volume V of the room, and

    b. inversely proportional to the sound absorption, i.e.

$$T \propto \frac{V}{A} \quad \text{or} \quad T = \frac{kV}{A}, \text{ where } k = 0.161, \text{ where } V \text{ is the m}^3 \text{ and } A \text{ is in m}^2.$$

Now, we proceed to derive the relation.

Energy Density

Let us consider that ds is a small element of plane wall AB, then the rate of energy received from the source at ds is obtained as $Eds \cdot \dfrac{C}{4}$ where C is the velocity of sound. Now, if α is the co-efficient of absorption of the wall AB, then, one finds

Rate of energy absorbed by $ds = EC\alpha \dfrac{ds}{4}$.

Obviously, the total absorption of all the surfaces of the wall where the sound is falling, obtained as

$$EC(\Sigma\alpha ds)/4 = ECA/4 \tag{6.1}$$

Here $A = \Sigma\alpha ds$, the total absorption of sound by all the surfaces of the wall where the sound falls.

Final Steady Value of E in Terms of P

If the rate of emission of energy from the source, i.e. the power output is P and V is the total volume of the room, then the total energy in the room at the instant when energy is E will be EV. Thus, we have

$$\text{Rate of growth of energy } = \frac{d}{dt}(EV) = \frac{VdE}{dt}$$

Now, rate of growth of energy in space = rate of supply of energy by source − rate of absorption by all the surfaces,

i.e.
$$\frac{VdE}{dt} = P - ECA/4 \tag{6.2}$$

Now, when steady state is reached,

$$\frac{dE}{dt} = 0$$

For steady energy E_m, we have
$$E_m = 4P/CA \qquad (6.3)$$

Therefore, Eq. (6.2) can be written as
$$\frac{dE}{dt} + \frac{ECA}{4V} = \frac{P}{V}$$

or
$$\frac{dE}{dt} + \alpha E = \frac{4P}{CA} \alpha \qquad (6.4)$$

where
$$\alpha = CA/4V \qquad (6.5)$$

Multiplying throughout by $\exp(\alpha t)$

$$\left(\frac{dE}{dt} + \alpha E \right) e^{\alpha t} = \frac{4P\alpha}{CA} e^{\alpha t}$$

or
$$\frac{d(Ee^{\alpha t})}{dt} = \frac{4P\alpha}{CA} e^{\alpha t}$$

Integrating, we get

$$Ee^{\alpha t} = \int \frac{4P\alpha}{CA} e^{\alpha t} = \frac{4P}{CA} + K \qquad (6.6a)$$

where k is constant of integration

(i) At $t = 0, E = 0$

Hence
$$K = -\frac{4P}{CA} \qquad (6.6b)$$

Now, we study the growth and decay of sound energy density in a hall or auditorium.

Growth

Substituting the value of K in Eq. (6.6b) in Eq. (6.6a)
$$E = 4P/CA(1 - e^{-\alpha t}), \text{ where } \alpha = CA/4P$$

At $t = \infty$, $\quad E = 4P/CA = E_m \quad (\because e^{-\infty} = 0)$
$$E = E_m(1 - e^{-\alpha t}) \qquad (6.7)$$

Figure 6.2 shows the exponential growth of energy with time t in accordance with Eq. (6.7). We note that E ultimately attains the value E_m.

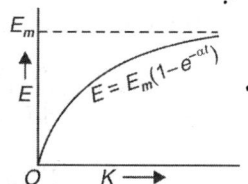

Fig. 6.2: Growth of sound with time in a hall

The sound gets reflected, absorbed and transmitted in the surroundings after it is transmitted from the speaker. When the absorption and transmission of sound is less, then most of the sound energy gets reflected back. In such a situation, the total energy in the hall will *increase linearly and attain the ultimate energy* in a very short time. However, due to the presence of absorbing materials in the hall sufficient absorption takes place and ultimately E attains the maximum value at $t = \infty$.

Decay

Suppose the source is cut-off when $E = E_m$, then $P = 0$ at $t = 0$, and now the decay of sound energy will take place. Here $P = 0$ at $t = 0$ and $E = E_m$. Thus, from Eq. (6.6), one obtains $E_m = K$, i.e.

$$E = E_m \exp(-\alpha t) \qquad (6.8)$$

Equation (6.8) reveals clearly the decay of sound energy density with time after the source is cut-off. Figure 6.3 shows the exponential decay of sound energy. Since $P = 0$, i.e. there is no transmission of power, so the sound will decay exponentially with time from E_m to O.

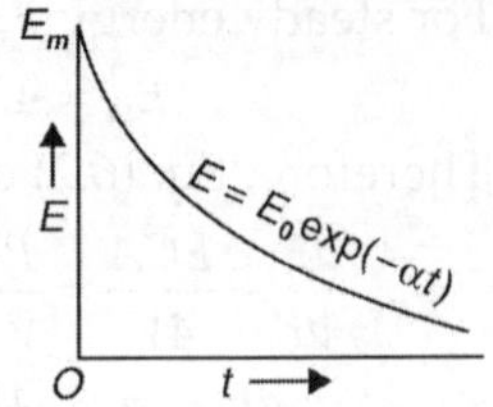

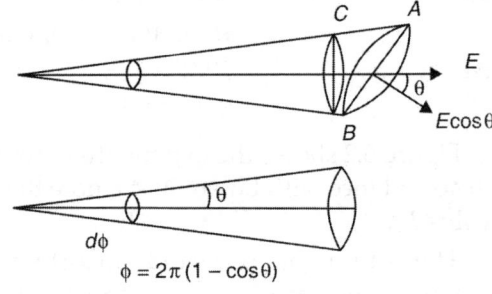

Fig. 6.3: Exponential decay of sound energy with time

Sabine's Reverbation Time Formula

Sabine developed the reverbation time formula to express the rise and fall of sound in an auditorium. The derivation is based upon the following main assumptions:

i. The average energy per unit volume is uniform. It is represented as σ.

ii. The energy is not lost in the auditorium. The energy is lost only due to the absorption of the material of the walls and ceiling. Energy is also lost due to escape through the window and ventilators. Both these factors are included in the *absorption* of energy.

Let us suppose that a source is producing sound continuously. This sound energy is propagated in all directions. Let σ represent the energy contained in a unit volume. Now, the energy contained in a *solid angle* $d\phi = \dfrac{\sigma d\phi}{4\pi}$.

Now, suppose that this energy be incident on a unit surface area of the wall at an angle θ. Let the velocity of sound be C, then the total energy falling per second on a unit surface area of the wall $= \left(\dfrac{\sigma d\phi}{4\pi}\right)(\cos\theta)C.$

Now, the total energy falling per second within a hemisphere $= \dfrac{\sigma C}{4\pi}\displaystyle\int \cos\theta\, d\phi.$

But $\quad \phi = 2\pi(1 - \cos\theta)$ [*see* Fig. 6.4]

or $\quad d\phi = 2\pi\sin\theta\, d\theta$

Now, substituting this value of $d\phi$, one obtains the total energy falling per second within a hemisphere

$$= \frac{\sigma C}{4\pi}\int_0^{\pi/2} 2\pi\sin\theta\cos\theta\, d\theta$$

$$= \frac{\sigma C}{2}\left[-\frac{\cos^2\theta}{2}\right]_0^{\pi/2} = \frac{\sigma C}{4}$$

$\phi = 2\pi(1 - \cos\theta)$

Fig. 6.4: Solid angle

Let us suppose that α is the absorption coefficient of the walls that refers to the fraction of the incident energy not reflected from the walls. The amount of energy absorbed per second per unit area $= \dfrac{\alpha\sigma C}{4}$. If A is the area of the walls and other absorbing materials including windows, ventilators, ceiling, etc. in a hall, the amount of energy absorbed per second $= \dfrac{A\alpha\sigma C}{4}$.

Let volume of the auditorium be V, then the total energy $= \sigma V$. The rate of increase of energy $= \dfrac{d}{dt}(V\sigma) = V\dfrac{d\sigma}{dt}.$

Let us suppose that the source supplies energy at the rate of Q units per second. Thus, the rate of increase of energy

$$= Q - \frac{A\alpha\sigma C}{4} \qquad (6.9)$$

Equating Eq. (6.8) and Eq. (6.9), one obtains

$$V\frac{d\sigma}{dt} = Q - \frac{A\alpha\sigma C}{4} \qquad (6.10)$$

Taking $\qquad \frac{A\alpha C}{4} = K,$

$$\frac{K}{V} = \beta \quad \text{and} \quad \frac{Q}{K} = \frac{4Q}{A\alpha C}$$

we obtain from Eq. (6.10)

$$V\frac{d\sigma}{dt} = Q - K\sigma$$

or $\qquad \dfrac{d\sigma}{dt} = \dfrac{Q}{V} - \dfrac{K}{V}\sigma \qquad (6.11)$

The general solution of Eq. (6.11) is

$$\sigma = B + be^{-\beta t} \qquad (6.12)$$

where $\qquad t = 0, \sigma = 0$

Therefore, from Eq. (6.12), we have

$$0 = B + b$$

or $\qquad b = -B$

$\therefore \qquad \sigma = B - Be^{-\beta t}$

$$= B[1 - e^{-\beta t}]$$

Now , substituting the values of B and β, we have

$$\sigma = \frac{4Q}{A\alpha C}\left[1 - e^{-\frac{A\alpha C}{4V}t}\right] \qquad (6.13)$$

Equation (6.13) represents the rise of average sound energy per unit time from the time the source commences to produce sound. The maximum value of average energy per unit volume is obtained as

$$\sigma_{max} = \frac{4Q}{A\alpha C} \qquad (6.14)$$

Similarly, after the source ceases to emit sound, the decay of the average energy per unit volume is obtained as

$$\sigma = \frac{4Q}{A\alpha C}\exp\left[-\frac{A\alpha C}{4V}t\right] \qquad (6.15)$$

$$\sigma = \sigma_{max}\exp\left[-\frac{A\alpha C}{4V}t\right] \qquad (6.16)$$

The factor $\dfrac{A\alpha C}{4V}$ gives the *reverbation time* in the hall or auditorium. If σ_0 represents the minimum audible intensity after a time t_1, then one obtains from Eq. (6.16)

$$\sigma_0 = \sigma_{max} \exp\left[-\frac{A\alpha C}{4V}t_1\right] \tag{6.17}$$

where t_1 is the time interval between the cutting off the sound and the time at which intensity falls below the minimum audible level. From Eq. (6.17), we have

$$\sigma_{max} = \sigma_0 \exp\left[\frac{A\alpha C}{4V}t_1\right]$$

Taking logarithms, one obtains

$$\log_e\left(\frac{\sigma_{max}}{\sigma_0}\right) = \frac{A\alpha C}{4V}t_1 \tag{6.18}$$

We may note that here α and σ_0 change with the frequency of sound.

Now, for calculating the reverbation time, a standard steady intensity is required. Sabine took the value $\dfrac{\sigma_{max}}{\sigma_0} = 10^6$. From Eq. (6.18), we have

$$\log_e\left(10^6\right) = \frac{A\alpha C}{4V}t_1$$

or $\qquad 2.303 \times 6 = \dfrac{A\alpha C}{4V}t_1$

Now, taking the velocity of sound approximately at room temperature as 350 m/s, we have

$$2.303 \times 6 = \frac{A\alpha \times 350}{4V}t_1$$

or $\qquad t_1 = \dfrac{2.303 \times 24V}{350 A\alpha}$

$$= \frac{0.158V}{A\alpha} \tag{6.19}$$

In general,

$$T = t_1 = \frac{0.158V}{\sum A\alpha} \tag{6.20}$$

We may note that the quantities appearing in Eq. (6.20) are represented in MKS units.

Equation (6.20) is *Sabine's equation* for reverbation time. Equation (6.20) is in general good agreement with experimental value obtained by Sabine.

According to Eq. (6.19), the reverbation time T or t_1 is (i) directly proportional to the volume of the hall or auditorium, (ii) inversely proportional to the area of ceiling walls, etc. and (iii) inversely proportional to the total absorption plus transmission through open surfaces. Experimentally it is found that the reverbation time of 1.03 s is most suitable for all rooms having approximately a volume of less than 350 cubic metres.

In order to decrease the reverbation time, the walls of the auditorium are usually covered with material having large absorption co-efficients. In good cinema halls, the area of the surfaces of the walls is increased to decrease the reverbation time.

Limitations of Sabine's Formula

i. Sabine's reverbation time formula is a special case of more general formula put forwarded by CF Eyring. We may note that Sabine's formula gives contradictory result in the case of a dead room. In the case of complete absorption, we have $a = 1$ and reverbation time T or t_1 should be 0, but according to Sabine's formula, we have $T = 0.161\,V/A$, where A is area in m^2.

ii. Sabine's formula does not give correct result for absorption co-efficient more than 0.2.

iii. In the derivation of Sabine's formula, it is assumed that there is uniform energy density and there is no loss of energy in the air.

However, this is not a practical reality.

Eyring's Formula

Eyring obtained a reverbation time formula on the assumption that sound images could be used to replace the walls of a room in calculating the rate of decay of sound energy after the sound source is cut-off. Eyring's formula reads as

$$T = \frac{KV}{-S\log_e(1-\bar{\alpha})}$$

where K is a constant, S is the total area of the absorbing surface, and $\bar{\alpha}$ is the average absorption co-efficient, defined as

$$\bar{\alpha} = \frac{\sum(\alpha\delta S)}{S} = \frac{A}{S}$$

Eyring's formula is applicable to *live* as well as *dead* halls. For a perfectly dead hall (open space), $\bar{\alpha} = 1$. Then the Eyring's formula yields

$$T = \frac{KV}{-S\log_e 0} = \frac{KV}{\infty} = 0$$

as expected.

For a *live* room or hall, the absorption co-efficient $\bar{a}$ is very small. Then one can write $\log_e(1-\bar{\alpha}) = -\bar{\alpha}$. Now, Eyring's formula takes the form

$$T = \frac{KV}{S\bar{\alpha}}$$

which is same as Sabine's relation. Obviously, Sabine's relation is a special case of Eyring's formula.

6.4 ABSORPTION CO-EFFICIENT

The absorption co-efficient (α) of a material is generally defined as the ratio of the sound energy absorbed by the surface to that of the total incident sound energy on the surface of the material, i.e.

$$\alpha = \frac{\text{Sound energy absorbed by the surface of the material}}{\text{Total sound energy incident on the surface}}$$

To compare the relative efficiency of different absorbing materials, it is essential to first select some standard of absorption in terms of which one can assess all the substances. Sabine selected a unit area of open window, as standard of absorption, because all the sound falling on the open window passes out and none is reflected back. Obviously, open window is ideal, i.e. perfect absorber of sound.

Thus, one can define the *absorption co-efficient (α) of a material as the reciprocal of its area which absorbs the same sound energy as absorbed by an unit area of an open window*. The unit of co-efficient of absorption is Sabine or Open Window Unit (OWU). For example, suppose 8 m² of certain carpet absorbs the same amount of energy, as absorbed by 1 m² of an open window. The co-efficient of absorption of carpet is $1/8 = 0.125$. We may note that the absorption co-efficient of a given material is different at different frequencies. Usually, it is higher at higher frequencies. The absorption co-efficient of some common materials calculated with a source of frequency 512 Hz are given in Table 6.1.

Table 6.1: Absorption co-efficient (α) of some common materials calculated with a source of frequency 512 Hz

Material	Absorption co-efficient (α)	Material	Absorption co-efficient (α)
Marble	0.01	Acoustic plaster	0.30
Glass	0.027	Acoustic felt	0.45
Common plasters	0.03	Fibre board	0.50
Concrete	0.17	Heavy curtains	0.50
Cork	0.23	Hair or Felt	0.58
Asbestos	0.26	Fibre glass	0.75
Carpet	0.30	Perforated cellulose fiber tiles	0.85

Measurement of Absorption Co-efficient

There are different methods for the determination of absorption co-efficient of materials as mentioned below.

Single Source Method

One can measure the absorption coefficient in terms of reverbation time. The reverbation time is first measured when the absorbing material is not inside the hall. Let its value of reverbation time be T_1. Thus,

$$1/T_1 = A/0.161V = \Sigma \alpha s/0.161V \tag{6.21}$$

where V is the volume of the hall, Σs is the inside surface area and $\bar{\alpha}$ is the average absorption co-efficient of reverbation chamber. Now, when an absorber is present inside the hall, one can measure absorbtion co-efficient in the following two ways:

1. *Absorber suspended inside the hall*: Let us consider an absorbing material, e.g. a stage screen, which is suspended inside the room. The reverbation time T_2 is now measured. We have

$$\frac{1}{T_2} = \frac{\Sigma \alpha s + 2\alpha_2 S_2}{0.161V} \tag{6.22}$$

where α_2 is the absorption co-efficient of the material of area S_2. This material is suspended inside the hall.

Now, since the absorbing material is suspended inside the hall and hence absorption will take place from both the sides of the screen. Obviously, the absorption by the material in this case $= 2\alpha_2 S_2$. Subtracting Eq. (6.21) from Eq. (6.22), we get

$$2\alpha_2 S_2 = 0.161V\left(\frac{1}{T_2} - \frac{1}{T_1}\right)$$

or $\qquad \alpha_2 = \frac{0.161V}{2S_2}\left(\frac{1}{T_2} - \frac{1}{T_1}\right)$ $\qquad\qquad$ (6.23)

From Eq. (6.23), knowing the values of T_1, T_2, S_2 and V, the value of a_2 can be obtained.

2. *Absorber spread on the floor or on the wall*: Let us consider that the absorber is spread on the floor of reverbation chamber, e.g. carpet or it may be fixed on the wall, e.g. window glass, curtain cloth. For such type of absorbing materials, absorption co-efficient can be determined in the following manner. Let reverbation time with such a material be T_3, whose area is S_3 and absorption co-efficient a_3, then we have

$$\frac{1}{T_3} = \frac{\sum \alpha s + \alpha_3 S_3 - \alpha S_3}{0.161V}$$ $\qquad\qquad$ (6.24)

Here, we have subtracted αS_3 because now the surface area S_3 of the floor or wall does not contribute to absorption which is included in $\sum \alpha s$, the total absorption by the chamber, and here it is $\alpha_3 S_3$ and not twice of that, because the absorption takes place from one side only.

Now, subtracting Eq. (6.21) from Eq. (6.24), we obtain

$$\alpha_3 = \frac{0.161V}{S_3}\left(\frac{1}{T_3} - \frac{1}{T_1}\right) + \alpha$$ $\qquad\qquad$ (6.25)

Knowing other quantities, one can determine absorption co-efficient α_3 of window glass or carpet.

Double Source Method

This method of measurement of absorption co-efficient requires two sound sources of powers P_1 and P_2. Let the reverbation times of two sound sources be T_1 and T_2. Actual powers of two sources are not necessarily to be known but the knowledge of their ratio will be sufficient. The loudspeaker with measured audio frequency current are used as the sources.

The steady energy densities maintained by two sound sources are

$$E_1 = \frac{4P_1}{AC} \quad \text{and} \quad E_2 = \frac{4P_2}{AC}$$

Let us consider that during the time of decay, they reach a value of bare inaudibility E_0 in times T_1 and T_2. We have

$$E_0 = \frac{4P_1}{AC}e^{-\alpha T_1} \quad \text{and} \quad E_0 = \frac{4P_2}{AC}e^{-\alpha T_2}$$

$\therefore \qquad \dfrac{P_2}{P_1} = e^{\alpha(T_1 - T_2)} \quad \text{where } \alpha = \dfrac{CA}{4V}$

Thus, $\dfrac{CA}{4V}(T_1 - T_2) = \log_e \dfrac{P_2}{P_1}$

or
$$A = \frac{4V \log_e (P_2/P_1)}{C(T_1 - T_2)} = \sum \alpha s$$

where A is the total absorption by the areas of the hall, and is given by $\sum \alpha s$.

The average absorption co-efficient of the hall is

$$A = \sum \alpha s = \frac{4V \log_e (P_2/P_1)}{CS(T_1 - T_2)} \qquad (6.26)$$

where S is the total internal surface area of the hall or auditorium.

Knowing the various quantities appearing on the RHS of Eq. (6.26), one can easily find out the average absorption coefficient A.

The experiment is carried out with empty chamber and then with various materials of different areas under test.

6.5 ACOUSTIC DESIGN

While constructing an acoustically good auditorium or hall a special consideration has to be made about reverbation time, absorption and reflections taking place inside. An acoustically good auditorium means, an auditorium in which every syllable or musical note reaches an audible level of loudness, at every point of the auditorium, and then quickly dies away to make room for the next group of notes or syllables. It is found that an acoustically good hall or auditorium should satisfy the following conditions:

    i. Adequate loudness,
    ii. Uniform distribution of sound, i.e. absence of echoes and focussing of sound,
    iii. The hall must be non-resonant,
    iv. Optimum reverbation,
    v. Exclusion of extraneous sound or noise,
    vi. Echelon effect.

Departure from these conditions makes the auditorium or hall defective. We shall now deal about these conditions.

Reverbation

When reverbation is large, there is overlapping of successive sounds, which results in loss of clarity in hearing in an auditorium. On the other side, if the reverbation is very small, the loudness is inadequate. This means that the time of reverbation for a hall or auditorium should neither be too large nor two small. Sabine's relation for standard time of reverbation is

$$T = 0.161V/A = 0.161V/\sum \alpha s$$

where A is the total absorption of the auditorium, V is the volume, $\sum \alpha$ is the average absorption co-efficient and s is the internal surface area of the hall.

On the basis of experimental observations it is found that rever-bation time depends upon the size of the hall, loudness of sound, and kind of music for which hall is used. The best reverbation time for a frequency of 512 Hz, is controlled by the following factors:

    i. To make the volume of reverbation time optimum, provision of windows and ventilators can be made and they can be opened or closed as per requirement,
    ii. Making use of heavy curtains with folds,
    iii. Walls of the auditorium can be decorated with pictures and maps,
    iv. Providing acoustic tiles and carpeting the floors in the auditorium,

v. Making full capacity of audience within the hall,

vi. The clothing worn by the persons in an auditorium also provides sound absorption. Obviously, the period of reverberation varies according to the number of persons present within the hall. To off set this variation, one can provide seats with cushions which themselves have a sufficiently large absorbent effect to make up when the seats are unoccupied.

Adequate Loudness

In an auditorium the sounds which reach an audience must be sufficiently loud. One can achieve this by placing large wooden boards near and behind the speaker. Then the sounds reflected from them reach the listener within 1/20 second of the direct sound and increase loudness without *creating confusion*. We may note that a hard plane wall behind the speaker and facing the audience in a hall also serves a very good reflecting surface. A *low* flat ceiling also plays a useful part in reflecting sound towards the audience. Large polished wooden reflecting surfaces immediately above the speakers in a hall are also quite helpful. If loud speakers are used, they should be fitted above the heads of the listeners and be pointed slightly downward.

Shape and Size of the Auditorium

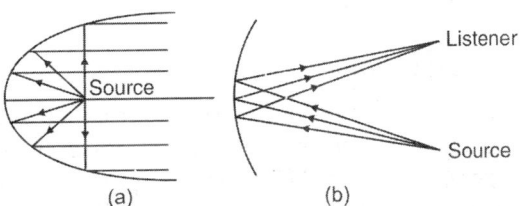

There should be uniform distribution of the voice of the speaker in an entire hall or auditorium. For this purpose, the walls and ceiling of the hall should not have curved surfaces because they focus the sound in some parts, while in some other parts no sound reaches at all. Normally, reflection from the plane surfaces is quite helpful in increasing the loudness of sound and for distributing it uniformly, whereas the curved

Fig. 6.5: (a) Parabolic reflection showing the uniform distribution of sound. (b) Reflection at the concave surface showing the undesirable concentration of sound.

surfaces can be troublesome. It is found that for a uniform distribution of sound intensity, it is better to have parabolic reflector and the use of concave, spherical or cylindrical reflectors should be discouraged as such types of surfaces give rise to undesirable focussing effects (Fig. 6.5). A paraboloidal ceiling (platform at the focus) is very useful in sending a uniform reflected beam of sound in the hall (Fig. 6.6).

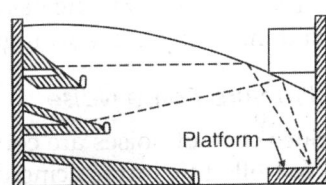

Fig. 6.6: Focussing of sound in a hall.

It is found that if the radius of curvature of the ceiling of the hall is made twice the height of the room, the bad focussing of the sound can be removed. In the case of halls of larger dimensions, the echoes due to the reflections from the back of the speaker can be minimised by the use of suitable absorbents. We may note that large size of hall results in the increase in reverbation time and therefore unnecessary large size should be avoided.

Absence of Echoes

It is found that the disturbing echoes are produced in a hall when the time interval between the sound received by the direct path and that by reflection from some walls or ceilings exceed about 1/10 second. It has been estimated that an interval of 1/20 second between direct and reflected sounds is the extreme limit of tolerance for speech and 1/15 second for music. If the reflected sound in a hall is arriving earlier than that, it helps in arising the loudness, while those arriving later produce echoes, and cause confusion. Echoes may be avoided by covering distant walls and high ceilings with absorbent materials.

Absence of Echelon Effect

It is found that a set of railings or staircase or any regular spacing of reflected surfaces may produce a musical note due to regular succession of echoes of the original sound to observer (Fig. 6.7). This effect is called *echelon effect* and this makes original sound to appear confused or unintelligible. Obviously, such types of surfaces should be avoided or be properly covered with heavy stair carpet.

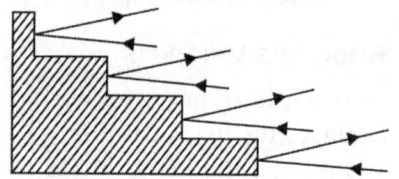

Fig. 6.7: Echelon effect

Freedom or Insulation from Noise

For good hearing, it is essential to take steps to guard against any noise reaching in the auditorium from outside or made by any instruments inside it. There are two types of noises (1) airborne noise (2) structure borne noise.

Airborne Noise

The noises travelling through air come in from outside through open windows, doors, etc. One can reduce this type of noise using double doors and *double or triple windows in separate windows*. Noise entering the hall through the ventilating ducts can be minimised by packing the ducts with hair felt baffles or bale of metal gauge. The noise travelling through the walls of the auditorium can be reduced by making the walls of layers of different materials, or by using double walls with air space between them. We may note that when sound passes from one medium to another, there is always a reflection of certain amount of energy that takes place at the boundary. One can also minimise the airborne noise by using heavy glasses in doors, windows and ventilators in the auditorium and making perfect arrangement of the auditorium.

Structure Borne Noise

Such type of noises are conveyed through the structure of the auditorium. They can be controlled by introducing discontinuity in the path of sound, e.g. noise from water pipe can be easily controlled by using rubber at the functions. Such type of noises can also be controlled by using double walls with an airspace between them.

Inside Noise

Noise is also produced inside the auditorium by machinery, typewriters, etc. One can reduce this type of noise by proper lubrication of machines, hanging curtains of absorbent material near the machines, and placing the typewriters on rubber pads or compressed cork so that the variations are prevented from being transmitted through the floor and walls.

Freedom from Resonance

Sometimes powerful tones of right frequency produced in the auditorium set the windowpanes, sections of wooden portions and the walls lacking in rigidity, or volume of air contained in small rooms in resonant vibrations causing unpleasant results. Resonant frequency of a big hall or auditorium is normally well below the audible range because it is proportional to $1/\sqrt{\text{volume}}$. Such resonant vibrations should be suitably damped. We may note that sometimes resonance can be useful. In order to improve the loudness of the hall, there are some halls famous for their acoustic properties which have a large area of resonant material in the form of wood panelling.

6.6 ACOUSTIC MATERIALS

Usually, common building materials absorb sound only to a small extent and therefore, to meet the acoustic requirements, materials with better sound absorption property are to be incorporated in the halls. Such acoustic materials having more capacity to absorb the incident sound are called *absorbent* or *acoustic materials*.

In general, acoustic materials are soft, porous and work on the basic principle that the sound waves penetrate into the pores and the sound energy gets converted into other forms of energy. The absorbing capacity of the material depends on its density, thickness and the frequency of the incident sound wave. One can broadly classify the acoustic materials into the following classes: i. Porous materials, ii. Cavity resonators, iii. Resonant panels, and iv. Composite materials.

ILLUSTRATIVE EXAMPLES

Example 1: A hall has dimensions $6 \times 4 \times 5$ cubic metres. Calculate: (i) mean free path of sound wave in the room (ii) the number of reflections made per second by the sound wave with the walls of the room. Given, velocity of sound in air = 350 m/s.

Solution:

i. One can define the mean free path of the sound waves as the average distance travelled by a sound wave through air between any two consecutive encounters with the walls of the room.

$$\therefore \text{Mean free path } (L) = \frac{4(\text{Volume of the room})}{\text{Total surface area}}$$

Volume of the room $= 6 \times 4 \times 5 = 120 \text{ m}^3$

Total surface area $= 2[6 \times 4 + 4 \times 5 + 6 \times 5] = 148 \text{ m}^3$

$$L = \frac{4 \times 120}{148} = 3.2 \text{ m}$$

ii. Number of reflections made per second

$$N = \frac{\text{Velocity of sound}}{\text{Mean free path}}$$

$$= \frac{350}{3.243} = 107.9$$

Example 2: Define sound absorption co-efficient of a material. A hall of volume 2000 m^3 is found to have a reverbation time of 2 s. If the area of sound absorbing surface be 800 m^2, calculate the average absorption co-efficient.

Solution: One can define the sound absorption co-efficient of a surface as the ratio of the sound energy absorbed by the surface to that falling on the surface. Since all the energy falling on an open window passes through it, the absorption co-efficient of open window is 1. Obviously, one can also define the absorption co-efficient of any surface as the ratio of the sound energy absorbed by the surface to that absorbed by the same area of an open window.

$$\text{Reverbation time, } T = \frac{0.161V}{A}$$

where V is volume in cubic metre and A is the total absorption of sound. If S be the total area of the sound absorbing surface and a the average absorption co-efficient, then $A = Sa$.

Given that $\quad V = 2000 \text{ m}^3, T = 2 \text{ s}, S = 800 \text{ m}^2$

$$\therefore \qquad\qquad T = \frac{0.161V}{Sa}$$

$$\therefore \qquad\qquad 2 = \frac{0.161 \times 2000}{800 \times a}$$

$$\text{or} \qquad\qquad a = \frac{0.161 \times 2000}{800 \times 2} = 0.2$$

Example 3: The volume of a room is 600 m^3. The wall area of the room is 220 m^2, the floor area is 120 m^2 and the ceiling area is 120 m^2. The average sound absorption co-efficient for the (i) walls is 0.03 (ii) ceiling is 0.80 (iii) floor is 0.06. Find the average reverbation time.

Solution: The average sound absorption co-efficient

$$a = \frac{\sum \alpha A}{\sum A}$$

$$= \frac{\alpha_1 A_1 + \alpha_2 A_2 + \alpha_3 A_3}{A_1 + A_2 + A_3}$$

Given $A_1 = 220 \text{ m}^2, \alpha_1 = 0.03$
$\qquad A_2 = 120 \text{ m}^2, \alpha_2 = 0.80$
$\qquad A_3 = 120 \text{ m}^2, \alpha_3 = 0.06$

$$\therefore \qquad\qquad a = \frac{0.3 \times 220 + 0.8 \times 120 + 0.06 \times 120}{220 + 120 + 120}$$

$$= \frac{109.8}{460} = 0.2389 \approx 0.24$$

Now, the total sound absorption of the room $= a\sum A = 0.24 \times 460 = 110.4$ metric Sabines

$$\text{Reverbation time, } T = \frac{0.161V}{a\sum A} = \frac{0.161 \times 600}{110.4}$$

$$= 0.86 \text{ s (approx.)}$$

Example 4: The reverbation time of a cubical hall of side 10 m is 0.7 s. If one of the walls is covered with felt, the reverbation time is reduced to 0.6 s. Show that the sound absorption co-efficient of the felt is 0.7.

Solution: Reverbation time $T = \dfrac{0.161V}{A} = \dfrac{0.161V}{\sum as} = \dfrac{0.161V}{a\sum s}$

where V is the volume of the hall in cubic metre and $A = a\Sigma S$, a being the absorption co-efficient and S the corresponding area. Here $T = 0.7$ s, $V = (10)^3$ m^3, A (for the four walls, ceiling and floor)

$$A = a \times (10)^2 \times 6 = 600a$$

$$0.7 = \frac{0.161 \times (10)^3}{a \times (10)^2 \times 6}$$

or $\qquad\qquad a = 0.38$

Obviously, a is the absorption co-efficient for the walls, ceiling and floor. If one of the walls is covered with felt whose absorption co-efficient is a' (say), then we have

or $\qquad\qquad A = a \times (10)^2 \times 5 + a' \times (10)^2$
$$= 0.38 \times (10)^2 \times 5 + a' \times (10)^2$$

Now, the reverbation time is reduced to 0.6 s. Again from Sabine's formula, we have

$$0.6 = \frac{0.161 \times (10)^3}{0.38 \times (10)^2 + 5 \times a' \times (10)^2}$$

$$= \frac{0.161 \times 10}{0.38 \times 5 + a'}$$

$\therefore \qquad\qquad a' = 0.76$

Example 5: Find the reverbation time of an office which has a volume of 2000 m^3 and a total sound absorption of 90 metric Sabine. What is the additional sound absorption required for an optimum reverbation time of 1.5 s?

Solution: We have Sabine's formula

$$T = \frac{0.161V}{A}$$

where V is the volume of the room in m^3 and A is the total absorption in metric Sabine. Now, substituting the given values in the above relation, we have

$$T = \frac{0.161 \times 2000}{90} = 3.6 \text{ s}$$

Now, the total absorption A for an optimum reverbation $T = 1.5$ s is given by

$$A = \frac{0.161V}{T} = \frac{0.161 \times 2000}{1.5}$$

$$= 216 \text{ metric Sabine}$$

$\therefore$ Additional sound absorption required

$$= 216 - 90 = 126 \text{ metric Sabine}$$

Example 6: The volume of an auditorium is 12000 m^3. Its reverbation time is 1.5 seconds. If the average absorption co-efficient of interior surfaces is 0.4 Sabine, find the area of interior surfaces.

Solution: Given, $V = 12000$ m^3, $T = 1.5$ s

$$\bar{a} = 0.4 \text{ Sabine}, S = ?$$

We know $\qquad T = \dfrac{0.167V}{\bar{a}S}$

or $\qquad\qquad S = \dfrac{0.167V}{\bar{a}T}$

$\therefore \qquad\qquad S = \dfrac{0.167 \times 12000}{0.4 \times 1.5} = \dfrac{2004}{0.6}$

$\therefore \qquad\qquad S = 3340$ m^2

Example 7: The volume of a room is 1500 m^3. The wall area of the room is 260 m^2, the floor area is 140 m^2 and the ceiling area is 140 m^2. The average sound absorption co-efficient for wall is 0.03, for ceiling is 0.8 and for the floor is 0.06. Calculate the average absorption co-efficient and the reverbation time.

Solution: Given, $V = 1500$ m^3, $a_1 = 0.03$ Sabine

$$a_2 = 0.8 \text{ Sabine}, a_3 = 0.06 \text{ Sabine}$$
$$S_1 = 260 \text{ m}^2, S_2 = 140 \text{ m}^2, S_3 = 140 \text{ m}^2$$
$$\bar{a} = ?, T = ?$$

Average absorption co-efficient is

$$\bar{a} = \dfrac{a_1 S_1 + a_2 S_2 + a_3 S_3}{S_1 + S_2 + S_3}$$

$$= \dfrac{0.03 \times 260 + 0.8 \times 140 + 0.06 \times 140}{260 + 140 + 140}$$

$$= \dfrac{7.8 + 112 + 8.4}{540}$$

$$\bar{a} = 0.2374 \text{ OWU}$$

$\therefore$ The total sound absorption of the room is

$$\bar{a}\, \Sigma S = 0.2374 \times 540 = 128.196 \text{ OWU m}^2.$$

Therefore, the reverbation time is

$$T = \dfrac{0.167V}{\bar{a}\Sigma S} = \dfrac{0.167 \times 1500}{128.196} = \dfrac{250.5}{128.196}$$

$$T = 1.9540 \text{ s}$$

Example 8: A hall has a volume of 12500 m^3 and reverbation time of 1.5 s. If 200 cushioned chairs are additionally placed in the hall, what will be the new reverbation time of the hall? The absorption of each chair is 1.0 OWU.

Solution: Given, $V = 12500$ m^3, $T_1 = 1.5$ s

$$a_2 S_2 = 200 \text{ Sabine m}^2,$$
$$\Sigma a_1 S_1 = ?, T_2 = ?$$

Let $T_1 = \dfrac{0.167V}{\Sigma a_1 S_1}$ be the reverbation time before placing cushioned chairs.

$$\Sigma a_1 S_1 = \frac{0.167 \times 12500}{1.5} = \frac{2087.5}{1.5}$$

$$\Sigma a_1 S_1 = 1391.66 \text{ Sabine m}^2$$

The reverbation time after placing the cushioned chairs be,

$$T_2 = \frac{0.167V}{\Sigma aS + a_1 S_2} = \frac{0.167 \times 12500}{1391.66 + 200}$$

$$= \frac{2087.5}{1591.66}$$

$$T_2 = 1.3115 \text{ s}$$

∴ The new reverbation time after placing the cushioned chairs is 1.3115 s.

Example 9: A hall has a volume of 2265 m³. Its total absorption is equivalent to 94.85 m² of open window. What will be the effect on reverbation time if audience fill the hall and thereby increases the absoption by another 94.85 m²?

Solution: Given, $V = 2265$ m³, $SaS = 94.85$,

$$\Sigma a_2 S_2 = 2 \times \Sigma aS = 189.7,\ T_1 = ?,\ T_2 = ?$$

Let T_1 be the reverbation time in the hall without audience

∴ $$T_1 = \frac{0.167V}{\Sigma a_1 S_1}$$

∴ $$T_1 = \frac{0.167 \times 2265}{94.85}$$

$$T_1 = 3.987 \text{ s}$$

Let T_2 be the reverbation time in the hall with audience

∴ $$T_2 = \frac{0.167V}{\Sigma a_2 S_2}$$

∴ $$T_2 = \frac{0.167 \times 2265}{189.7} = \frac{378.255}{189.7}$$

$$T_2 = 1.993 \text{ s}$$

Thus, the reverbation reduces to half of its initial value when the audience fill the hall.

REVIEW QUESTIONS AND PROBLEMS

1. What are the characteristics of an acoustically perfect hall or auditorium? What is the role of reverbation time? What steps would you take to improve the acoustics of a hall?

2. Discuss the various acoustic defects commonly found in larger halls. How are these defects removed?

3. Deduce Sabine's formula for the reverbation time of a hall or auditorium. Explain why does it fall in the case of dead room.

4. Investigate the growth of sound in hall, and then deduce Sabine's reverbation formula. Explain why does it fall in the case of dead room. Explain how the above result can be used to determine the sound absorption co-efficient of a piece of felt.

5. Mention some effects of reverbation in daily life.

6. Give the theory of growth and decay of sound in a live room. Find the reverbation time. Why the theory fails in a dead room?

7. What are the acoustic requirements of a good auditorium? Discuss the theory of reverbation.

   What is the order of magnitude of optimum reverbation?

8. How absorption co-efficient of a material is measured?

9. Enumerate the features that an auditorium should have for good acoustics. How can these features be incorporated in the design of an auditorium?

10. What remedial measures would you suggest for the following defects in an existing auditorium?
    i. Uneven distribution of sound
    ii. Unintelligible audibility of speech.

11. The reverbation time for an empty hall is 1.5 s and 1 s when a curtain cloth of 20 m^2 is suspended at the centre of the hall. The dimensions of the hall are 10 m × 8 m × 6 m. Show that the absorption co-efficient for the curtain cloth is about 0.6 Sabines.

12. A hall has dimensions 20 m × 15 m × 5 m. The reverbation time is 3.5 s. Determine the total absorption of its surface.   [*Ans.* 67.7 metric Sabines]

13. The reverbation time for an empty hall of size 20 m × 15 m × 10 m is 3.5 s. Calculate (a) the average absorption co-efficient of the hall, (b) area of the wall that should be covered by curtains so as to reduce the reverbation time to 2.5 s. The absorption coefficient of curtain cloth is 0.5.   [*Ans.* (a) 0.106 OWU (b) 140.0 m^2]

14. A hall has a volume of 7500 m^3. It is required to have reverbation time of 1.5 s. Determine total absorption of the hall. If the total absorption of the hall is increased, what will be its effect?   [*Ans.* 7500 metric Sabine]

SHORT ANSWER QUESTIONS

1. For an acoustically perfect hall, is it essential that 'quality' of the sound must remain unaltered?

   Ans. Yes.

2. How an acoustically perfect hall have marked effects on speech and music being delivered there in the hall?

   Ans. Speech delivered is quite intelligible.

3. How echelon effect can be avoided?

   Ans. By using a stair carpet.

4. Why an echo cannot be heard if the distance between the source and the obstacle is less then 17 m?   [V = 340 m/s]

Ans. If a sound source be situated at a distance d from a reflecting obstacle, then the time-interval t for the sound to go and return is given by

$$t = \frac{2d}{V} = \frac{2 \times 17\,m}{340\,m/s} = 1/10\,s$$

Obviously, if d is less than 17 m, then the time-interval t will be less than 1/10 s, and the echo cannot be heard.

5. What steps will you take to adjust the reverbation time to optimum value?

Ans. i. The windows and ventilators will be kept open.

ii. Heavy curtains will be hung.

iii. The floor will be covered with carpets or tapestries.

iv. The walls and ceiling will be lined with absorbent material such as felt, celotex, fibre-board, glass wool, rock wool, mineral wool, etc.

v. We will provide acoustic tiles.

vi. The clothing worn by the persons in the hall also provides sound absorption. Therefore, the period of reverbation varies according to the number of persons present. To offset this variation, we will provide the seats with cushions which themselves have a sufficiently large absorbent effect to make-up when the seats are unoccupied.

OBJECTIVE QUESTIONS

1. In an acoustically perfect hall, the sound heard must be sufficiently .... in every part of the hall.   [loud]

2. The undue persistence of sound of the speaker or the singer in an acoustic hall is called ....   [reverbation]

3. The Sabine relation for the period of reverbation in seconds is .... $\left[T = \dfrac{0.161V}{A}\right]$

4. For adequate loudness in a hall, loudspeaker should be fitted .... the heads of the listeners and pointed slightly ....   [above, downward]

5. The echoes in an acoustic hall may be avoided by covering distant walls and high ceilings with .... materials.   [absorbent]

6. Reverbation arises due to .... reflection from the various surfaces in a hall.
                                                                                      [multiple]

7. Reverbation time of a hall is .... on its volume and nature of the .... surface.
                                                                         [dependent, absorbing]

8. Sound waves are .... waves and exhibit all .... phenomena.
                                                                         [longitudinal, wave]

MULTIPLE CHOICE QUESTIONS

1. The unit of intensity of sound is
   (a) Wm^{-2} (b) Wm^{-1} (c) Wm^2 (d) Jm^{-2}

2. The minimum sound intensity which a human ear can sense is called
   (a) zero point energy   (b) threshold energy
   (c) vibrational energy   (d) rotational energy

3. A change in sound intensity level of 1 dB alters the intensity by
   (a) 74%   (b) 36%   (c) 26%   (d) None of the above

4. The intensity of a source of sound is increased 20 times its value. The intensity level increase as

(a) 1.301 dB   (b) 0.1301 dB   (c) 13.01 dB   (d) 7.01 dB

[Hint: $L = 10\log\dfrac{I}{I_0}$

$$L_1 = 10\log\dfrac{I}{I_0} = 10\log I - 10\log I_0$$

$$L_2 = 10\log\left(\dfrac{20I}{I_0}\right) = 10\log 20 + 10\log I - 10\log I_0$$

Subtracting gives

$$L_2 - L_1 = 10\log 20 = 10 \times 1.3010 = 13.01\,pdB$$

5. The ratio of intensity I and I_0 of a sound in a heavy traffic is $10^{-6}/10^{-12}$. The intensity level in dB is

(a) 6 dB   (b) 30 dB   (c) 60 dB   (d) 120 dB

[Hint: $I_L = 10\log_{10}\left(\dfrac{I}{I_0}\right) = 10\log_{10}\left(\dfrac{10^{-6}}{10^{-12}}\right)$

$$= 10\log_{10}(10^6) = 10 \times 6\ dB = 60\ dB]$$

6. The time taken by the sound to fall below the minimum audibility level after the source stopped sounding is called

(a) periodic time   (b) reverbation time   (c) decay time   (d) growth time

ANSWERS

1. (a)   2. (b)   3. (c)   4. (c)   5. (c)   6. (b)

7

Ultrasonics

7.1 INTRODUCTION

Ultrasonic waves are sound waves of frequency greater than that of audible sound. These waves are produced by an object vibrating at a frequency higher than the human ear can hear, i.e. greater than 20000 cycles/s = 20 kHz/s. On the basis of frequency range we have

Audio frequency range	– 20 Hz to 20 kHz
Radio frequency range	– 550 kHz to 22 MHz
TV frequency range	– 47 MHz to 230 MHz
Above audio range	– Ultrasonic (above 20 kHz)
Below audio range	– Infrasonic (below 20 Hz)

With the help of modern techniques, one can produce ultrasonic waves of frequency upto 20 million Hz, which has a wavelength of 10^{-8} m (~x-ray wavelength).

Ultrasonic waves are highly energetic and have extremely short wavelength because of its high frequency and energy. Our ear is not sensitive to ultrasonic waves although some animals like dogs, bats, and some kinds of birds show response to these waves. These waves can be produced by the high frequency vibrations of a quartz crystal under an alternating fields (piezoelectric effect) or by the vibrations of a ferromagnetic rod under an alternating magnetic field (magnetostriction effect). Because of small wavelength and high energy, the ultrasonic waves find use especially in the field of medicine and in various industries. Due to this ultrasonic waves have great promises in future.

The longitudinal waves whose frequency lies below 20 Hz are called *infrasonic* waves. Human ear is not sensitive to these waves. These waves are produced by large vibrating bodies such as during an earthquake.

We may note that *supersonic* also means of very high frequency. However, we must differentiate ultrasonic from supersonic. Supersonic effects are produced by objects that travel through a medium at a faster speed than the waves they generate. Supersonic is essentially confined to aeroplanes and missiles, which fly through air at a speed of sound in air.

7.2 PRODUCTION OF ULTRASONIC WAVES

There are number of ways by which ultrasonic waves can be generated. Usually, the method to be employed depends upon the power output necessary and the frequency range to be covered. Generation of mechanical type waves such as tuning forks or Galton's whistle can be used up to 10,000 cycles/s.

Similar to sonic range, ultrasonic range requires a source of energy. A device which transfers energy from one system to another, generally with the change of the form of energy mainly from acoustical to mechanical or *vice versa* is known as *transducer*. We may note that a medium is also necessary for transmission of ultrasonic waves.

Galton's Whistle

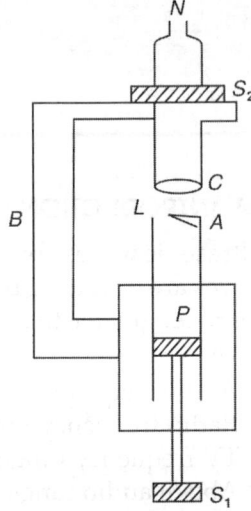

Galton devised a miniature organ pipe in the form of a whistle to determine the limit of audibility. It consists of a closed end air column A whose length can be adjusted with the help of a moveable piston. With the help of a screw S_1, the piston P can be moved to the desired position. The open end of the pipe A is fitted with a lip L (Fig. 7.1). With the help of the screw S_2, the position of the pipe C can be adjusted. One can adjust the gap between the ends of A and C with the help of the screw S_2 as shown in Fig. 7.1.

An air blast is blown through the nozzle at the top and the blast of air coming out of C strikes against the lip L. Due to this the column of air in the pipe is set into vibration. Now, adjusting the length of air column in A, it is brought to the resonance position. The resonance frequency depends on the length and diameter of the pipe A. Now, if l is the length of the air column in A, x is the end correction, then the wavelength is obtained as

Fig. 7.1: Galton's whistle

$$\lambda = 4(l + x) \tag{7.1}$$

The frequency of sound

$$n = \frac{V}{\lambda} = \frac{V}{4(l+x)} \tag{7.2}$$

One can produce frequencies of the order of 30,000 Hz with the help of whistle.

Magnetostriction Oscillator

Joule in 1847 observed that a rod of ferromagnetic substance such as nickel undergoes a change in length when magnetic field is applied along the length. The increase in length is proportional to the applied magnetic field. This increase in length is greater in the case of nickel in comparison to other ferromagnetic substances. We may note that the increase in length is independent of the direction of the magnetic field acting. If the magnetic field is alternating and applied parallel to the length of the nickel rod or tube, as magnetic field grows the length increases and the vibration sets up in the rod with a frequency double that of the frequency of the magnetic field. However, if a permanent steady polarizing field of suitable strength is applied from direct current supply, and a changing alternating current supply giving rise to alternating field is superimposed on it, the longitudinal vibration of the nickel rod will set-up, having frequency of vibration same as the frequency of the alternating field. However, resonance will occur when the frequency of alternation of the magnetic field be equal to the natural frequency of the bar which depends on the elastic constant and the dimension of the bar. The amplitude of vibration will be large. The rod vibrates with ultrasonic frequency and will emit ultrasonic waves.

The velocity of ultrasonic waves in a longitudinal bar is given by

$$c = \sqrt{Y/\rho} \qquad (7.3)$$

where, Y is Young's modulus and ρ is density of the material. If l denotes the length of the bar, the fundamental wavelength of the bar is $= 2l$ and hence the frequency $f = \dfrac{c}{2l}$.

Thus, in general $f = \dfrac{sc}{2l} = \dfrac{s}{2l}\left(\sqrt{\dfrac{Y}{\rho}}\right)$ or $f = \dfrac{s}{2l}\sqrt{Y/\rho}$, where s is the order of harmonic. One can easily see from the above relation that at a frequency of 20000 c/s, the length of the nickel rod that will resonate will be of the order of m and moreover, this length will be smaller and smaller with higher frequencies so that the length will become very small at a frequency greater than 60,000 c/s; hence the usual range of a magnetostriction ultrasonic oscillator will be from 5000 c/s to 60,000 c/s. We may note that magnetostriction has been used in laboratories to generate ultrasonic signals of frequency 2 M c/s but at higher frequencies the output is small that it cannot be utilized for any practical purposes.

An arrangement using magnetostriction for producing ultrasonic waves is shown in Fig. 7.2. A short rod XY of nickel is placed in a solenoid, fed by DC supply, which produces a steady polarizing magnetic field. L_1 and L_2 are two coils wound on the ends of the rod and are included in the grid- and anode-circuit respectively of a triode valve. The frequency of the oscillating anode circuit is adjusted with the variable capacitor C. When the frequency equals the natural frequency of the rod, then the longitudinal oscillations of the rod are maintained and ultrasonics are produced in the surrounding medium.

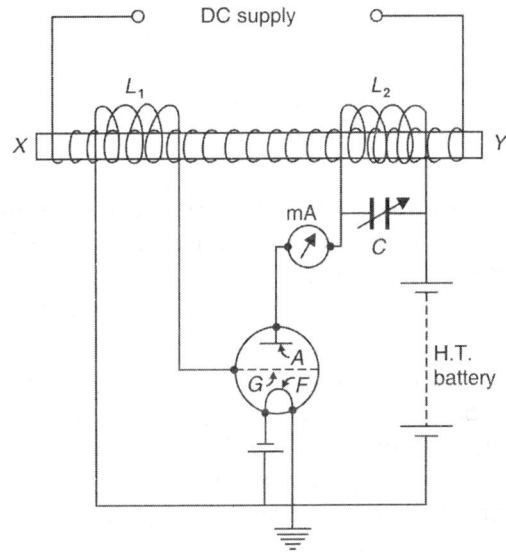

Fig. 7.2: Magnetostriction oscillator

The rod is already magnetised by the steady polarising magnetic field. The periodic variation in the anode current passing in the coil L_2 causes a variation in magnetisation and a consequent variation in the length of the rod. Now, this variation in length causes a variation in the magnetic flux through the grid coil L_1. Obviously, by the converse magnetostriction effect, an induced emf is set-up in coil L_1. This emf acts on the grid, and produces an amplified current variation in coil L_2. Hence, the oscillations of the rod are maintained. When the oscillations start, the milliammeter (mA) will show an alteration. To make this alteration maximum, the capacitor C is adjusted, because at this stage the frequency of the rod and the vibrations of the rod will be most vigorous. A nickel rod of length 10 cm, gives out ultrasonic waves of frequency 25000 c/s. The oscillation frequency f of the nickel rod is given by

$$f = \dfrac{1}{2\pi\sqrt{LC}} \qquad (7.4)$$

When this frequency matches with the natural frequency of the rod $f = \dfrac{P}{2L}\sqrt{\dfrac{E}{D}}$, where L is the length of the rod, E is Young's modulus, D the density of the rod materials and P is the harmonic mode 1, 2, 3, etc. resonance will occur. With the help of this method, one can obtain frequencies of the ultrasonic waves up to 3×10^5 Hz. However, at higher frequencies the output is so small that it cannot be utilized for any practical purposes.

To secure maximum transfer of energy the *transducer* housing is fitted close to the medium. A common type of powerful magneto-striction transducer is shown in Fig. 7.3. A number of nickel tubing act as a driving element, each having a length equal to one quarter of the wavelength of the sound to be radiated. From Fig. 7.3, we can see that one end of these tubing is free and the other is embedded in a circular sheet plate P whose resonant frequency is very nearly equal to that of the nickel tube. Nickel tubes are the driving elements. Each nickel tube is surrounded by its own driving coil C and all are driven in phase by the current from a very powerful oscillator. The permanent magnets M mounted inside the water tight housing containing the tubes supply the optimum polarising field. The alternating force exerted on the plate by the reaction to the stress in the nickel tubes are transmitted by the plate.

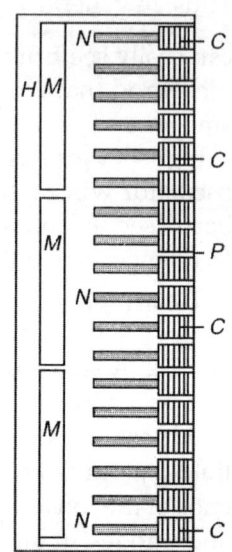

Fig. 7.3: Transducer

The power handling capacity of this type of transducer is higher. It has a sharper resonance curve which helps in reducing the disturbing effect of the background noise. However, it is difficult to produce optimum polarising field in this type of instrument.

Since the magnetostrictive effect is reversible and hence the same instrument can be used as a *receiver*. Sound waves impinging on the plate P set it into vibrations, causing the Ni tubes to vibrate in magnetic field. The resulting alternating currents are induced in the surrounding coils which may be amplified for detection. We may note that the frequency range of magnetostriction is 5–60 kc/s beyond which rods of very small lengths are required and handling becomes difficult. Usually, 25 kc/s are used for obtaining good results.

The main advantages of magnetostriction method are:
  i. The method is simple and its cost of construction is low.
 ii. At low ultrasonic frequency, large power output is possible without damage or risk to the oscillator circuit, even under temporary overload.
iii. In comparison to piezoelectric transducers, magnetostriction transducers have certain advantages. Magnetostriction transducers possess large power handling capacity and sharp resonance curve.

The main drawbacks of this method are:
  i. One finds difficulty in imposing permanent polarising magnetisation of the Ni rod initially.
 ii. One cannot use this method for very high frequency range, i.e. there is a limitation of frequency.
iii. Greater dependence of frequency on temperature and *breadth of resonance curve* causes changes in elastic constants of ferromagnetic substances with degree of magnetisation.

Piezoelectric Method

J and P Curie brothers in 1880 discovered that electric charges are developed on two opposite faces of quartz, rochelle salt crystal cut in a perpendicular way when two opposite faces are subjected to pressure or compression perpendicularly. The sign of the charge developed on two opposite faces will change with the change in tension or compression. The electric charge developed is proportional to the amount of pressure or tension. Conversely, we can say that if a properly cut piezoelectric crystal is placed in an electric field, compression or extension, depending on the nature of electric field occurs, the amount of compression or extension being proportional to the potential difference between the faces.

Hence, when a crystal (quartz, rochelle salt or tourmaline) is subjected to an alternating potential difference, it is set into elastic vibrations, i.e. it contracts and expands periodically and sets up mechanical vibrations in any acoustic medium in which it is placed. The frequency of the vibrations is within the ultrasonic range. Ordinarily the amplitude is very small. If the frequency of the electrical oscillations coincides with one of the natural frequency of the crystal, which are of the order of 250 to 10,000 kc/s a large amplitude of vibrations results. This is adopted in the construction of piezoelectric transducer and the phenomenon is utilized to produce ultrasonic waves.

Quartz Crystal

Natural quartz crystal occurs in the shape of a prism of six sides with a pyramid attached to each end (Fig. 7.4a). The line joining the two opposite vertices of the pyramids is defined as the *optic axis* of the crystal. If the opposite corners of the crystal are joined together, the line is known as X-axis or also called as the *electrical axis* of the crystal. Figures 7.4b, c and d represent the sections of the crystal perpendicular to Z-axis. Line perpendicular to Z-axis and joining two opposite corners of hexagonal slice of the crystal

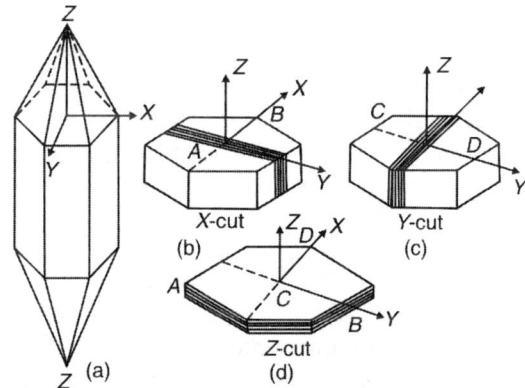

Fig. 7.4: Quartz crystal

as shown in Fig. 7.4b is the *electric axis* known as X-axis. A line represented by CD (Fig. 7.4c) joining the *opposite faces* of the hexagonal slice and perpendicular both to x-and z-axes represents Y-axis or *mechanical axis*.

One can obtain Y-cut crystal by cutting the hexagonal slice along X-axis, i.e. along the corners, such that the largest force is perpendicular to Y-axis and Z-cut crystal is obtained by cutting the crystal perpendicular to Z-axis. X-cut crystal is obtained by cutting the hexagonal slice along Y-axis such that the largest face is perpendicular to X-axis. These crystals are generally used for piezoelectric oscillators. Faces of the crystal are silvered to ensure proper electrical contact. The alternate electric field is applied by the metal plates pressing the faces, along X-axis. The crystal vibrates in the direction of Y-axis. If the frequency of the electrical oscillations coincides with one of the natural frequencies of the crystal, which are of the order of 250 to 10,000 kc/s, a large amplitude of vibrations results.

Pressure along the direction of X-axis produces the charges on two surfaces normal to this axis but tension will produce opposite charges, opposite to what is produced when pressure is applied.

Tension along the direction of Y-axis charges two surfaces normal to X-axis as is produced when pressure is applied along X-axis and pressure along the direction of Y-axis charges oppositely two opposite surfaces normal to X-axis, as the charge is produced when tension is applied.

We may note that the pressure or tension along Z-axis produces no effect at all.

Electric field along X-axis only causes the crystal to expand along X-axis and to contract along Y-axis, reversal of the field produces the reversal of the effect. The deformation along Y-axis will produce longitudinal vibration and mechanical deformation along X-axis will produce surface vibration which is generally used as a source.

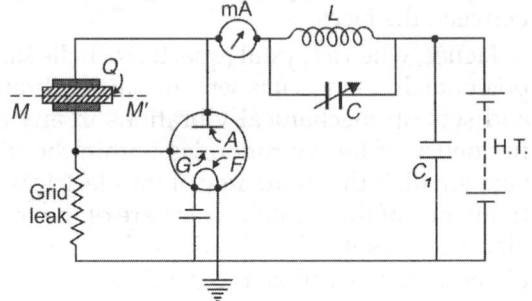

The necessary arrangement is shown in Fig. 7.5. A quartz slab Q is placed between two metal plates which are connected to the anode A and the grid G of a triode valve as shown in Fig. 7.5. Also connected to the anode is an inductance–capacitance (L–C) circuit and a high tension (H.T.) battery shunted by a bypass capacitor C_1. The bypass capacitor C_1 prevents high frequency current from

Fig. 7.5: Piezoelectric oscillator

passing through the battery. The variable capacitor C in the anode circuit is then adjusted to produce electrical oscillations of frequency equal to that of the modes of vibrations of the crystal.

A small alternating current is set-up in the anode circuit of the triode valve. At some instant, the anode plate A is more negative and grid G is more positive. Since the anode plate A and the grid G are connected to the opposite plates of Q, these plates are oppositely charged. The charges reverse in sign when anode A becomes more positive and grid G more negative. Obviously, an alternating potential difference is applied on the opposite faces of the crystal. It sets the crystal into elastic vibrations of the frequency of AC supply. When the resonance occur between this frequency and the natural frequency of the plate, which depends upon dimension, the amplitude of vibrations will be large. We may note that such vibratory system is used as standard of frequency in electrical circuit.

The simplest form of the circuit is mostly the circuit as shown in Fig. 7.5 to maintain the oscillation of triode valve. Obviously, the crystal is thrown into resonant mechanical oscillations along MM' generating ultrasonic waves in the surrounding medium.

It is found that the ultrasonic generator delivers maximum power when it is operated at the fundamental frequency of the crystal. In order to generate higher frequency ultrasonic waves, the L–C circuit is made to oscillate at a frequency equal to one of the harmonics of the crystal. However, only odd harmonics of the crystal are used because piezoelectric effect can occur only when opposite charges appear on the electrodes.

The distribution of pressure and charge in the thickness of a crystal oscillating at three harmonics are shown in Fig. 7.6. It is

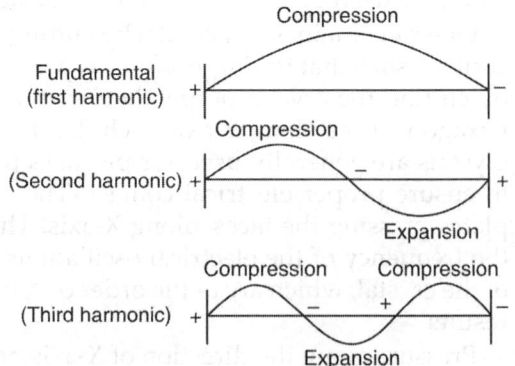

Fig. 7.6: Distribution of pressure and charge in the thickness of a crystal at three harmonics

found that the effect will occur in the *fundamental and third harmonic, and not in the second harmonic.*

Design of the Ultrasonic Crystal

One may cut the crystal along the X-axis or along the Y-axis. It is called the *x-cut* if it is cut along the x-axis and *y-cut* if it is cut along the Y-axis. Usually the x-cut is used since it generates longitudinal or *L*-waves. For the production of shear waves, y-cut crystals are also used. However, for the propagation of this type of waves through a liquid or a gas, the medium must have a shear elasticity. In case of an x-cut crystal, the longitudinal waves will occur along the thickness dimension of the crystal. If the crystal vibrates at its natural frequency, one finds that the thickness (t) of the crystal is

$$t = \lambda/2$$

where λ is the wavelength of ultrasonic waves. For a quartz crystal, the velocity of wave propagation is given by

$$c = \sqrt{\frac{E}{\rho}} = \sqrt{\frac{770 \times 10^9}{2.654}} = 5500 \text{ m/s}$$

If n is frequency of the wave, we have

$$n = \frac{c}{\lambda} = \frac{540 \times 10^3}{2t} = \frac{2700}{t} \text{kc/s}$$

If t is expressed in millimetres, usually, the generated frequency is given by

$$n = \frac{2870}{t(\text{mm})} \text{kc/s}$$

The Frequency of Vibration

When the frequency of the alternating voltage is equal to the natural frequency of the vibration of the crystal or its simple higher multiples, the crystal is thrown into resonant vibrations and the amplitude will be large. We may note that these vibrations are longitudinal in nature. The frequency of vibrations can be calculated from

$$n = \frac{p}{2l} \sqrt{\frac{E}{\rho}}$$

where $p = 1, 2, 3, \ldots$, etc., E is the elasticity and ρ is the density of the crystal. The velocity of longitudinal wave propagation is (since it vibrates in odd harmonics, then $p = 1, 3, 5, \ldots$, etc.)

$$c = \sqrt{\frac{E}{\rho}}$$
$$= 5.5 \times 10^3 \text{ m/s for quartz}$$

For a crystal of length 0.05 m, the frequency for the first mode of vibration will be

$$n = \frac{1 \times 5.5 \times 10^3}{2 \times 0.05} = 5.5 \times 10^4 \text{ Hz}$$

The modes of frequency are simple integral multiples of 5.5×10^4 Hz. One can produce frequency upto 150,000 kHz by tourmaline crystal.

Piezoelectric oscillator has the following advantages:

i. Piezoelectric oscillator has to be used for higher frequency range,
ii. Size of piezoelectric oscillator is small and they are economical,
iii. When used in detector, these oscillators are very sensitive,
iv. The waveform is good.

However, power handling capacity of piezoelectric oscillator is low. This is the disadvantage of these oscillators. We may also note that intense ultrasonic radiations has a disruptive effect on liquids by causing bubbles to be formed.

7.3 DETECTION OF ULTRASONICS

One cannot directly detect the ultrasonic waves, although some animals, specially the bat, can do so. Ultrasonic waves propagated through a medium can be detected indirectly by the following methods:

Piezoelectric Detector

The ultrasonic waves when fall on one pair of the faces of a quartz crystal, varying electric charges are produced on the other perpendicular faces. Though they are very small but can be amplified and detected with the help of some suitable means.

Kundt's Tube Method

One can detect ultrasonic waves with the help of Kundt's tube. When ultrasonic waves of relatively large wavelength, almost comparable to audible sound, were allowed to pass through Kundt's tube, then lycopodium powder sprinkled inside the tube, collects in the form of heaps at nodal points and is blown off at antinodal points. The average distance between two adjacent heaps is equal to half the wavelength. One cannot use this method if the wavelength of ultrasonic waves is very small, i.e. less than a few millimetres. In the case of liquid medium, powdered coke is used instead of lycopodium powder to detect the position of nodes.

Acoustic Diffraction Method

According to Brillouin, when an ultrasonic wave passes through a liquid, the density of liquid varies from layer to layer due to periodic variation of pressure and the structure of the liquid thus agitated can be simulated to that of a light diffraction, i.e. if under this condition monochromatic light is passed through the liquid at right angles to the wave, then liquid behaves as diffraction grating. Moreover, this grating behaves in the same way as a ruled grating. Consequently, if a beam of monochromatic light be incident on such a simulated grating, there should be visible the diffraction pattern in the transmitted light. Such a grating is known as *acoustic grating*. The grating element is equal to the wavelength of the ultrasonic waves. Experimental arrangement for acoustic diffraction is shown in Fig. 7.7. Light from a sodium lamp S is condensed by a lens L on the slit S_1 of a collimator C which renders it parallel. The parallel beam is allowed to pass through a glass cell filled with the experimental liquid. A quartz

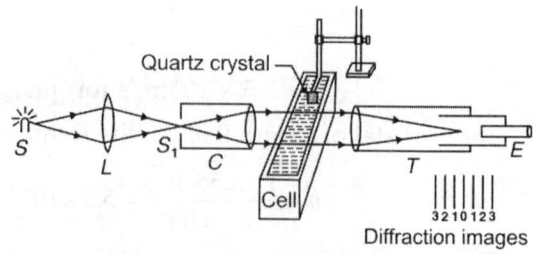

Fig. 7.7: Acoustic diffraction method

crystal fed by a radio frequency oscillator and thus generating ultrasonic waves is suspended in the liquid such that the waves are propagated in the liquid at right angles to the beam of light. The waves get reflected back from the wall of the cell and a stationary wave pattern is set-up in the liquid. The light beam passing through this pattern, i.e. acoustic grating is diffracted. The diffracted beam of light is viewed through the telescope. It consists of a central maximum and principal maxima on either side. If θ_n be the angle of diffraction for the nth order maximum, then

$$d \sin \theta_n = n\lambda \tag{7.5}$$

Here $n = 1, 2, 3...$ etc., λ is the wavelength of sodium light and d is the distance between two adjacent nodal or antinodal planes. Knowing n, θ_n and λ, the value of d can be calculated using Eq. (7.5). Now, if λ_a is the wavelength of the ultrasonic waves through the medium, we have

$$d = \frac{\lambda_a}{2}$$

or $$\lambda_a = 2d \tag{7.6}$$

If the resonant frequency of the piezoelectric crystal oscillator is N, the velocity of the ultrasonic waves can be determined from the following relation

$$V = N\lambda_a = 2Nd \tag{7.7}$$

We may note that the theory of this method is not so simple as deduced above. This method is useful in measuring the velocities of ultrasonic waves through liquids and gases at various temperatures.

Thermal Detectors

For the detection of ultrasonic waves, thermal detectors method is most commonly used. A fine platinum wire is used in this method. This wire is moved through the medium. At the positions of nodes, due to alternate compression and rarefaction, adiabatic changes in temperature take place and the resistance of the platinum wire changes with respect to time. One can detect this with the help of Callender and Garrifith's bridge arrangement. At the positions of antinodes, the temperature remains constant and the resistance of the platinum wire remains constant. The undisturbed balanced position of the bridge indicate this situation.

Sensitive Flame Method

Along the medium a narrow sensitive flame is moved. The flame is steady at the position of the antinodes. The flame flickers at the position of node because there is change in pressure. In this way, one can find the positions of nodes and antinodes in a medium. The average distance between two adjacent nodes is equal to half the wavelength. Knowing the value of the frequency of the ultrasonic wave, one can calculate the velocity of the ultrasonic wave through the medium.

Smoke Method

When one introduces light, solid particles or liquid drops in the field of a sound wave, the particles take up the motion, the amplitude of which is given by Konig's following relation:

$$\frac{y_1}{y_2} = \sqrt{1 + a^2} \quad \text{where} \quad a = \frac{2}{8} \frac{\rho_1}{\rho_0 \lambda b^2}$$

where y_1 and y_2 are amplitudes of the particles when set into vibration and that of sound wave respectively, b and ρ are respectively the radius and density of detectors, ρ_0 the density of the gas and λ the wavelength of the sound wave. The method then demands in introducing small particles in the field to photograph them with long-time of exposure and then measure the amplitude of the motion of the particles *which appear as streaks in the negative*. The above equation reveals that particles of very small radius must be introduced in the sound field. However, these limitations restrict the use of this method for detection of ultrasonic waves having wavelength smaller than 0.1 m. We may note that this method is not only useful for the detection but also for the measurement of the amplitude of sound wave.

7.4 PROPERTIES OF ULTRASONICS

i. Ultrasonic waves have a large energy content, i.e. they are highly energetic.

ii. The speed of propagation of ultrasonics depends upon their frequency, i.e. it increases with increase in frequency.

iii. Because of small wavelength, these waves show negligible diffraction and can be transmitted over long distances without any appreciable loss of energy. This is why these waves have been used in determining the depth of ocean by echo-sounding.

iv. When a plane stationary ultrasonic wave is set-up in a liquid, a structure is developed in which the density of the liquid varies from layer to layer along the direction of propagation of the waves. This structure within the liquid can diffract light in the same way as the structure of a crystal diffract x-rays.

v. Intense ultrasonic radiations has a disruptive effect on liquid by causing bubbles to be formed.

7.5 APPLICATIONS OF ULTRASONICS

Ultrasonics is a very useful tool in the hands of physicists and chemists because from the nature of the absorption and dispersion of ultrasonics in liquids and gases much information can be obtained regarding the structure of molecules. Scientists have obtained very useful data regarding the superconducting state by propagation of ultrasonics through liquid helium. Besides these theoretical considerations, ultrasonics have found numerous applications in the following fields: (a) communication (b) industry (c) medical and biological fields (d) scientific research.

We shall consider some of these applications here.

Communication

Ultrasonic Signalling

As the wavelength of ultrasonics is very small, it is possible to produce a short beam of ultrasonics without diffraction. The high frequency sound waves may be readily formed into a beam and set in the desired direction. It is found that the beam of sound gets more and more narrow as the radius of the plate of the generator is increased in relation to the wavelength of radiated sound. A plate of 3.5 cm diameter is found to be sufficient to confine the sound within an angle of 5 degrees at 600 kc/s. We may note that such narrow beams can never be attained in the audible frequency range. This low power source can radiate very high intensity of ultrasonic sound waves, which can travel several kilometres in water prior being absorbed.

Detection of Submarines, Icebergs and Other Objects

Ultrasonics can be used to detect the reflecting objects like a submerged submarine and a hidden iceberg. For this purpose ultrasonic transmitters have been developed. A typical submarine ultrasonic transmitter uses piezoelectric resonator which generates powerful ultrasonic waves of frequency about 40 kHz. These submarine ultrasonic transmitter finds following uses:

Ultrasonic signalling

It is used for signalling from ship-to-ship especially in submerged submarines and detecting a hidden iceberg.

Recently, ultrasonic microscope has been invented. It is used to detect concealed objects. The frequency is very high so that the wavelength is of the order of the wavelength of visible light.

Depth of sea

One can determine the depth of the sea, position of a submerged submarine, position of a ship, etc. with the help of submarine ultrasonic transmitter. This uses the *echo principle*. The ultrasonic waves transmitted by the quartz crystal in ultrasonic transmitter are directed towards the bed of the sea and these waves are reflected back from the bed and the echo is detected by the crystal itself. We may note that the working principle is the same as that of a conventional pulsed radar. At short intervals, the pulses of ultrasonic signals are sent out at short intervals from a piezoelectric transmitter which is typically energised at a frequency of about 40 kc/s. As stated above, the returned echo is received in the ultrasonic receiver, which also uses piezo crystal followed by an amplifier and the time interval between the transmitted and received signals is recorded by time measuring instrument.

Let t be the time interval between the transmission of the ultrasonic wave and receipt of the echo. The velocity of sound in sea water is

$$v = v_0 + 1.45 + 4.21T - 0.037T^2$$

where v is the velocity of sound at $t°C$ in sea water (m/s), v_0 is the velocity of sound at $0°C$ in water (m/s) = 1510 m/s, S is the salinity (gm/litre), T is the temperature of the sea water in $°C$. Knowing v, the velocity of sound waves through sea water, one can obtain the depth of the sea

$$h = \frac{vt}{2}$$

The method can also be used to detect the presence and depth of submarines, rocks, etc. from the surface of the water. The instrument directly calibrated to show the depth of sea is called a *fathometer* and *echometer*. This method of determining the depth of sea, the direction and distance of a submarine or iceberg is called *Sound Naviagation and Ranging* (SONAR).

Industrial Applications

Ultrasonic Flaw Detector

The strength of a material plays a significant role in most of the engineering materials. If there are defects in the material, the strength of the material gets reduced.

The defects in materials may be as large as cracks or as tiny as cavities produced during casting. The fine internal cracks and flaws in metals act as good sound reflectors of ultrasonics whose wavelength is small compared to the size of the crack or the flaw. Hence they can be detected and located by echo sounding techniques. A part of the surface of the metal is polished and a small ultrasonic generator is placed upon it which sends a beam into the specimen. The beam will be reflected at the far surface and also at the crack or flaw if any, as shown in Fig. 7.8a. A receiver is also placed near the ultrasonic generator, which picks up the reflected echoes, which are amplified and displayed along the time base of cathode ray oscilloscope (CRO). The time

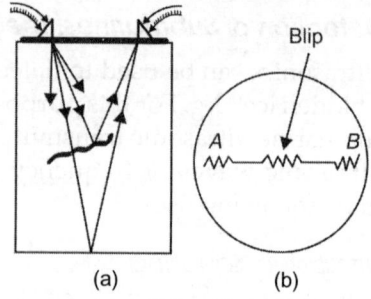

Fig. 7.8: Ultrasonic flaw detection.

base of CRO shows a blip A in the beginning corresponding to the signal received directly, and a blip B at the end due to the signal reflected from the far surface. Any signal due to flaw appears between A and B as shown in Fig. 7.8b. The interval between the first and the last blip gives a length scale by which the exact location of the flaw can be known as *low power application*.

Non-homogeneties

One can use ultrasonic waves for the study of non-homogeneities in a medium such as metal or plastic. For visual presentation one can convert the ultrasonic waves coming from medium into corresponding light waves.

Thickness of the Gauge

One can use ultrasonic waves for measuring the thickness of boiler, atomic pile structure where one side is not accessible by the process of reflection of ultrasonic wave as low power application.

Cavitation

When one places an ultrasonic transducer in a liquid, it produces standing wave formation and it is observed that these result in the development and implosion of bubbles. When subjected to powerful ultrasonic wave, pressure develops at some points in a liquid, such that excessive stress breaks a part of the liquid and produces hollow bubbles. However, these bubbles collapse soon and at the instant of implosion, the pressure around the bubbles becomes very high (~ several hundreds of atmosphere). This extremely high pressure results in emulsification. Interestingly, the cavitation bubbles formed by ultrasonic vibration hamper the propagation of ultrasonic waves and produces noise. This action of ultrasonic waves has been successfully used in various industrial devices described as follows as *high power application of ultrasonic waves*:

Soldering and metal cutting

One can use the cavitation action of ultrasonic waves for soldering, metal cutting and drilling processes in metals.

Preparation of uniform alloy

The process of cavitation has been used successfully for dispersion of metals in molten materials to obtain uniform alloying. By the use of ultrasonic waves it has become possible to disperse different metals such as Na, K, Hg, Pb, Zn, Cu, etc. and also their fusible alloys in oil, water and alcohol.

Preparation of very stable emulsion

The process of cavitation has been used for the preparation of very stable emulsion of two immiscible liquids such as oil and water.

Ultrasonic cleaning device

One can clear contaminated articles by irradiating ultrasonically in sequences in suitable solvents. Nowadays, one uses this technique for washing textiles. The large variation of acoustic pressure actually breaks off the contaminated particles from the surface in the ultrasonic cleaning.

Scientific Applications

Photographic Emulsion

The homogeneity and stability of photographic emulsions are improved by the application of ultrasonic waves. Thorough dis-persion of the dye in the emulsion under the influence of ultrasonic waves increases colour sensitivity of photographic emulsions.

Coagulation and Crystallisation

Using ultrasonics, one can bring the particles of a suspended liquid quite close to each other and thus coagulation may take place which helps in accelerating the rate of crystallisation.

Structure of Matter

Ultrasonics can be used for the study of structure of matter. This is a technique of ultrasonics displaying smaller and smaller in homogeneities in metal. It is reported that a limit of improvement is reached when the wavelength of the ultrasonic wave becomes comparable to the size of the crystal grain of the test materials for the frequency range of 1–20 Mc/s. Then due to the scattering of the ultrasonic waves at the boundaries of the crystal grain, multiple reflections and interferences are caused.

Medical and Biological Applications

Ultrasonography have a large number of applications in the field of medicine. Some important applications are as follows:

Neuralgic Pain

Ultrasonic waves are useful for relieving neuralgic and rheumatic pains. The affected portion of the body is exposed to ultrasonic waves. The waves produce a soothing massage action and relieves pain.

Detection of Abnormal Growth

Cerebral ventricles are explored by ultrasonic waves for locating extraordinary abnormal growth in the brain.

Arthritis

Ultrasonic waves are used to relieve pain due to arthritis. Here a small metal head, vibrating with a frequency of more than 10^6 Hz is moved over the skin of the patient. These vibrations after passing through the tissues, produce a deep massage action and the patient gets relieved of the pain.

Contracted Fingers

To restore the contracted fingers, ultrasonic waves are used. Ultrasonic waves are also used to loosen up the scar tissues in various parts of the human body.

Dental Cutting

Ultrasonic waves are used by dentists for the proper extraction of broken teeth. These waves have been found very useful by dentists because: (i) they make the cutting almost painless, (ii) they cut the hard material very easily, and (iii) they do not require any additional mechanical device for cutting purpose.

Bloodless Surgery

Ultrasonic waves are focussed with the help of a sharp instrument and the tissues are destroyed without any loss of blood. Doctors have used such instruments for conducting bloodless brain operations.

Sterilization

Ultrasonic waves can destroy unicellular organisms. Under the action of ultrasonic waves bacteria perish. Ultrasonic waves are also used in the sterilization of water and milk.

Biological Effects

When some small animals like rats, frogs, fishes, etc. are exposed to high intensity ultrasonic waves, they become lame or killed. This is the only destructive application of ultrasonic waves.

Ultrasonic waves are finding more and more practical applications in various fields. Active research work is still in progress to study the effect of ultrasonic waves in various fields.

7.6 PZT (LEAD ZIRCONATE TITANATE BASE) CERAMICS

There are number of materials, e.g. barium titanate, lead metaniobate, lead titanate, lead zirconate, as well as materials on lead zirconate titanate base (PZT) resemble barium titanate. These are used as sintered ceramic materials because it is not possible to produce large single crystal from them. These materials exhibit piezoelectric property, some of them are ferroelectric also. We may note that in contrast to quartz, lithium sulphate and other neutral piezoelectric crystals, ceramics are giving their piezoelectric properties by polarization.

One can form ceramic piezoelectric crystals in different shapes as per requirement. The ground, raw material mixed with binders, is moulded as per requirement by pressing and sintering, above 1000°C and then get shaped accurately by grinding as per requirement. We may note that PZT ceramics are white to yellowish materials of lower hardness and resistance than quartz and one can get their dull surfaces to be silver plated by baking process.

PZT ceramic material is first heated to 350° C (curie temperature) which is higher than the curie temperature 120°C for $BaTiO_3$, and then the material is permitted to cool off, a direct voltage of few thousand volts/cm thickness is applied. Due to this voltage, the small elementary crystals in the ceramic material, which oriented randomly earlier, now align along one axis and get frozen. Thus, the PZT ceramic materials become polarized along one axis, provided the material is not reheated close to the curie temperature due to the alignment along one axis, the PZT ceramic materials thus remain piezoelectric in the frozen state.

One can use ceramic piezoelectric materials for obtaining transducers with curved surfaces by sintering and grinding, at every point, normal to the surface during polarization. We may note that for quartz crystal in transducer ground with a curve, is not possible. However, if the transducer ground is excessively concave or convex, then these faces no longer contribute to the radiation of longitudinal waves. This helps in chewing the considerable concentrations of the sound field along the axis of the cylinder by using PZT ceramic piezoelectric over natural crystal. Lead metaniobate and lithium sulphate are the best for non-destructive testing.

7.7 ULTRASONIC TRANSDUCER

These are actually the energy converters. These convert the energy from one transmission system where it is activated to second transmission system either in the same form or in another form. This can be used with electronic amplifier for development of intense acoustic fields in solutions and materials at frequency from few thousand Hz to several megahertz. Magnetostriction bar and the piezoelectric crystal are the common forms of ultrasonic transducers, both of which have been found useful for converting electric energy to mechanical acoustic energy.

ILLUSTRATIVE EXAMPLES

Example 1: Determine the frequency of the first and second modes of vibration for a quartz crystal of piezoelectric oscillator. The velocity of longitudinal waves in quartz crystal is 5.5×10^3 m/s. Given, thickness of quartz crystal is 0.05 m.

Solution: The distance between the two faces of the crystal (thickness t) in the lowest mode of vibration will be $\lambda/2$. Thus

$$\lambda = 2t = 2 \times 0.05 = 0.1 \text{ m}$$

∴ The lowest frequency

$$n_1 = \frac{v}{\lambda} = \frac{5.5 \times 10^3 \text{ m/s}}{0.1 \text{ m}} = 5.5 \times 10^4 \text{ Hz}$$

In the second mode, the frequency would be

$$n_2 = 2n_1 = 11 \times 10^4 \text{ Hz}$$

Example 2: A quartz crystal of thickness 0.001 m is vibrating at resonance. Find the fundamental frequency. Given Y for quartz $= 7.9 \times 10^{10}$ N/m^2 and ρ for quartz $= 2650$ kg/m^3.

Solution: For longitudinal vibration, we have

$$v = \sqrt{Y/\rho}$$
$$Y = 7.9 \times 10^{10} \text{ N/m}^2$$
$$r = 2650 \text{ kg/m}^3$$

∴

$$v = \sqrt{\frac{7.9 \times 10^{10}}{2650}} = 5460 \text{ m/s}$$

For the fundamental mode of variation, we have

Thickness $(t) = \lambda/2$

∴

$$\lambda = 2t = 2 \times 0.001 = 0.002 \text{ m}$$

Frequency $n = v/\lambda = \dfrac{5460}{0.002}$

$$= 2730,000 \text{ Hz} = 2730 \text{ kHz}$$

Example 3: An ultrasonic beam is used to determine the thickness of a steel plate. It was found that the difference in two adjacent harmonic frequencies is 50 kHz. The velocity of sound in steel is 5000 m/s. Show that the thickness of the steel plate is 0.05 m/s.

Solution: The velocity of ultrasonic wave

$$v = 2Nd$$

where N is the fundamental frequency and d is the thickness

$$\therefore \qquad N = \frac{v}{2d}$$

We know that the harmonic frequencies are the multiples of the fundamental frequencies, we have

$$N_{(n)} - N_{(n-1)} = \frac{v}{2d}$$

$$v = 50,000 \text{ Hz}$$
$$= 5000 \text{ m/s}$$

$$\therefore \qquad d = \frac{v}{2\left[N_{(n)} - N_{(n-1)}\right]}$$

$$= \frac{5000}{2 \times 50,000}$$

$$= 0.05 \text{ m}$$

Example 4: Calculate the capacitance to produce ultrasonic waves of 10^6 Hz with an inductance of 1 henry.

Solution: We have

$$n = \frac{1}{2\pi}\sqrt{\frac{1}{LC}}$$

or

$$C = \frac{1}{4\pi^2 n^2 L}$$

$$n = 10^6/s$$
$$L = 1 \text{ henry}$$

$$= \frac{1}{4 \times (3.14)^2 \times (10^6)^2 \times 1}$$

$$= 0.025 \times 10^{-12} \text{ farad}$$
$$= 0.025 \text{ } \mu\text{F}$$

Example 5: A quartz crystal of thickness 0.001 m radiates ultrasonic waves of frequency 20 kHz and intensity 5×10^4 W/m² into water. Calculate (i) maximum acceleration (ii) maximum displacement. Velocity of sound in water = 1480 m/s and density of water = 10^3 kg/m³. Density of crystal = 2650 kg/m³.

Solution: We have

Sound intensity $(I) = \dfrac{p^2}{2\rho v}$ W/m², where P is the sound pressure, r is the density and v is the velocity of sound.

Now, sound pressure

$$P = \sqrt{2I\rho v}$$

$$I = 5 \times 10^4 \, \text{W/m}^2$$

$$\rho = 10^3 \, \text{kg/m}^3$$

$$v = 1480 \, \text{m/s}$$

$$\therefore \quad P = \sqrt{2 \times 5 \times 10^4 \times 10^3 \times 1480}$$

$$= 3.85 \times 10^5 \, \text{N/m}^2$$

Force = Pressure × Area

$$= \text{Mass} \times \text{acceleration}$$

Here $M \rightarrow$ Mass of the crystal = Volume × density

$$= A t D$$

i. Maximum acceleration

$$a = \frac{P}{Dt} = \frac{3.85 \times 10^5}{2.65 \times 10^3 \times 10^{-3}} = 1.45 \times 10^5 \, \text{m/s}^2$$

ii. Maximum displacement

$$y = \frac{a}{\omega^2} = \frac{1.45 \times 10^5}{(2\pi f)^2}$$

where $f = 20 \, \text{kHz} = 20{,}000 \, \text{Hz}$

$$\therefore \quad y = \frac{1.45 \times 10^5}{(2\pi \times 20{,}000)^2} = 9.4 \times 10^{-6} \, \text{m}$$

Example 6: Show that the natural frequency of 40 mm length of a pure iron rod is 49.75 kHz. Given the density of pure iron is $2.75 \times 10^3 \, \text{kg/m}^3$ and $Y = 115 \times 10^9 \, \text{N/m}^2$.

Solution:

$$n = \frac{1}{2L}\left(\frac{Y}{\rho}\right)^{1/2}$$

$$= \frac{1}{2 \times 40 \times 10^{-3} \, \text{m}}\left(\frac{115 \times 10^9 \, \text{N/m}^2}{7.25 \times 10^3 \, \text{kg/m}^3}\right)^{1/2}$$

$$= 12.5 \times \left(15.86 \times 10^6\right)^{1/2} \frac{1}{\text{m}}\left(\frac{\text{N m}^3}{\text{m}^2 \text{kg}}\right)^{1/2}$$

$$= 49750\frac{1}{\text{m}}\left(\frac{\text{kg} - \text{m}}{\text{s}^2} \times \frac{\text{m}}{\text{kg}}\right) = 49.75\text{kHz} \cdot$$

Example 7: Find the frequency of the first and second modes of vibration for a quartz crystal of piezoelectric oscillator. The velocity of longitudinal waves in quartz crystal is $5.5 \times 10^3 \, \text{ms}^{-1}$. Thickness of quartz crystal is 0.05 m.

Solution: Given, $v = 5.5 \times 10^3 \, \text{ms}^{-1}$; $t = 0.05 \, \text{m}$;

$$n_1 = ?; n_2 = ?$$

In the lowest mode of vibration, the distance between the two faces of the crystal of thickness t will be $\lambda/2$.

Therefore, $\qquad t = \dfrac{\lambda}{2}$

or $\qquad\qquad \lambda = 2t = 2 \times 0.05$

$\qquad\qquad\qquad \lambda = 0.1$ m

Therefore, the frequency in the first mode of vibration

$$v_1 = \frac{v}{\lambda} = \frac{5.5 \times 10^3}{0.1}$$

$$v_1 = 5.5 \times 10^4 \text{ Hz}$$

The frequency in the second mode of vibration is

$$v_2 = 2v_1 = 2 \times 5.5 \times 10^4$$

$\therefore \qquad\qquad v_2 = 110 \times 10^3$ Hz

Example 8: An ultrasonic source of 0.09 MHz sends down a pulse towards the seabed which returns after 0.55 sec. The velocity of sound in water is 1800 m/s. Calculate the depth of the sea and wavelength of pulse.

Solution: Given, $f = 0.09$ MHz $= 0.09 \times 10^6$ Hz;

$\qquad\qquad t = 0.55$ sec; $v = 1800$ ms^{-1}

Depth of the sea = ?; $\lambda_u = ?$

Depth of the sea, $d = \dfrac{vt}{2} = \dfrac{1800 \times 0.55}{2}$

$\therefore \qquad\qquad d = 495$ m.

The wavelength of the ultrasonic pulse is

$$\lambda_u = \frac{v}{f} = \frac{1800}{0.09 \times 10^6}$$

$\therefore \qquad\qquad \lambda_u = 0.02$ m.

Example 9: Calculate the frequency to which piezoelectric oscillator circuit should be tunned so that a piezoelectric crystal of thickness 0.1 cm vibrates in its fundamental mode to generate ultrasonic waves. (Young's modulus and density of material of crystal are 80 GPa and 2654 kg·m^{-3}).

Solution: Given, $E = 80$ GPa $= 80 \times 10^9$ Pa; $\rho = 2654$ kgm^{-3};

$\qquad\qquad t = 0.1$ cm $= 0.1 \times 10^{-2}$ m

The frequency of vibration is given by

$$f = \frac{P}{2t}\sqrt{\frac{E}{\rho}} = \frac{1}{2 \times 0.1 \times 10^{-2}}\sqrt{\frac{80 \times 10^9}{2654}}$$

$$= \frac{5490.28}{2 \times 10^{-3}}$$

$$f = 2.7451 \times 10^6 \text{ Hz}$$

REVIEW QUESTIONS AND PROBLEMS

1. Distinguish between audible, infrasonic and ultrasonic waves. Mention one source of each class.

2. What are ultrasonic waves? Describe a method for their production.

3. How are ultrasonics detected? Explain the magnetostriction method of producing ultrasonic energy and hence compare with piezoelectric method.

4. What is piezoelectric effect? Explain the generation of ultrasonics by piezoelectric generator. Show that only odd harmonics exist in this generator.

5. What is magnetostriction effect? Explain how is it used to design an ultrasonic generator. Give its working.

6. Describe a laboratory method to determine the velocity of ultrasonic waves in liquids.

7. Discuss the advantages and disadvantages of magnetostriction and piezoelectric methods in production of ultrasonic waves.

8. What are the various applications of the ultrasonic waves? Describe the non-destructive application in detail.

9. Bats, who are blind, fly about avoiding obstacles. Explain.

   [**Hint:** Bats have the natural power of emitting and detecting ultrasonic waves. When they fly, they emit continuously ultrasonic waves. These ultrasonic waves get reflected from the obstacles which happens to come in the path of flying bats. The reflected waves are immediately detected by the bats who change their path avoiding the obstacle].

10. What are PZT ceramics? How do these help in the construction of ultrasonic transducers?

11. Draw circuit diagram for production of ultrasonic waves using piezoelectric effect. Explain its working.

12. Draw circuit diagram of magnetostriction oscillator. Explain its working.

13. Discuss briefly the scientific application of ultrasonics.

14. Write short notes on the following:
    i. Piezoelectric effect
    ii. Magnetostriction effect
    iii. Echo sounding
    iv. Cavitation
    v. PZT ceramics
    vi. Ultrasonic transducers

15. A piezoelectric x-cut quartz crystal plate has a thickness of 1.6 mm. If the velocity of propagation of sound waves along the x-direction is 5760 m/s, calculate the fundamental frequency of the crystal. [**Ans.** 1.8 MHz]

16. A quartz crystal of thickness 0.005 m is vibrating at resonance. Calculate the fundamental frequency. Given Y for quartz is 7.9×10^{10} N/m^2 and ρ for quartz is 2650 kg/m^3. [**Ans.** 5.46×10^5 Hz]

17. Ultrasonic pulse echo method is employed to detect possible defect in steel bar of thickness 40 cm. If the pulse arrival times are 30 and 80 microseconds, find the distance of the defect. [**Ans.** 15 cm]

[**Hint.** Let x be the distance of the defect from the end of the bar at which the pulse enters the bar. The pulse arriving first is the one reflected from the defect and that arriving later on is the one reflected from the other end.

Obviously, the pulse takes 30 microseconds to cover a distance of $2x$, and 80 microseconds to cover a distance 80 cm.

Hence $30 = \dfrac{2x}{v}$ and $80 = \dfrac{80}{v}$.

Solving, one obtains $x = 15$ cm]

18. Longitudinal standing waves are set-up in a quartz plate with antinodes at opposite faces. The fundamental frequency of vibration is given by

$$N = \frac{2.87 \times 10^5}{t}$$

where N is in cycles/s and t is the thickness of the plate in cm. Compute (a) Young's modulus of the quartz plate (b) the thickness of the plate required for a frequency of 1200 kc/s. The density of quartz is 2.66×10^3 kg/m^3.

[**Ans.** $Y = 8.76 \times 10^{10}$ N/m^2, $t = 0.24$ cm]

19. An ultrasonic source of 0.07 MHz sends down a pulse towards the seabed, which returns after 0.61 s. The velocity of sound in sea water is 1700 m/s. Show that the depth of sea is 552.5 m and the wavelength of pulse is 0.0243 m.

[**Hint.** $2d = vt = 1700 \times 0.65$

∴ $d = 552.5$ m

$v = n\lambda$ ∴ $\lambda = v/n = 1700/0.07 \times 10^6$

            $= 2.43 \times 10^{-2}$ m

            $= 0.0243$ m]

SHORT ANSWER QUESTIONS

1. What are infrasonic waves?

Ans. Those longitudinal waves whose frequency lies below 20 Hz are called *infrasonic waves*. These waves are produced by large vibrating bodies such as during an earthquake and our ear is not sensitive to listen these waves.

2. What are hypersonic waves?

Ans. Longitudinal waves of frequency higher than 10^8 Hz are called *hypersonic waves*.

3. When does the ultrasonic generator delivers maximum power in piezoelectric method of production of ultrasonics?

Ans. The ultrasonic generator delivers maximum power when it is operated at the fundamental frequency of the crystal. In order to generate higher frequency ultrasonic waves, the L–C circuit is made to oscillate at a frequency equal to one of the harmonics of the crystal.

4. What happens when ultrasonic waves pass through a substance?

Ans. Intense heat is produced.

5. Can a glass rod oscillating with ultrasonic frequency lose a hole through steel, diamond?

Ans. Yes.

OBJECTIVE QUESTIONS

1. Ultrasonic waves are used to detect tumours, breast cancer and also growth of .... can be studied.   [foetus]
2. Weak ultrasonic waves are able to produce good .... of two immissible liquids, so it is used for the manufacture of good ....   [emulsion, paints]
3. The speed of propagation of ultrasonics increases with increasing .... [frequency]
4. Intense ultrasonic radiation has a disruptive effect on liquid by causing .... to be formed.   [bubbles]
5. Bats, who are blind, fly about avoiding ....   [obstacles]
6. Magnetostriction consists in changing of length of a ferromagnetic material subjected to a ....   [magnetic field]
7. Sound waves having frequencies above 20 kHz are inaudible and are called .... waves.   [ultrasonic]
8. The phenomenon of magnetostriction and piezoelectric effect are used to produce .... waves.   [ultrasonic]
9. Certain crystals undergo periodic mechanical deformation when subjected to AC voltage. This is knows as ....
                                                                       [piezoelectric effect]
10. The materials that exhibit changes in their mechanical dimen-sions, create pressure variations in the surrounding air. These pressure variations of high frequency constitute ....   [bubbles]

MULTIPLE CHOICE QUESTIONS

1. The frequency of ultrasonic sound wave is
   (a) greater than 5 kHz   (b) greater than 10 kHz
   (c) greater than 20 kHz   (d) less than 20 kHz
2. When a ferromagnetic material in the form of a rod is subjected to an alternating magnetic field, the rod undergoes alternate contractions and expansions at a frequency equal to the frequency of the applied magnetic field. This phenomenon is called
   (a) hysteresis   (b) meissner effect
   (c) magnetostriction effect   (d) piezoelectric effect
3. When pressure is applied to one pair of opposite faces of crystals like quartz, tourmaline, rochelle salt, etc. cut with their faces perpendicular to its optic axis, equal and opposite charges appear across its other faces. This phenomenon is known as
   (a) meissner effect   (b) superconductivity
   (c) magnetostriction effect   (d) piezoelectric effect
4. If an alternating voltage is applied to one pair of opposite faces of the crystal, alternatively mechanical contractions and expansions are produced in the crystal and the cystal starts vibrating. This phenomenon is known as
   (a) piezoelectric effect   (b) inverse piezoelectric effect
   (c) magnetostriction effect   (d) meissner effect
5. Fathometer or echometer is a device which is directly calibrated to determine the
   (a) height of a satellite   (b) depth of the sea
   (c) frequency of a tuning fork   (d) flaws and cracks in metals

6. When ultrasonic waves propagate in a liquid medium, the alternating compressions and rarefactions vary the density of the medium. This change in density results in the variation of

(a) density of the liquid
(b) temperature of the liquid
(c) pressure of the liquid
(d) refractive index of the liquid

ANSWERS

1. (c)   2. (c)   3. (d)   4. (b)   5. (b)   6. (d)

Appendices

APPENDIX I

SURFACE WAVES IN A LIQUID

Waves on the surface of a liquid are examples of wave motion in one direction. These are the most familiar kind of waves, e.g. the waves, we observe on the ocean and lakes or simply when we drop a stone in a pond. A study of its exact nature is complicated yet interesting.

Let us consider the water contained in a long rectangular tank of depth h as shown in Fig. A1.1(a). To make treatment simple, let us consider a sinusoidal disturbance is created over the surface along the length of the tank by some source (e.g. wind blowing over the surface or by dipping a vibrating rod) which excites water wave. In this way, all the waver in the tank is a transverse harmonic disturbance of surface as shown in Fig. A1.1(b).

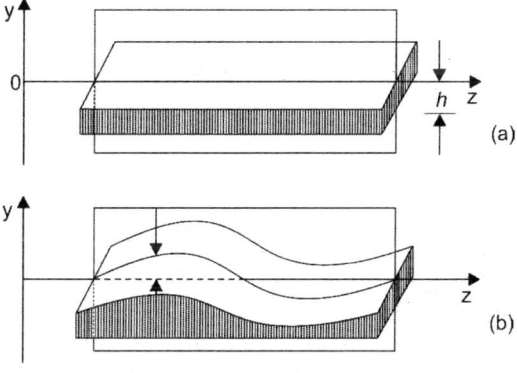

Fig. A1.1

Here, we will consider the motion confined to Y–Z plane which is parallel to the walls of tank. We are justified in our assumption because of the symmetry of the problem. Z-axis coincides with the equilibrium level of water surface. Let the instantaneous displacement of the surface be represented by ξ. In order to study the wave motion in liquid, we will make some assumption. The main assumption is that the viscosity of water (liquid) is zero, so that the damping effects and induction of rotational motion due to viscosity are neglected.

When the displacement of a point on the surface is downwards, the water beneath that point must move out sideways into a region where the surface movement is upwards making $\xi > 0$. This is due to the fact that the liquid (water) under consideration is assumed to be incompressible. This causes the displacement ξ as a function of the vertical coordinates y as well as the horizontal coordinates z, as the disturbance propagates along z-direction. Because of wave motion, a particle located at (y, z) will have displacement defined by $\xi_y(y, z, t)$ and $\xi_z(y, z, t)$ as measured from point (y, z) (Fig. A1.2). Since the disturbance on the surface is assumed to be harmonic, the displacements ξ_1 and ξ_2 also vary harmonically. This means that the motion of the water particle at any point is super-position of two orthogonal vibrations—one along Y-axis and other along Z-axis. The nature of the propagating harmonic wave is represented by the term $e^{i(\omega t - kz)}$, where ω is angular frequency and k is wave vector.

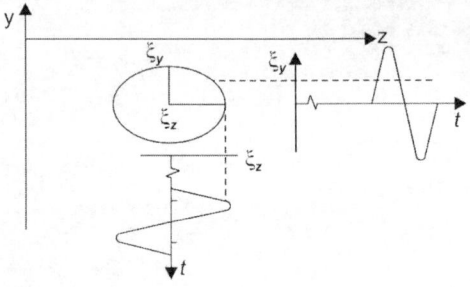

Fig. A1.2

The phase difference between the displacements ξ_y and ξ_z, can be understood as follows. Let us consider the case when displacement $\xi_y = 0$ at $y = 0$. At that point ξ_y has a maximum slope at the surface and the maximum rearrangement of water must take place there. As a result of the displacement ξ_z must be maximum. It naturally follows that the phase difference between displacements ξ_y and ξ_z must be $\pi/2$.

The last and final assumption that we will make is that the amplitude of the displacement depends only on the depth of water (liquid) alone and not on the other coordinates. Keeping these assumptions in mind, we can write the equations for horizontal and vertical displacements as follows:

$$\xi_y = A_y(y) \cos(\omega t - kz) \quad \text{and} \quad \xi_z = A_z(y) \sin(\omega t - kz) \tag{A1.1}$$

This gives us

$$\frac{\xi_y^2}{A_y^2} + \frac{\xi_z^2}{A_z^2} = 1$$

This is an equation of an ellipse. This means that the path of each water particle participating in the wave motion is an ellipse centered at its equilibrium position (Fig. A1.2).

In case of fluid motion, the measurement of velocity is much easier than the measurement of displacement. As such, the velocities associated with the displacement ξ_y and ξ_z can be easily equated. Then:

$$V_z = \frac{\partial \xi_z}{\partial t} = \omega A_z(Y) \cos(\omega t - kz)$$

and

$$V_y = \frac{\partial \xi_y}{\partial t} = - \omega A_y(Y) \sin(\omega t - kz) \tag{A1.2}$$

With the assumptions that the small change in amplitude at the points (y, z) and $(y + \xi_y, z + \xi_z)$ are neglected, the negative sign merely shows the phase difference of $\pi/2$ between the two.

Let us now find a connecting equation in terms of measurable qualities for evaluation of A_y and A_z.

1. The mass of water in small volume element, anywhere inside the water should remain unchanged as water is incompressible [Fig. A1.3(a)] and this is mathematically

expressed as $\Delta \cdot v = 0$, i.e. divergence of vector point function is zero and $v_x = 0$, which gives

$$\frac{\partial v_y}{\partial y} + \frac{\partial v_z}{\partial z} = 0 \tag{A1.3}$$

2. The assumption that viscosity of water being zero implies that there are no shear forces across the different layers. In such a case, the work done by viscous forces over a closed loop $ABCD$ [Fig. A1.3(b)] must be zero. The use of Stokes theorem allows to express the line integral in terms of the surface integral. Thus

$$\oint v \cdot dl = \int_s (\Delta \times v) \cdot ds = 0 \tag{A1.4}$$

This result implies that each vector component of the elemental integral is zero, i.e.

$$\left(\frac{\partial v_z}{\partial y} - \frac{\partial v_y}{\partial z}\right) \Delta y \, \Delta z = 0 \quad \text{or} \quad \frac{\partial v_z}{\partial y} = \frac{\partial v_y}{\partial z} \tag{A1.5}$$

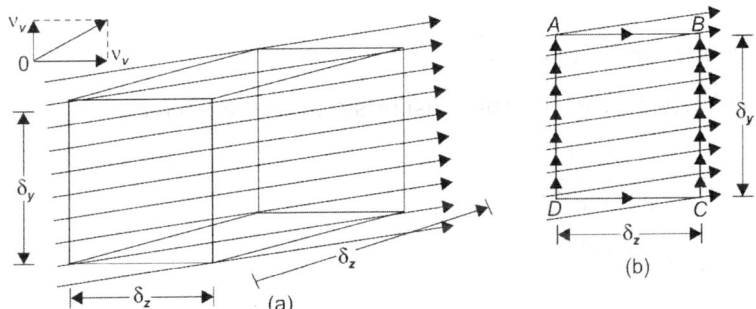

Fig. A1.3

Equation (A1.3) gives $\dfrac{\partial v_y}{\partial y} = -\dfrac{\partial v_z}{\partial z}$ and Eq. (A1.2) on differentiation gives

$$\frac{\partial v_z}{\partial z} = + \omega k A_z \sin(\omega t - kz)$$

and

$$\frac{\partial v_y}{\partial y} = - \omega \frac{\partial A_y}{\partial y} \sin(\omega t - kz)$$

This yields

$$\frac{\partial A_y}{\partial y} = k A_z \tag{A1.6}$$

Again differentiation of Eq. (A1.2) gives

$$\frac{\partial v_z}{\partial y} = \omega \frac{\partial A_y}{\partial y} \cos(\omega t - kz)$$

and

$$\frac{\partial v_y}{\partial z} = k \omega A_y(Y) \cos(\omega t - kz)$$

and using Eq. (A1.5), we have

$$\frac{\partial A_z}{\partial y} = k A_y \tag{A1.7}$$

differentiating Eq. (A1.6) w.r.t. y, we have

$$\frac{\partial^2 A_y(Y)}{\partial y^2} = k\frac{\partial A_z}{\partial y} = k^2 A_y \tag{A1.8}$$

and the solution of this equation is

$$A_y(Y) = \alpha e^{ky} + \beta e^{-ky} \tag{A1.9}$$

where α and β are unknown constants to be determined from the boundary conditions.

We know that the top layer of water corresponds to $y = 0$, where we have sinusoidal wave, i.e.

$$\xi_y(y,z,t) = \xi_y(0,z,t) = A\cos(\omega t - kx)$$

If we assume $A_y(0) = A$, then we have

$$A = \alpha + \beta \tag{A1.10}$$

Further at the bottom of the tank, i.e. $y = -H$, the vertical displacement is zero. This gives

$$A_y(-h) = \alpha e^{-kH} + \beta e^{kH} = 0 \tag{A1.11}$$

Solving Eqs (A1.10 and A1.11) for constants α and β, we have

$$\alpha = \frac{Ae^{kH}}{e^{kH} - e^{-kH}}$$

$$\beta = -\frac{Ae^{-kH}}{e^{kH} - e^{-kH}}$$

$$\therefore \qquad A_y(y) = A\frac{e^{k(H+y)} - e^{-k(H+y)}}{e^{kH} - e^{-kH}}$$

or $\qquad A_y(y) = A\frac{\sin h\{k(H+y)\}}{\sin h k H}$

Similarly, we can prove

$$A_z(Y) = A\frac{\cos h[k(H+y)]}{\sin h(kH)} \tag{A1.12a}$$

where $A_y(Y)$ and $A_z(Y)$ define the two axis of the elliptical path of water particles with their ratio given by

$$\frac{A_y(Y)}{A_z(Y)} = \tan h[k(H+y)] \le 1 \tag{A1.13}$$

Thus, the elliptical path is in the form of an ellipse flattened horizontally. As the depth H increases, the vertical amplitude decreases much faster until at the bottom-most layer where the motion is just along a horizontal line, the water just rocks the ground forward and backward.

The conclusion to be drawn from these solutions is that for the wave propagation ξ_z and ξ_y differs in phase by 90° with former lagging behind and each water particle describes an ellipse in clockwise sense as viewed in positive X-direction (Fig. A1.4). Further as depth increases, the flattening of the ellipse sets in and ultimately the vertical amplitude becomes zero at the bottom.

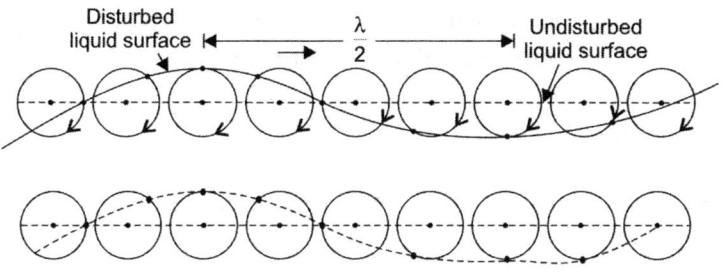

Fig. A1.4

DISPERSION RELATION

The undisturbed surface of liquid is plane and horizontal. A disturbance of the surface produces a displacement of all molecules directly underneath the surface (Fig. A1.4). Each volume element of the liquid describes a closed path. The amplitude of the horizontal and the vertical displacements of a volume element of a fluid varies in general with the depth. Of course, the molecules at the bottom do not suffer any vertical displacement, since they cannot separate from the bottom. On the surface of the liquid certain forces enter into play, in addition to that due to the atmospheric pressure. One force is due to the surface tension of the liquid, which gives an upward force on an element of the surface, similar to that found in the case of a string. Another force is the weight of the liquid above the undisturbed level.

To understand the effect of these forces on the wave motion, we will make use of Bernoulli's theorem, which is applicable to the steady state flow of liquids, i.e., the flow in which motion does not depend upon time. The water waves discussed above are not in steady state. However, the steady state conditions can be achieved by describing the dynamics of water waves in coordinate system moving with a velocity (ω/k), same as that of the motion of water waves along the Z-axis. In the moving coordinate system, a water particle with the coordinate z in the stationary system has coordinate $z' = z + (\omega t/k)$. The velocity components using equation (2) are

$$v_y = \xi_y = \omega A_y(Y)\sin kz'$$

and
$$v_z = \xi_z = \omega A_z(Y)\cos kz' \qquad (A1.14)$$

It is clear from Eq. (A1.14) that $\xi_y(v_y)$ and $\xi_y(v_z)$ are independent of time and hence represent the steady state components. The surface of the water appears stationary in this moving coordinate system. Bernoulli's theorem says that total energy per unit mass at every point lying on a streamline must be the same. The total energy per unit mass is given by

$$E = \frac{P}{\rho} + \frac{v^2}{2} + U \qquad (A1.15)$$

where P is the pressure, v is the velocity of the water particles, ρ is the density and U is the potential energy per unit mass. The pressure P at any point on surface is due to the atmospheric pressure, which is constant and the surface tension. As the wave disturbance travels, the surface of the water is subjected to deviation from equilibrium configuration and the pressure due to the surface tension will be acting on curved boundaries. The approximate value of the curvature as it develops along the flow direction Z, is given by

$\left(\dfrac{\partial^2 \xi_y}{\partial z^2}\right)$. If $\left(\dfrac{\partial^2 \xi_y}{\partial z^2}\right)$ is positive, the surface is concave upwards and for a negative, the surface is convex upward. On the water surface positive curvature results in the reduction of pressure. Hence the total pressure at any point is given by

$$P = P_a - \sigma\left(\frac{\partial^2 \xi_y}{\partial z^2}\right)$$

where P_a is atmospheric pressure and σ the surface tension. For a sinusoidal surface wave, in the moving system, we have

$$\xi_y = A_y \cos kz'$$

and
$$\left(\frac{\partial^2 \xi_y}{\partial z'^2}\right) = \left(\frac{\partial^2 \xi_y}{\partial z^2}\right) = -k^2 A_y \cos kz'$$

The total pressure at any point is given by

$$P = P_a + \sigma k^2 A_y \cos kz' \tag{A1.16}$$

The term $\frac{1}{2}v^2$ is given by:

$$\frac{v^2}{2} = \frac{1}{2}\left[v_y^2(0, z') + v_z^2(0, z')\right]$$

Since the surface corresponds to $y = 0$, Eq. (A1.14) gives

$$\frac{v^2}{2} = \frac{1}{2}\left[\omega^2 A_y^2 \sin^2 kz' + \left(\omega A_z \cos kz' - \frac{\omega}{k}\right)^2\right]$$

$$= \frac{1}{2}\left[\omega^2 A_y^2 \sin^2 kz' + \omega^2 A_z^2 \cos^2 kz' + \frac{\omega^2}{k^2} - \frac{2\omega^2}{k} A_z \cos kz'\right]$$

But we know that

$$\frac{A_y}{A_z} = \tan h[k(H + y)] = \tan h(kH) \quad \text{for} \quad y = 0$$

$$\therefore \quad \frac{1}{2}v^2 = \frac{1}{2}\left[\omega^2 A_y^2 \sin^2 kz' + \omega^2 A_y^2 \cot h^2(kH) \times \cos^2 kz' + \frac{\omega^2}{k^2} - \frac{2\omega^2}{k^2} A_y \cot h(kH)\cos kz'\right]$$

Neglecting first two terms as they are very small for small amplitude (A_y) waves, we have

$$\frac{1}{2}v^2 = \frac{\omega^2}{2k^2} - \frac{\omega^2}{k} A_y \cot h(kH)\cos kz' \tag{A1.17}$$

The potential energy per unit mass in a gravitational field is given by

$$U = g\xi_y = gA_y \cos kz' \tag{A1.18}$$

where g is the acceleration due to gravity and the potential energy is due to the weight of water lifted upwards. Here gravity tries to restore the sinusoidal surface to the equilibrium plane $y = 0$. Substituting the values, we get for the total energy as

$$E = \frac{P_a}{\rho} + \sigma k^2 A_y \cos kz' + \frac{\omega^2}{2k^2} - \frac{\omega^2}{k} A_y \cot h(kH)\cos kz' + gA_y \cos kz' \tag{A1.19}$$

This energy should be same at all points z' and hence is independent of z'. This implies that the co-efficient of $\cos kz'$ should be zero. This gives

$$\sigma k^2 A_{ij} - \frac{\omega^2}{k} A_y \cot h(kH) + gA_z = 0$$

or

$$\frac{\omega^2}{k} = \frac{g + \sigma k^2}{\cot h(kH)}$$

or

$$\omega^2 = (s^2 k^3 + gk) \tan h(kH) \qquad \text{(A1.20)}$$

Equation (A1.20) gives the dispersion relation for water waves. This dispersion relation is valid for the incompressible and non-viscous fluids as discussed earlier. Phase velocity is given by

$$v = \frac{\omega}{k} = \sqrt{\left(\frac{\sigma k}{\rho} + \frac{g}{k}\right) \tan h(kH)} \qquad \text{(A1.21)}$$

The most interesting aspect of Eq. (A1.21) is that the velocity of propagation depends on the wavelength, a situation never encountered in wave motion. The plot of phase velocity for water waves as a function of wavelength is shown in Fig. A1.5.

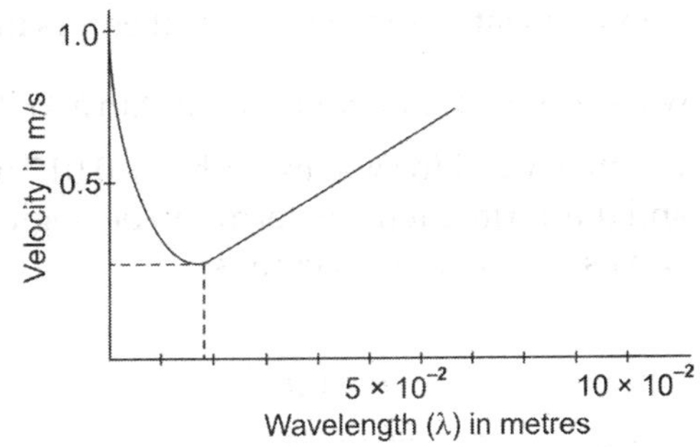

RIPPLES OR SHALLOW WATER WAVES (OSCILLATORY WAVES)

The dispersion relation given by Eq. (A1.20) contains terms due to gravity and surface tension. Either the gravity or the surface tension becomes significant depends on the

Fig. A1.5

wavelength $\left(k = \dfrac{2\pi}{\lambda}\right)$ of the water wave. The two terms are equal when

$$\frac{\sigma k}{\rho} = \frac{g}{k} \quad \text{or} \quad k = \sqrt{\frac{g\rho}{\sigma}}$$

Under this condition, wavelength

$$\lambda = 2\pi \sqrt{\frac{\sigma}{g\rho}}$$

For water $\rho = 10^3$ kg/m^3, $\sigma = 0.073$ N/m and $g = 10$ m/s^2 giving $\lambda = 0.017$ m.

Thus, the water waves with $\lambda < 0.017$ m are governed mainly by surface tension while gravity is the governing parameter for waves with $\lambda > 0.017$ m.

The surface tension governed waves constitute ripples in which water surface is curved more and $kH \gg 1$ as $k \left(= \dfrac{2\pi}{\lambda}\right)$ is very large. This condition is readily realized when H is not very small. For $kH \gg \tan h(kH) \approx 1$ and the dispersion is given by

$$\omega = \sqrt{\frac{\sigma k^3}{\rho}}$$

and the phase velocity

$$v = \frac{\omega}{k} = \sqrt{\frac{\sigma k}{\rho}} = \sqrt{\frac{2\pi\sigma}{\lambda\rho}} \tag{A1.22}$$

and the group velocity is given by

$$v_g = \frac{d\omega}{dk} = \frac{3}{2}\left(\frac{\sigma k}{\rho}\right)^{\frac{1}{2}} = \frac{3}{2}v \tag{A1.23}$$

It is important to note that the group velocity is greater than the phase velocity and that the individual waves appear to travel backwards through the wave group. Further the phase velocity increases as the wavelength decreases and the dispersion is seen to be anomalous.

GRAVITY WAVES

Depth of water is important as it enters into the dispersion relation as parameter. For deep water $kH \gg 1$ and we may put $\tan h(kH) \approx 1$. For smaller depths $\tan h(kH) \approx kh - \frac{1}{3}(kH)^3$, which is valid for waves with $\lambda > 0.017$ m. These waves are mainly controlled by gravity and the surface tension term can be neglected. When $kH \gg 1$ as H is usually large for these waves, the relation becomes:

$$\omega = (gk)^{\frac{1}{2}}$$

and the phase velocity is given by

$$v = \frac{\omega}{k} = \left(\frac{g}{k}\right)^{\frac{1}{2}}$$

and group velocity

$$v_g = \frac{d\omega}{dk} = \frac{1}{2}\sqrt{\frac{g}{k}} \tag{A1.24}$$

Thus

$$v_g = \frac{1}{2}v \tag{A1.25}$$

This is normal dispersion. Waves in most of the fluids are gravity waves. A familiar example is the ocean waves. An ocean wave of wavelength $\lambda = 100$ m travels with a velocity of 10 m/s but the energy of the wave travels at only half the velocity.

It is generally the storm that cause *ocean waves* to rush towards the shore. In fact, it is even possible to predict the distance of the origin of an ocean storm from the shore. If τ is the time interval between two successive waves reaching the shore, we have

$$\tau = \frac{\lambda}{v} = \sqrt{\frac{2\pi\lambda}{g}} \tag{A1.26}$$

where v is the phase velocity of the wave. If the storm breaks at time t_0 at a distance L inside the ocean, then the time taken by the storm disturbance to reach the shore is given by

$$t = t_0 + \frac{L}{v} = t_0 + \frac{L\tau}{\lambda} \tag{A1.27}$$

Eliminating λ from Eqs (A1.26 and A1.27), we have

$$\tau = \frac{2\pi L}{g(t - t_0)}$$

and the time variation of τ is given by

$$\frac{-d\tau}{dt} = \frac{2\pi L}{g(t - t_0)^2} = \frac{g\tau^2}{2\pi L}$$

Thus, by measuring τ and $-\left(\dfrac{d\tau}{dt}\right)$, L can be determined.

The waves in shallow water are nearly longitudinal. When $kH \ll 1$, we use $\tan h(kH) \approx kH - \dfrac{1}{3}(kH)^3$. This gives dispersion relation as

$$\omega^2 = gHk^2\left(1 - \frac{H^2 k^2}{3}\right)$$

or

$$v_g = \frac{\omega}{k} = \sqrt{gH}\left(1 - \frac{H^2 k^2}{3}\right)^{\frac{1}{2}}$$

$$\therefore \qquad \omega = k\sqrt{gH}\left(1 - \frac{H^2 k^2}{3}\right)^{\frac{1}{2}}$$

$$= k\sqrt{gH}\left(1 - \frac{H^2 k^2}{6}\right)$$

$$\therefore \qquad \omega = kv(1 - dk^2)$$

with $\qquad v = \sqrt{gH} \quad \text{and} \quad d = \dfrac{H^2}{6}$ \qquad (A1.28)

Thus, we see that v_g is less than v when $d > 0$ and $v_g = v = \sqrt{gH}$ for small dk^2.

In shallow water (for $kH \ll 1$), we have $\sin h(kH) \approx kH$ and $\sin h[k(H + y)] \approx k(H + y)$ and $\cos h[k(H + y)] = 1$. Thus,

$$\xi_y \approx A\left(1 + \frac{y}{H}\right)\cos(\omega t - kz)$$

$$\xi_z \approx A\left(\frac{1}{kH}\right)\sin(\omega t - kz) \tag{A1.29}$$

It is seen that the horizontal amplitude does not depend upon the vertical position but it is very large due to the term $1/kH$ where $kH \ll 1$. Thus, the shallow water waves are almost longitudinal. To have a physical insight into the wave motion, we define the longitudinal velocity as

$$\xi_y(z,t) = \left(\frac{\omega A'}{kH}\right)\sin(\omega t - kz) = \frac{\omega}{kH}\xi(z,t) = \sqrt{\frac{g}{H}}\,\xi$$

We have already discussed the relation between velocity and amplitude. We observe

that $\sqrt{\dfrac{g}{H}}$ is proportional to the characteristic impedance of the medium.

A familiar example is that of the waves at shore. The approaching waves slow down as H decreases gradually. To conserve energy, the amplitude of the wave increases while approaching the shore. This wave ultimately breaks into surf, at which stage the analysis of small amplitude motion is no longer valid. The waves of longer wavelength, generated in deep sea due to earthquakes approach the coast. The dispersion is nominal as $kH << 1$ and the energy gets concentrated into mighty waves rests with devastating consequences on reaching the shore.

LIMITING VALUES OF PHASE VELOCITY

The general expression for the velocity of propagation of surface waves in a liquid is

$$v = \sqrt{\left(\frac{g}{k} + \frac{\sigma k}{\rho}\right)\tan h(hk)} = \sqrt{\left(\frac{g\lambda}{2\pi} + \frac{2\pi}{\lambda}\frac{\sigma}{\rho}\right)\tan h\left(\frac{2\pi}{\lambda}H\right)}$$

when depth H is very large compared to the wavelength, i.e. $\dfrac{2\pi H}{\lambda} \gg 1$ the value of

$\tan h\left(\dfrac{2\pi H}{\lambda}\right) \approx 1$ and replacing $\tan h(Hk)$ by 1 in the expression for velocity, we have

$$v = \sqrt{\left(\frac{g\lambda}{2\pi} + \frac{2\pi\sigma}{\lambda\rho}\right)}$$

On the other hand, when the depth H is very small compared with the wavelength λ,

the quantity $Hk = \dfrac{2\pi H}{\lambda} \ll 1$ and using approximation $\tan h(x) \approx x$, we can write

$\tan h\left(\dfrac{2\pi H}{\lambda}\right) \approx \dfrac{2\pi H}{\lambda}$. Thus,

$$v = \sqrt{\left(\frac{g\lambda}{2\pi} + \frac{2\pi\sigma}{\lambda\rho}\right)\frac{2\pi H}{\lambda}} = \sqrt{gh + \frac{4\pi^2 H\sigma}{\lambda^2\rho}}.$$

Since we are assuming relatively large value of λ, as such neglecting the term $\dfrac{4\pi^2 H\sigma}{\lambda^2\rho}$,

we have

$$v = \sqrt{gH}$$

In these circumstances, the velocity is independent of the wavelength.

Example A1.1: If a fairly large stone of diameter D is dropped into deep water, most of the energy will go into gravity waves whose wavelength is roughly $2D$, show that the approximate radius of the ring of disturbed water is $\sqrt{\dfrac{gDt^2}{\pi}}$, where t denotes the time interval after the stone was dropped.

Solution: Velocity of the deep water waves is given by

$$v = \left(\frac{g\lambda}{2\pi}\right)^{\frac{1}{2}} = \left(\frac{g2D}{2\pi}\right) = \sqrt{\frac{gD}{\pi}}.$$

In a time t sec after dropping the stone, the disturbance travels a distance vt which represents the radius of the disturbed water around the centre of disturbances

$$vt = \sqrt{\frac{gD}{\pi}}\, t = \sqrt{\frac{gDt^2}{\pi}}$$

Example A1.2: For a canal of 5 m depth, estimate the phase velocity and group velocity for wavelength (a) 0.01 m, (b) 1 m, (c) 100 m.

Solution: (a) A wavelength of 0.01 m (< 0.017) represents ripples. The phase velocity is

$\sqrt{\dfrac{2\pi\sigma}{\lambda\rho}}$, where surface tension $\sigma = 0.073$ N/m, for water

$\therefore$
$$v = \sqrt{\frac{2\pi \times 0.073}{1000 \times 0.01}} = 0.21\,\text{m/s}$$

Group velocity

$$v_g = \frac{3}{2}v = \frac{3}{2} \times 0.21 = 0.315 \text{ m/s}$$

(b) A wavelength of 1 m (> 0.017) corresponds to gravity waves and the phase velocity is given by

$$v = \sqrt{\frac{g\lambda}{2\pi}} = \sqrt{\frac{9.8 \times 1}{2\pi}} = 1.25 \text{ m/s}$$

and group velocity

$$v_g = \frac{1}{2}v = \frac{1}{2} \times 1.25 = 0.62 \text{ m/s}$$

(c) A wavelength of 100 m in a canal of depth 5 m represents gravity waves in shallow water. For these waves

$$kH = \frac{2\pi}{100} \times 5 = 0.314 < 1$$

Under such conditions $\tan h(kH) = kH - \dfrac{1}{3}(kH)^3 + \dots$

The dispersion relation becomes

$$\omega^2 = gH\left(1 - \frac{k^2H^2}{3}\right)k^2$$

The phase velocity

$$v = \frac{\omega}{k} = \sqrt{gH\left(1 - \frac{k^2 H^2}{3}\right)} = \sqrt{gH}\left(1 - \frac{k^2 H^2}{6}\right)$$

and the group velocity

$$v_g = \frac{d\omega}{dk} = \sqrt{gH}\left(1 - \frac{k^2 H^2}{2}\right)$$

where kH being very small and can be neglected in both the expressions. Thus

$$v_g = v = \sqrt{gH}$$

i.e., there is no dispersion

$$\therefore \qquad v = v_g = \sqrt{gH} = \sqrt{5 \times 9.8} = 7 \text{ m/s}$$

APPENDIX **II**

DIFFRACTION

Diffraction is the phenomenon characteristic of wave motion. Diffraction is observed when a wave is distorted by an obstacle which has dimensions comparable to the wavelength of the wave. The obstacle may be a screen with a small opening or slit which allows only a small portion of the incident wavefront to pass. The obstacle may also be a small object, such as a wire or a disc that blocks the passage of a small portion of the wave front.

If a stream of particles fall on a screen, which has small opening, only those falling on the opening will be transmitted and allowed to continue their motion undisturbed (Fig. A2.1). The others will be either stopped or will bounce back. Conversely, if an object is placed in a stream of particles, it will block those particles falling on it, but the remaining particles will continue their motion undisturbed. However, we know from common experience, specially for the case of sound waves and surface waves in water, that waves behave in a different way and that they extend around the obstacles interposed in their path. as shown in Fig. A2.2. This effect becomes more and more noticeable as the dimensions of the slits or the size of the obstalces approach the wavelength of the waves. One cannot usually observe diffraction of light with the naked eye, since most of the objects interposed in a beam of light are much larger than the wavelength of light whose magnitude is of the order of 5×10^{-7} m.

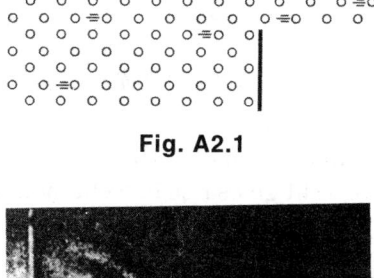

Fig. A2.1

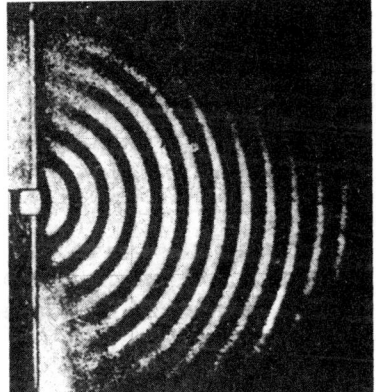

Fig. A2.2

HUYGENS–FRESNEL PRINCIPLE

The effect of obstruction on an advancing wave front can be explained by the Huygens–Fresnel principle. As per this principle, a given wave front H–B–G–I (Fig. A2.3) becomes responsible for the development of a similar wave front D–C–E–F at a later time. This is so because each point such as *bbbb* gives rise to a secondary wavelet K–L and the net effect of these wavelets emanating from every point on H–B–G–I is to reinforce along D–C–E–F, which is the tangent to all of them at a given instant. When a travelling wave falls on a slit, the original wave front at the slit acts like a source of secondary wavelets. The disturbance beyond the obstacle is a linear superposition of the waves spreading out from the secondary sources.

In wave propagation, every point has a similar displacement and transmits its energy to *neighbouring points*. Hence each point of the wave front can be treated as a source of secondary waves with the same characteristics as the original wave. This concept of wave propagation becomes relevant

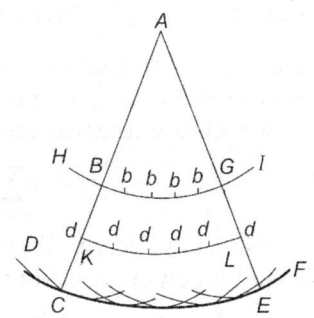

Fig. A2.3

when we deal with the diffraction and interference of waves. As the medium of wave propagation is homogeneous, the obstruction along the path of a wave can stop it from propagating into a shadow region, since the medium is a coupled system and once the disturbance is initiated, the waves spread in all directions. Figure A2.4(a and b) show the generation of Huygen's wavelets to strengthen the concept of secondary wave generation. In Fig. A2.4(a), the secondary wave generation at narrow aperture due to circular primary wave is shown. Whereas Fig. A2.4(b) depicts the stationary wave generation at narrow aperture due to straight primary waves. These are illustrated with ripple waves on the water surface.

(a)

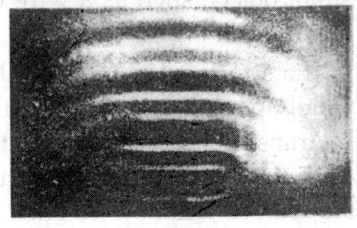

(b)

Fig. A2.4

DIFFRACTION FROM A SINGLE SLIT

The width of the slits used for the study of diffraction cannot be made arbitrarily narrow. The finite aperture of a slit gives rise to a characteristic interference pattern in the shadow region. If a plane wave front such as A–B–C–D (Fig. A2.5) is moving towards a point P, every part of wave ABCD acts as a source of secondary waves which eventually reach the point P. The time taken by a disturbance from a point such as Q to reach P is a little longer than the time taken by it from a point O. This difference in time causes a spatial path difference resulting in phase difference. The coherent addition of wavelets coming from a secondary source at the slit region leads to a peculiar pattern of interference known as *diffraction*. This diffraction pattern depends on the size and shape of the geometric configuration of the opening.

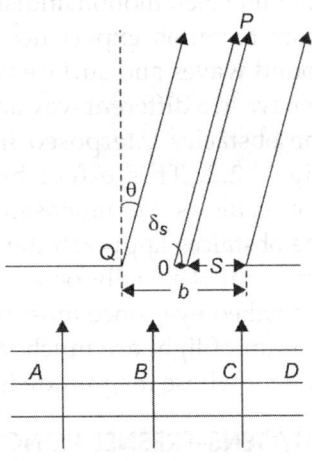

Fig. A2.5

Because of the finite dimensions of the slit, we consider the slit width b as divided into small line element δs (along the Y-axis). The path difference at P due to elemental slit (δs) is (δs) sin θ, which corresponds to a phase difference of $\delta = \dfrac{2\pi}{\lambda}$ (path difference) $= \dfrac{2\pi}{\lambda}(\delta s)\sin\theta$. If we divide the slit opening b into N equal parts as $N = \dfrac{b}{(\delta s)}$, then the total disturbance reaching P is the sum of N wavelets with relative phase difference of δ. The wavelets emanating from the wave front falling on the finite slit add coherently at P. The linear superposition of these N wavelets with successive phase difference δ gives the resultant displacement at P as

$$y_p = A\sum_{n=0} \cos(\omega t - n\delta + \phi_0) \tag{A2.1}$$

We have already discussed the superposition of N simple harmonic motions and the result when expressed in the complex representation is given by

$$z = \sum_{n=0}^{N-1} Z_n = Ae^{i(\omega t + \phi_0)}\left(\sum_{n=0}^{N-1} \cdot e^{in\delta}\right)$$

$$\therefore \qquad Re(z) = y_p(t) = A \frac{\sin\left(\dfrac{N\delta}{2}\right)}{\sin\left(\dfrac{\delta}{2}\right)} \tag{A2.2}$$

as $\delta_s \to 0$, $N \to \infty$, the phase diagram (discussed already) becomes a smooth curve with its arc length representing the resultant time-independent amplitude.

$$R = A \frac{\sin\left(\dfrac{N\delta}{2}\right)}{\sin\left(\dfrac{\delta}{2}\right)}$$

Let us put $\dfrac{N\delta}{2} = \alpha$ and $\sin\dfrac{\delta}{2} \approx \dfrac{\delta}{2} = \dfrac{\alpha}{N}$ as δ is very small. Thus,

$$R = A \frac{\sin\left(\dfrac{N\delta}{2}\right)}{\sin\left(\dfrac{\delta}{2}\right)} = NA\frac{\sin\alpha}{\alpha} = A_0 \frac{\sin\alpha}{\alpha} \tag{A2.3}$$

with $\qquad \alpha = \dfrac{N\delta}{2} = \dfrac{N}{2}\left(\dfrac{2\pi}{\lambda}b\sin\theta\right) = \dfrac{N\pi b}{\lambda}\sin\theta \tag{A2.4}$

and $\qquad A_0 = NA.$

The expression R is known as the Bessel function of order zero. The general behaviour is shown in Fig. A2.6. It is seen that the amplitude varies as a function of diffraction angle θ which defines the parameter α. Fig. A2.6(a) depicts the variation of wave amplitude with direction in a single slit diffraction and Fig. A2.6(b) gives magnitude variation of the diffracted wave amplitude.

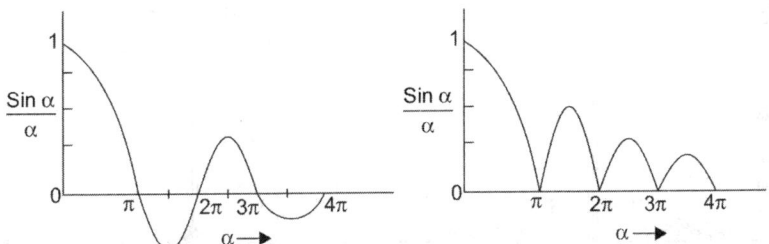

Fig. A2.6

The first zero occurs at a location such that the path difference for the secondary waves from the two edges of the slit is of one wavelength. This is understood if we consider the slit to be made of two continuous slit each of width $\lambda/2$. The path difference between the wavelets starting from the centres of these two parts is then $\lambda/2$ leading to the destructive interference, i.e., the resulting amplitude in that direction is zero. This result is important as the interference effets are observable even with a single slit. At a position close to the slit, we see the effects of slit width more directly as a strong disturbance exists over this region. Fig. A2.7(a), (b) and (c) show the relative intensity in single slit diffraction for different ratio b/λ and (c) and (d) show single slit diffraction of sound waves.

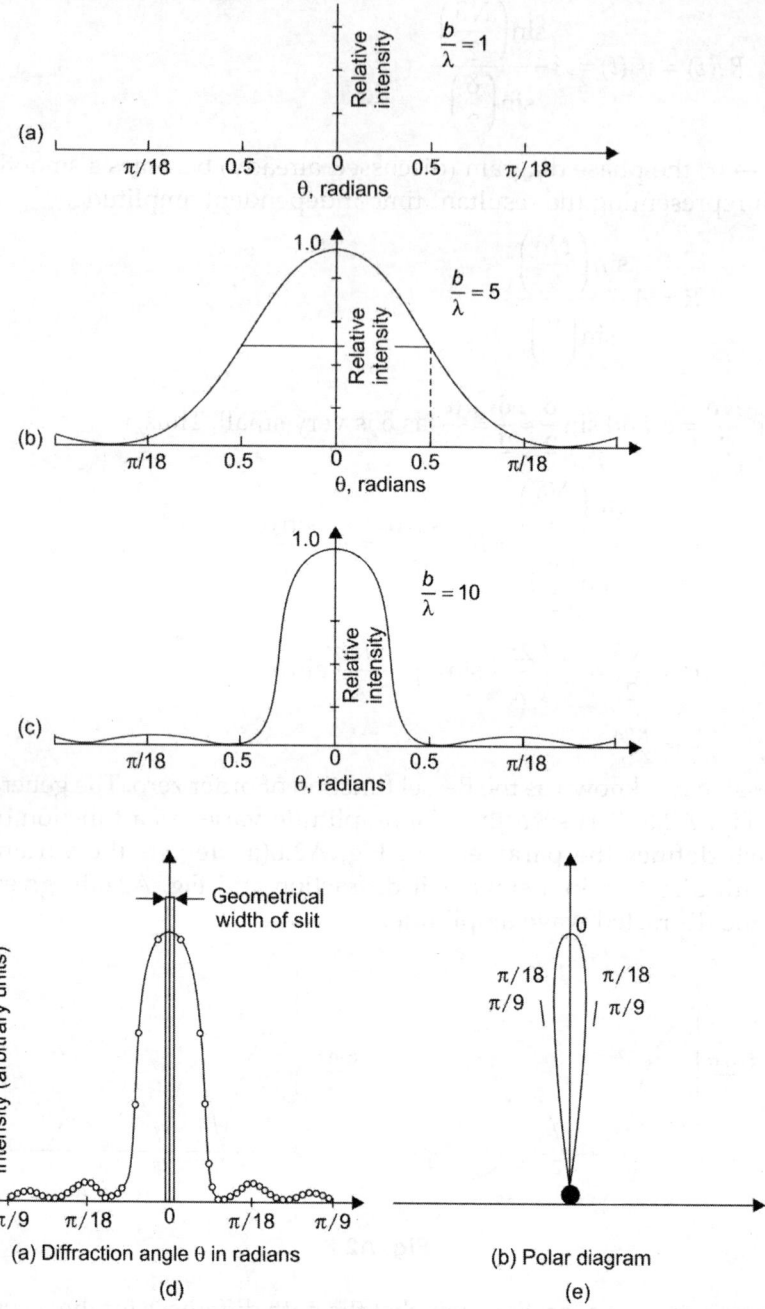

Fig. A2.7

The central maximum has a spread of θ_m defined by $\sin \theta_m = b/\lambda$ giving a linear spread of the diffraction peak at $\pm D \tan \theta_m$, D being the distance from the slit to the screen.

The criteria to describe this spread due to diffraction is that $D \tan \theta \gg$ the slit width b. Therefore, when $b \gg \lambda$, we have $\sin \theta_m = \lambda/b$, the condition $D \tan \theta_m \gg b^2$ means $D\lambda \gg b^2$ as $\tan \theta_m \approx \theta_m = \lambda/b$.

From this analysis, it follows that a single slit can also give rise to a diffraction pattern with a system of nodal lines. Equation (A2.2) gives that the condition for diffraction minima is the same as that for interference maxima ($\Delta = d \sin \theta = n\lambda$) between two slits separated by a distance that has been taken as b in diffraction as d in interference. This observation becomes more important when we consider the effects of diffraction due to double slit interference through slits with finite widths.

In Fig. A2.8 explains diffraction due to an opaque object of linear dimensions. If $AB = b$, the path difference $\Delta = b \sin \theta$. If $\Delta = b \sin \theta = n\lambda$, we can hear sound at P. If we take $n = 1$ and $\theta = 90°$, we get $\lambda = b$. Taking treble A as 440 Hz and velocity of sound as 330 m/s, $\lambda = \dfrac{330}{440} = 0.75 = b$. Hence a sound wave travelling from left to right around an obstacle of 0.75 m width, a person standing at a point such

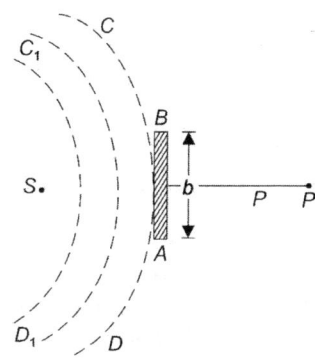

as P in Fig. A2.8 can hear the sound of frequency corresponding to treble A. Ordinary frequencies of speech ranges from 440 Hz to 8800 Hz. For diffraction to occur at frequency of 8800 Hz, the width of AB in Fig. A2.8 should be $b = \lambda = 0.0375$ m. Note that large opening cause hindrance to waves being heard in the shadow region. The consequence of diffraction of sound is that a shadow region exists for short values of wavelength, whereas for large values of wavelengths it is almost non-existent.

Fig. A2.8

If s is a source of sound (Fig. A2.8), it is not possible to avoid the sound by standing behind an obstacle. In the figure, CD is a wave of sound starting from S and AB is an obstacle in its path. The secondary waves from BC and AD still reach P and as they reinforce each other, the waves reach P as if AB did not exist.

For sound of frequency 256 Hz, we have $\lambda = 1.289$ m. Simple estimation shows that the physical extension of the obstacle should be 5.5 m to enable a dead zone. That is why, the geometrical shadow in an auditorium is not a shadow region for sound waves of large wavelengths, since the obstacles are not usually over 6 m wide. However, for a frequency of 10 kHz, we have $\lambda = 0.03$ m. Hence the quality of the music which constitute a large number of frequencies will be affected due to frequency dependence of the diffraction pattern. This is generally ignored in public places, although effects are perceptible. These obstructions become important in the high quality recording and reproduction of music.

DIFFRACTION GRATING OR MULTIPLE (FINITE) SLIT INTERFERENCE

The above discussion can be extended to an array of N equally spaced slits along the Y-axis, each with a slit width b separated by a distance d. Under the simplifying assumption of each slit being of very small width, the total amplitude at P is obtained by the linear superposition of secondary waves starting from the slits. Since all the slits are driven by the same wave front, the relative phase difference between any two successive waves reaching P is $\left(\dfrac{2\pi}{\lambda}\right) d \sin \theta$. Thus, the resultant displacement is given by

$$Z = A \sum_{n=0}^{N-1} \cdot e^{i(\omega t - N\delta + \phi_0)}$$

$$\therefore \quad Re(z) = A\frac{\sin\left(\dfrac{N\delta}{2}\right)}{\left(\dfrac{\delta}{2}\right)}\cos\left[\omega t - (N-1)\frac{\delta}{2} + \phi_0\right] \tag{A2.5}$$

Consideration of the finite slit-width introduces diffraction effects common to all slits. Therefore, the intensity which is proportional to the square of the amplitude is given by

$$I(\theta) = I_0\left(\frac{\sin\alpha}{\alpha}\right)^2\left\{\frac{\sin\left(\dfrac{N\delta}{2}\right)}{\left(\dfrac{\delta}{2}\right)}\right\}^2 \tag{A2.6}$$

where $\alpha = \left(\dfrac{\pi}{\lambda}b\sin\theta\right)$ and $\delta = \left(\dfrac{2\pi}{\lambda}\right)d\sin\theta$. The important aspect of this analysis is that the intensity pattern of N-slits interference is modified by the slit diffraction function given by $\left(\dfrac{\sin\alpha}{\alpha}\right)^2$. When $d\sin\theta = m\lambda$ (m being an integer), the interference factor in the intensity expression has zero in the numerator with the denominator denoting maximum intensity, this is because around $d\sin\theta = m\lambda$, the factor

$$\frac{\sin^2\left(\dfrac{N\delta}{2}\right)}{\sin^2\left(\dfrac{\delta}{2}\right)} \rightarrow \frac{(N\delta)^2}{(\delta)^2} \rightarrow N^2$$

makes the intensity as large as N^2 times the normal single slit intensity. For each value of m thus have one principal maximum. The magnitude of these maxima is modified by $\left(\dfrac{\sin\alpha}{\alpha}\right)$, the slit diffraction term. Further, we also have $(N-1)$ zeroes in the intensity function $I(\theta)$ in between the two principal maxima which occur when the factor in the numerator of the intensity function $I(\theta)$ is zero, whereas the factor in the denominator is finite. These zeroes occur at

$$d\sin\theta = \frac{\lambda}{N}, \frac{2\lambda}{N}, \frac{3\lambda}{N}, \cdots, \frac{(N-1)\lambda}{N}.$$

Subsidiary maxima of $(N-2)$ will be of very small intensity as there is no multiplying factor of N^2 at these angles. Figure A2.9 illustrates a few examples of diffraction patterns under different conditions. If d and θ are known, this method of N slits diffraction is the best way to evaluate the unknown wavelengths.

Example A2.1: The horn of loud speaker is in the form of a rectangle of 1 m length and 0.5 m width, which of these is more effective to keep the long side vertical or

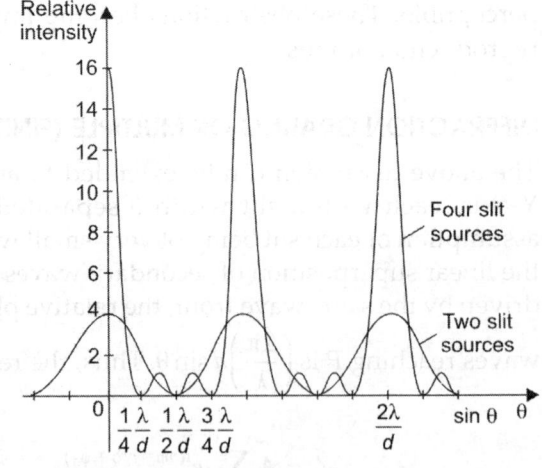

Fig. A2.8

to keep the short side vertical. For greater audience coverage out of doors, calculate the diffraction angle in each case.

Solution: Velocity $v = v\lambda$, $v = 330$ m/s and $\nu = 4000$ Hz

$$\therefore \qquad \lambda = \frac{330}{4000} = 0.0825$$

Since $d \sin \theta = n\lambda$, we take $n = 1$. Hence θ for $d = 1$ m

$$1 \times \sin \theta = \frac{330}{4000}$$

or $\qquad \sin \theta = 0.0825 \quad$ or $\quad \theta = \dfrac{\pi}{40}$ radians

for $d = 0.5 \qquad 0.5 \sin \theta = \dfrac{330}{4000} \quad$ or $\quad \theta \approx \dfrac{\pi}{20}$ radians

Hence for greater audience coverage, it is more effective to keep the short side horizontal.

Example A.2: A circular plate of diameter 0.5 m is used as a transducer in a submarine sonar set up, the plate has a frequency 1 MHz. Assuming the speed of sound in sea water to be 1500 m/s, calculate the angle between the normal to the transducer and first minimum.

Solution: $\quad b \sin \theta = m\lambda$

we have: $\ 0.5 \sin \theta = \dfrac{1 \times 1500}{1 \times 10^6}$

or $\qquad\qquad \theta = 0°10'$ arc

Suggested Readings

1. Kakani SL and Hemrajani C. *Mechanics*, 3rd ed, 2014, Viva Books, New Delhi-2.
2. Kakani SL and Kakani S. *Applied Physics*, 1st ed, 2014, Viva Books, New Delhi-2.
3. Kakani SL and Kakani S. *Engineering Physics*, 3rd ed, 2015, CBS Publishers & Distributors Pvt Ltd, New Delhi-2.
4. Pain HJ. *The Physics of Vibration and Waves*, 6th ed, 2005, Wiley.
5. Kakani SL and Kakani S. *Modern Physics*, 2nd ed, 2014, Viva Books, New Delhi-2.
6. French, AP. *Vibration and Waves*, The MIT Introductory Physics Series, Thomas Nelson and Sons, UK.
7. Bajaj NK. *The Physics of Waves and Oscillations*, Tata McGraw Hill, New Delhi.
8. Puri SP. *Vibrations and Waves*, Tata McGraw Hill, New Delhi.

Index